Plant Biotechnology

Plant Biotechnology
Current and Future Applications of Genetically Modified Crops

Edited by

NIGEL G. HALFORD

Crop Performance and Improvement, Rothamsted Research, UK

John Wiley & Sons, Ltd

Other Wiley Editorial Offices

John Wiley & Sons Inc., 111 River Street, Hoboken, NJ 07030, USA

Jossey-Bass, 989 Market Street, San Francisco, CA 94103-1741, USA

Wiley-VCH Verlag GmbH, Boschstr. 12, D-69469 Weinheim, Germany

John Wiley & Sons Australia Ltd, 42 McDougall Street, Milton, Queensland 4064, Australia

John Wiley & Sons (Asia) Pte Ltd, 2 Clementi Loop #02-01, Jin Xing Distripark, Singapore 129809

John Wiley & Sons Canada Ltd, 22 Worcester Road, Etobicoke, Ontario, Canada M9W 1L1

Wiley also publishes its books in a variety of electronic formats. Some content that appears in print may not be available in
electronic books.

Library of Congress Cataloging-in-Publication Data
Plant biotechnology : current and future uses of genetically modified crops /
editor, Nigel Halford.
    p.   cm.
   ISBN-13 978-0-470-02181-1
   ISBN-10 0-470-02181-0
   1. Plant biotechnology.   I. Halford, N. G. (Nigel G.)
   SB106.B56P582 2006
   631. 5′233–dc22   2005024912

British Library Cataloguing-in-Publication Data

A catalogue record for this book is available from the British Library

ISBN-10 0-470-02181-0
ISBN-13 978-0-470-02181-1

Typeset in 10/12pt Times by Thomson Press (India) Ltd., New Delhi, India
Printed and bound in Great Britain by Antony Rowe Ltd, Chippenham, Wiltshire
This book is printed on acid-free paper responsibly manufactured from sustainable forestry
in which at least two trees are planted for each one used for paper production.

Contents

List of Contributors

B. Baker   Department of Molecular and Cell Biology, University of Cape Town, Private Bag, Rondebosch 7701, South Africa

Frédéric Beaudoin   Crop Performance and Improvement, Rothamsted Research, Harpenden, Hertfordshire AL5 2JQ, UK

K. Berry   Plant and Invertebrate Ecology, Rothamsted Research, Harpenden, Hertfordshire AL5 2JQ, UK

Eduardo Blumwald   Department of Plant Sciences, University of California, One Shields Ave, Davis, CA 95616, USA

Michael M. Burrell   Department of Animal and Plant Sciences, University of Sheffield, Sheffield S10 2TN, UK

M. Carmen Cañizares   John Innes Centre, Colney Lane, Norwich NR4 7UH, UK

N. Conrad   Department of Molecular and Cell Biology, University of Cape Town, Private Bag, Rondebosch 7701, South Africa

E.J. Davis   Department of Molecular and Cell Biology, University of Cape Town, Private Bag, Rondebosch 7701, South Africa

K. Govender   Department of Molecular and Cell Biology, University of Cape Town, Private Bag, Rondebosch 7701, South Africa

Anil Grover   Department of Plant Molecular Biology, University of Delhi South Campus, Benito Juarez Road, New Delhi 110021, India

Nigel G. Halford   Crop Performance and Improvement, Rothamsted Research, Harpenden, Hertfordshire AL5 2JQ, UK

Karin Herbers   BASF Plant Science GmbH, Agricultural Centre, D-67117, Limburgerhof, Germany

R. Iyer   Department of Molecular and Cell Biology, University of Cape Town, Private Bag, Rondebosch 7701, South Africa

John Jenkins   Institute of Food Research, Norwich NR4 7UA, UK

Huw D. Jones   Crop Performance and Improvement, Rothamsted Research, Harpenden, Hertfordshire AL5 2JQ, UK

George P. Lomonossoff   John Innes Centre, Colney Lane, Norwich NR4 7UH, UK

P.J.W. Lutman   Plant and Invertebrate Ecology, Rothamsted Research, Harpenden, Hertfordshire AL5 2JQ, UK

A.T. Maredza   Department of Molecular and Cell Biology, University of Cape Town, Private Bag, Rondebosch 7701, South Africa

Louise V. Michaelson   Crop Performance and Improvement, Rothamsted Research, Harpenden, Hertfordshire AL5 2JQ, UK

E.N. Clare Mills   Institute of Food Research, Norwich NR4 7UA, UK

S.G. Mundree   Department of Molecular and Cell Biology, University of Cape Town, Private Bag, Rondebosch 7701, South Africa

Johnathan A. Napier   Crop Performance and Improvement, Rothamsted Research, Harpenden, Hertfordshire AL5 2JQ, UK

Liz Nicholson   John Innes Centre, Colney Lane, Norwich NR4 7UH, UK

Dietrich Rein   BASF Plant Science Holding GmbH, Building 444, D-67117 Limburgerhof, Germany

Sujatha Sankula   National Center for Food and Agricultural Policy, 1616 P Street NW, First Floor, Washington, DC 20036, USA

Olga Sayanova   Crop Performance and Improvement, Rothamsted Research, Harpenden, Hertfordshire AL5 2JQ, UK

Peter R. Shewry   Crop Performance and Improvement, Rothamsted Research, Harpenden, Hertfordshire AL5 2JQ, UK

Maarten Stuiver   BASF Plant Science GmbH, Agricultural Center, Carl Bosch Strasse 64, 67117 Limburgerhof, Germany

J.A. Thomson   Department of Molecular and Cell Biology, University of Cape Town, Private Bag, Rondebosch 7701, South Africa

Qingzhong Xue   College of Agriculture and Biotechnology, Zhejiang University, People's Republic of China

Xianyin Zhang   College of Agriculture and Biotechnology, Zhejiang University, People's Republic of China

Yuhua Zhang   Crop Performance and Improvement, Rothamsted Research, Harpenden, Hertfordshire AL5 2JQ, UK

Preface

The beginning of the 20th century saw the rediscovery of Gregor Mendel's work on the inheritance of phenotypic traits in plants. Mendel's work laid the foundations of modern, scientific plant breeding by enabling plant breeders to predict how traits brought into breeding lines would be inherited, and what had to be done to ensure that the lines would breed true. As a result, scientific plant breeding from the early part of the 20th century onwards brought huge increases in crop yield, without which current human population levels would already be unsustainable.

In the following decades, science made great strides in the elucidation of the molecular processes that underpin inheritance; genes, the units of inheritance, were linked with proteins, DNA was shown to be the material of inheritance, the structure of DNA was resolved, DNA polymerases, ligases and restriction enzymes were discovered, recombinant DNA molecules were created and techniques for determining the nucleotide sequence of a DNA molecule were developed.

Plant scientists were quick to exploit the new tools for manipulating DNA molecules and also made the astounding discovery that a naturally occurring bacterium, *Agrobacterium tumefaciens*, actually inserted a piece of its own DNA into that of a plant cell during its normal infection process. As a result, by the mid-1980s everything was in place to allow foreign genes to be introduced into crop plants and scientists began to predict a second green revolution in which crop yield and quality would be improved dramatically using this new technology.

All plant breeding involves the alteration of plant genes, whether it is through the crossing of different varieties, the introduction of a novel gene into the gene pool of a crop species, perhaps from a wild relative, or the artificial induction of random mutations through chemical or radiation mutagenesis. However, the term 'genetic modification' was used solely to describe the new technique of artificially inserting a single gene or small group of genes into the DNA of an organism; organisms carrying foreign genes were termed genetically modified or GM. Another decade passed before the first GM crops became available for commercial use. Since then, genetic modification has become an established technique in plant breeding around the world and, in 2004, GM crops were grown on 81 million hectares in 17 countries.

With the first decade of GM crop cultivation drawing to a close, it seemed appropriate to assess the successes and failures that have marked that decade, and the prospects of new GM crop varieties reaching the market in the coming 10 years. This book is intended for students who are studying plant biotechnology at degree level and for specialists in academia and industry. It covers the impact of GM crop cultivation in two leading countries in the commercial application of plant biotechnology, the USA and China, and the advances being made in the use of genetic modification to increase crop resistance to biotic and abiotic stresses, improve the processing and nutritional value of crop products and enable plants to be used for novel purposes such as vaccine production.

GM crop production is, of course, one of the most controversial issues of our time, and two aspects of GM crops that have worried the public the most, the inadvertent synthesis of antigens and the risk of gene flow between GM and non-GM crops and wild relatives, are covered. Governments, particularly in Europe, have responded to public concern over these issues by introducing rafts of regulations to control GM crop production and use. I have discussed these in the last chapter.

I am delighted to have been able to bring together leading specialists in different topics to write the individual chapters, enabling the book to cover the subject comprehensively and in depth; I owe a debt of gratitude to all the authors who contributed.

Nigel G. Halford

PART I
THE CURRENT SITUATION

1.1

From Primitive Selection to Genetic Modification, Ten Thousand Years of Plant Breeding

Nigel G. Halford

Crop Performance and Improvement, Rothamsted Research, Harpenden, Hertfordshire, AL5 2JQ, United Kingdom

Introduction

In the mid-1990s plant biotechnology burst onto the scene in world agriculture, beginning a second 'green revolution' and precipitating one of the great public debates of our time. Approximately a decade later, this book describes the impact of genetically modified (GM) crops on world agriculture, recent advances in the technology and the areas of research from which the next generation of GM crops is likely to emerge, as well as addresses the issues of safety and regulation that have dogged the technology, particularly in Europe.

This chapter defines exactly what GM crops are (in other words, what distinguishes them from other crops) and describes the GM crops that are currently in commercial use. It covers the traits of herbicide tolerance, insect resistance, virus resistance, increased shelf life and modified oil profile, as well as the genes used to impart them. It also chronicles the uptake of GM crop varieties around the world from their widespread introduction in 1996 to the present day, contrasting the situation in the Americas, Australia and Asia with that in Western Europe.

First, it is necessary to put the advent of plant genetic modification into the context of a long history of advances in plant breeding and genetics.

Plant Biotechnology. Edited by Nigel Halford.
© 2006 John Wiley & Sons, Ltd.

Early Plant Breeding

Arguably the most important event in human history occurred approximately 10 000 years ago when people in what is now called the Middle East began to domesticate crops and livestock, and adopt a sedentary way of life based on farming rather than a nomadic one based on hunting and gathering. Ultimately this led to the growth of villages, towns and cities, and provided the stability and time for people to think, experiment, invent and innovate. Technological advancement, which had barely progressed at all for half a million years, accelerated enormously (Figure 1.1.1). The great civilizations of ancient Mesopotamia (Assyria, Sumeria and Babylon) and Egypt arose within a few thousand years, laying the foundation of modern civilization.

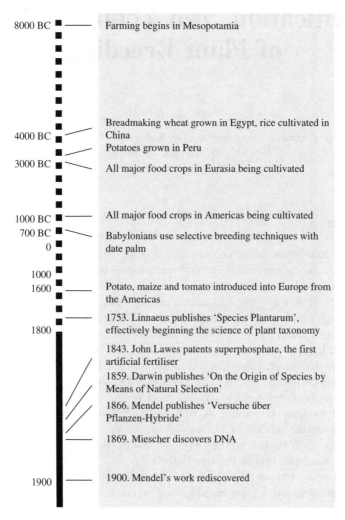

8000 BC — Farming begins in Mesopotamia

4000 BC — Breadmaking wheat grown in Egypt, rice cultivated in China
Potatoes grown in Peru

3000 BC — All major food crops in Eurasia being cultivated

1000 BC — All major food crops in Americas being cultivated

700 BC — Babylonians use selective breeding techniques with date palm

0

1000

1600 — Potato, maize and tomato introduced into Europe from the Americas

1753. Linnaeus publishes 'Species Plantarum', effectively beginning the science of plant taxonomy

1800

1843. John Lawes patents superphosphate, the first artificial fertiliser

1859. Darwin publishes 'On the Origin of Species by Means of Natural Selection'

1866. Mendel publishes 'Versuche über Pflanzen-Hybride'

1869. Miescher discovers DNA

1900 — 1900. Mendel's work rediscovered

Figure 1.1.1   *Timeline showing some of the major landmarks in the development of agriculture and plant breeding.*

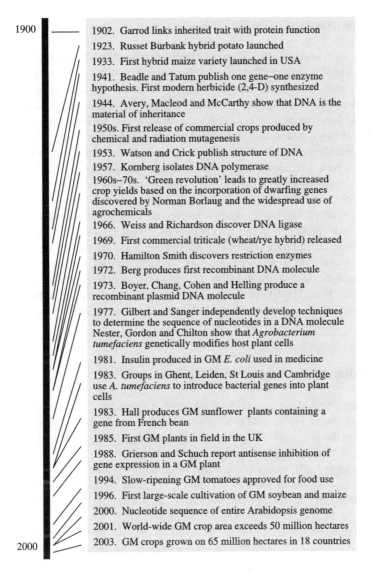

1900

1902. Garrod links inherited trait with protein function

1923. Russet Burbank hybrid potato launched

1933. First hybrid maize variety launched in USA

1941. Beadle and Tatum publish one gene–one enzyme hypothesis. First modern herbicide (2,4-D) synthesized

1944. Avery, Macleod and McCarthy show that DNA is the material of inheritance

1950s. First release of commercial crops produced by chemical and radiation mutagenesis

1953. Watson and Crick publish structure of DNA

1957. Kornberg isolates DNA polymerase

1960s–70s. 'Green revolution' leads to greatly increased crop yields based on the incorporation of dwarfing genes discovered by Norman Borlaug and the widespread use of agrochemicals

1966. Weiss and Richardson discover DNA ligase

1969. First commercial triticale (wheat/rye hybrid) released

1970. Hamilton Smith discovers restriction enzymes

1972. Berg produces first recombinant DNA molecule

1973. Boyer, Chang, Cohen and Helling produce a recombinant plasmid DNA molecule

1977. Gilbert and Sanger independently develop techniques to determine the sequence of nucleotides in a DNA molecule Nester, Gordon and Chilton show that *Agrobacterium tumefaciens* genetically modifies host plant cells

1981. Insulin produced in GM *E. coli* used in medicine

1983. Groups in Ghent, Leiden, St Louis and Cambridge use *A. tumefaciens* to introduce bacterial genes into plant cells

1983. Hall produces GM sunflower plants containing a gene from French bean

1985. First GM plants in field in the UK

1988. Grierson and Schuch report antisense inhibition of gene expression in a GM plant

1994. Slow-ripening GM tomatoes approved for food use

1996. First large-scale cultivation of GM soybean and maize

2000. Nucleotide sequence of entire Arabidopsis genome

2001. World-wide GM crop area exceeds 50 million hectares

2000

2003. GM crops grown on 65 million hectares in 18 countries

Figure 1.1.1   *(Continued)*

The crop species responsible for this change was probably wheat. Certainly by 6000 years ago, wheat was being baked into leavened bread in Egypt in much the same way as it is today. Farming was also developing in South America and China, with potato and rice, respectively, being the predominant cultivated crops.

It is probable that crop improvement began as soon as farming did. At first, such improvement may well have occurred unconsciously through the harvesting and growing of the most vigorous individuals from highly variable populations, but then became more systematic. For example, there is evidence that the Ancient Babylonians bred for certain characteristics in palm trees by selecting male trees with which to pollinate female trees.

Over time such practices had dramatic effects on crop characteristics. For example, the wheat grain found in Ancient Egyptian tombs is much more similar to modern wheat than to its wild relatives. Indeed, breadmaking wheat arose through hybridization events between different wheat species that only occurred in agriculture; there is no wild equivalent. It first appeared within cultivation, probably in Mesopotamia between 10 000 and 6000 years ago, and its use spread westwards into Europe.

Another excellent example of the effects of simple selection is the cabbage family of vegetables, which includes kale, cabbage itself, cauliflower, broccoli and Brussels sprouts. The wild relative of the cabbage family was first domesticated in the Mediterranean region of Europe approximately 7000 years ago. Through selective breeding over many centuries, the plants became larger and leafier, until a plant very similar to modern kale was produced in the 5th century BC. By the 1st century AD, cabbage had appeared, characterized by a cluster of tender young leaves at the top of the plant. In the 15th century, cauliflower was produced in Southern Europe by selecting plants with large, edible flowering heads and broccoli was produced in a similar fashion in Italy about a century later. Finally, Brussels sprouts were bred in Belgium in the 18th century, with large buds along the stem. All of these very different vegetables are variants of the same species, *Brassica oleracea*.

The Founding of the Science of Genetics

The examples above show how crop plants were improved by farmers who for millennia knew nothing about the scientific basis of what they were doing. Modern, systematic plant breeding did not come about until the science of genetics was established as a result of the work of Charles Darwin and Gregor Mendel.

Darwin is regarded by many as the father of modern genetics but it was Mendel's work that showed how Darwin's theories on natural selection could work. Ironically the two men never met and Darwin died unaware of Mendel's findings. Darwin's seminal book, 'On the Origin of Species by Means of Natural Selection', was published in 1859. In it, Darwin described the theory of evolution based on the principle of natural selection. The theory was proposed independently at approximately the same time by Alfred Russell Wallace, but it was Darwin's meticulous accumulation of evidence collected over decades that gave weight to the hypothesis.

In simple terms, Darwin's theory of evolution proposed that the diversity of life on Earth had arisen through the adaptation of species to different and changing environments, leading to the extinction of some species and the appearance of others. Species that were similar had arisen from a recent common ancestor. This process was driven by natural selection, in which individuals competed with each other and those best fitted for their environment would be most likely to survive, reproduce and pass on their characteristics to the next generation. If the environment changed or a species colonized a new environment, different characteristics would be selected, leading to change and eventually to the evolution of a new species.

Natural selection (or artificial selection, for that matter) can only work because individuals within a species are not all the same; individuals differ or show variation. Darwin and his contemporaries believed that traits present in two parents would be mixed in the offspring so that they would always be intermediate between the two parents. This posed a problem for Darwin's theory of evolution because it would have the effect of reducing variation with every successive generation, leaving nothing for selection to work on.

The solution to the problem was provided by Gregor Mendel, a monk at the Augustinian monastery in Brno. In 1857, Mendel began experimenting with pea plants, noting different characteristics such as height, seed color and pod shape. He observed that offspring sometimes, but not always, showed the same characteristics as their parents. In his first experiments, he showed that short and tall plants bred true, the short having short offspring and the tall having tall offspring, but that when he crossed short and tall plants all of the offspring were tall. He crossed the offspring again and the short characteristic reappeared in about a quarter of the next generation.

Mendel concluded that characteristics were passed from one generation to the next in pairs, one from each parent, and that some characteristics were dominant over others. Crucially, this meant that variation was not lost from one generation to the next. Whether the offspring of two parents resembled one parent or were an intermediate between the two, they inherited a single unit of inheritance from each parent. These units were reshuffled in every generation and traits could reappear. Although Mendel did not use the term, units of inheritance subsequently became known as genes.

Mendel's findings were published by the Association for Natural Research in 1866, under the title 'Versuche über Pflanzen-Hybride', but were ignored until the beginning of the next century as the work of an amateur. Later they became known as the Mendelian Laws and the foundation of modern plant breeding.

The Elucidation of the Molecular Basis of Genetics

The pace of discovery accelerated greatly in the 20th century (Figure 1.1.1) and gradually the molecular bases for the laws of genetics were uncovered. In 1902, Sir Archibald Garrod found that sufferers of an inherited disease, alkaptonuria, lacked an enzyme that breaks down the reddening agent, alkapton, and therefore excreted dark red urine. This was the first time that a link had been made between a genetic trait and the activity of a protein. The significance of Garrod's work was only recognized decades later when George Beadle and Edward Tatum showed that a genetic mutation in the fungus, *Neurospora crassa*, affected the synthesis of a single enzyme required to make an essential nutrient. Beadle and Tatum published the one gene–one enzyme hypothesis in 1941 (Beadle and Tatum, 1941) and were subsequently awarded a Nobel Prize. The hypothesis was essentially correct, with the exception that some proteins are made up of more than one subunit and the subunits may be encoded by different genes.

Underpinning the laws of genetics and evolution, which have now been established, is the ability of organisms to pass on the instructions for growth and development to their offspring. The obvious question was in what substance was this information carried, in other words what was the genetic material. Deoxyribonucleic acid (DNA) was identified as this substance in 1944 by Oswald Avery, Colin MacLeod and Maclyn McCarty. Their conclusive experiment showed that the transfer of a DNA molecule from one strain of a bacterium, *Streptomyces pneumoniae*, to another changed its characteristics (Avery *et al.*, 1944).

DNA was first discovered in 1869 by Friedrich Miescher but its structure was not determined for another 84 years. The breakthrough was made by James Watson and Francis Crick in 1953 (Watson and Crick, 1953). They came up with their model after analyzing X-ray crystallographs produced by Rosalind Franklin and Maurice Wilkins, but it is fair to say that Watson and Crick made an intellectual leap that Franklin and Wilkins had failed to make. Watson, Crick and Wilkins were awarded a Nobel Prize; tragically, Franklin missed out because she died before the prize was awarded and it is not awarded posthumously.

The structure of DNA is so elegant that it has become iconic. The molecule consists of two strands (it is said to be double-stranded); each strand is made up of units of deoxyribose (a type of sugar) with an organic base attached, linked by phosphate groups. Each unit is called a nucleotide and there are four kinds, each with a different organic base: adenine, cytosine, guanine or thymine. These are often represented as A, C, G and T. The two strands run in opposite directions and are coiled into a double helix structure, the two strands linked together by hydrogen bonds between opposing bases. The separation distance of the two strands means that the bases on opposing strands occur in pairs (base pairs) that will fit: adenine on one strand always paired with thymine on the other, and cytosine always paired with guanine. This means that the sequence of bases on one strand determines the sequence on the other (they are said to be the reverse and complement of each other), an important factor when the molecule is being duplicated. If double-stranded DNA is unraveled to form two single strands, each strand can act as a template for the synthesis of a complementary chain and two replicas of the original double-stranded molecule are created. Information is encoded within DNA as the sequence of nucleotides in the chain, a four-letter language in which all the instructions for life on Earth are written.

Information encoded within the DNA molecule determines the structure of a protein through the process of gene expression. The first part of this process is called transcription, in which a molecule related to DNA called ribonucleic acid (RNA) is synthesized using the DNA molecule as a template. Like DNA, RNA consists of a sugar–phosphate backbone along which are attached organic bases, but the RNA molecule consists of a single strand, not two, and the base thymine is replaced with uracil (U). The sequence of nucleotides on the newly synthesized RNA molecule is determined by the sequence of bases on the DNA template.

The RNA molecule is processed and transported to protein complexes called ribosomes where protein synthesis occurs; this is called translation. Proteins consist of chains of amino acids and the amino acid sequence is specified by the sequence of nucleotides in the RNA molecule, each amino acid in the protein being represented by a triplet of nucleotides called a codon. It is the sequence of amino acids in the protein that determines its function and properties, and ultimately it is the protein structure and function that determines the characteristics of an organism.

This link from DNA to RNA to proteins explains the observations of Garrod and underpins the one gene–one enzyme hypothesis of Beadle and Tatum. Furthermore, the processes of evolution and the changes in plants and animals brought about by selective breeding can be seen to result from changes (mutations) in the DNA sequence that lead to variations between individuals and traits that are selected.

DNA molecules can be huge and in plants, animals and fungi they are wrapped around proteins to form structures called chromosomes. In humans, they are organized into 23 pairs of chromosomes, each chromosome containing a DNA molecule ranging from 50 to 250 million base pairs so that 23 individual chromosomes (one from each pair, making up the genome) comprise a total of approximately 3 billion base pairs. If this length of DNA were stretched out it would be several centimetres long, yet it has to be coiled and packaged to fit into a cell. In comparison, the rice genome contains only 466 million base pairs on 12 chromosomes, while that of Arabidopsis, a plant widely used as a model in plant genetics, contains approximately 126 million base pairs on five chromosomes. The maize genome contains 2.6 billion base pairs on 10 chromosomes, while that of wheat is estimated to contain more than 16 billion base pairs on seven chromosomes.

Distributed unevenly along these huge DNA molecules are genes, just below 30 000 in Arabidopsis, 30 000–40 000 in humans and 45 000–56 000 in rice. Genes can be over a million base pairs long but are usually much smaller, averaging about 3000 base pairs. In fact, they make up a small proportion of the total genome; the rest (often referred to as 'junk DNA') appears to have no function and its amount varies greatly between different species, hence the great disparity in genome size between quite closely related species such as rice and wheat.

There is no structure marking the beginning and end of a gene. Rather, the units of heredity described by Mendel can be defined simply as functional units within a DNA molecule. Perhaps the most readily recognizable part of the gene is that containing the information for the sequence of amino acids in the protein that the gene encodes. This part of the gene is called the coding region and at least it has a definite beginning and end, although it is usually split into sections called exons interspersed with non-coding regions called introns. A gene also contains information that determines when, where and in response to what the gene is active. This information is usually contained in regions of the DNA 'upstream' of the coding region in what is called the gene promoter, but it can be in regions downstream of the coding region or within introns. The region 'downstream' of the coding region also contains information for the correct processing of the RNA molecule that is transcribed from the gene and is called the gene terminator.

Genes that are active throughout an organism all the time are referred to as constitutive or house-keeping genes. Other genes are active only in certain organs, tissues or cell types, while some are active during specific developmental stages or become active in response to a particular stimulus. In the case of plants, genes respond to many stimuli, including light, temperature, frost, grazing, disease, shading and nutritional status.

The Manipulation of DNA and Genes

Once DNA had been identified as the genetic material and its structure described, studies on the properties of DNA itself and the enzymes present in cells, which work on it, began in earnest. DNA polymerase, an enzyme that synthesizes DNA, was isolated by Arthur Kornberg in 1955 (Lehman *et al.*, 1958); DNA ligase, an enzyme that 'glues' two ends of DNA together, was isolated by Bernard Weiss and Charles Richardson in 1966 (Weiss and Richardson, 1967); a restriction endonuclease (also known as restriction enzyme), an enzyme that recognizes specific short sequences of base pairs in a DNA molecule and cuts the molecule at that point, was characterized by Hamilton Smith in 1970 (Smith and Wilcox, 1970). Both Kornberg and Smith received Nobel Prizes.

The molecular tools for repairing DNA, cutting it at specific places and sticking its pieces together in a test tube to make new molecules were now available. They were used by Paul Berg in 1972 to construct a DNA molecule by cutting viral and bacterial DNA sequences with restriction enzymes and then recombining them (Jackson *et al.*, 1972); he received a Nobel Prize in 1980. A year after Berg's experiment, Stanley Cohen, Annie Chang, Herbert Boyer and Robert Helling demonstrated that DNA which had been cut with a restriction enzyme could be recombined with small, self-replicating DNA molecules from bacteria called plasmids (Cohen *et al.*, 1973). The new plasmid could then be reintroduced into bacterial cells and would replicate. If the bacterial cells were cultured, each cell carrying copies of the recombinant plasmid, large amounts of plasmid DNA with the new piece of DNA inserted in it could be isolated from the culture. This enabled a section of DNA from any species to be cloned and bulked up in bacteria to generate enough of it to work on. This process is often called gene cloning. The bacterium of choice for this purpose is usually *Escherichia coli* (*E. coli*). This is a human gut bacterium, although the strains used in the laboratory have been disabled so that they are not pathogenic.

The ability to clone genes underpinned the molecular analysis of gene structure and function. Some people regarded this as a new branch of science and called it molecular biology. Its commercial exploitation was termed biotechnology and the first example of this was in the pharmaceutical industry; insulin produced from a modified human gene in *E. coli* was approved by the Food and Drug Administration of the USA in 1981.

Two other advances are worthy of note: in 1977, Walter Gilbert and Fred Sanger separately developed methods for determining the sequence of nucleotides in a DNA molecule (Maxam and Gilbert, 1977; Sanger *et al.*, 1977), and in 1983, Kary Mullis invented a method called the polymerase chain reaction (PCR) by which short sections of DNA could be bulked up (amplified) without cloning in bacteria (Mullis and Faloona, 1987). All three received a Nobel Prize. The methods for determining the nucleotide sequence of a DNA molecule were developed and automated to such an extent by the early 1990s that projects were initiated to obtain the nucleotide sequence of entire genomes. A first draft of the nucleotide sequence of the human genome was published in 2001. The first plant genome sequence was that of Arabidopsis, which was published in 2000, and the first crop plant genome sequence to be published was that of rice in 2002.

Modern Plant Breeding

The practice of planting different variants of the same crop in adjacent plots to promote the production of hybrid seed is used widely today by farmers and plant breeders, and has probably been practiced for millennia. It is done to exploit hybrid vigor, the phenomenon of a hybrid outperforming both of its parents. Hybrid vigor occurs because the ongoing process of genetic change by mutation leads to the existence of different forms of the same genes within a population. These different forms are called alleles, and the crossing of two parent lines with different characteristics results in a hybrid population with different combinations of alleles (genotypes) from the two parents. Some of these combinations are advantageous.

When Mendel's work on the inheritance of characteristics and the genetics of plant hybrids was rediscovered around 1900, plant breeding through the crossing of plants with different genotypes had a sound scientific basis. Plant breeders now understood what would happen to a genetic trait when it was crossed into a breeding line and how to produce a true-breeding line (a variety) in which that trait and other characteristics would be present in every individual in every generation. That is not to say that the process is simple; the fact that plants have several tens of thousands of genes which can be mixed in a myriad of combinations when a cross is made can make the outcome unpredictable. Furthermore, desirable traits may be linked with undesirable ones, usually as a result of being close together on the same chromosome.

Despite these difficulties, plant breeders have been incredibly successful at improving crop yield and it is just as well that they have. At the end of the 18th century, Reverend Thomas Malthus wrote in his 'Essay on the Principle of Population' that food supply could not keep up with rising population growth (Malthus, 1798). At that time, world population was approximately 1 billion. In 1999, the world population reached 6 billion, and yet famine remains relatively rare and localized and arises through extreme climate conditions combined with government incompetence and/or war, rather than inadequate crop plant performance.

An example of the dramatic increases in crop yield that have been achieved is that of wheat grown in the United Kingdom. It has increased approximately tenfold over the last 800 years, with more than half that increase coming since 1900. Similar increases have been achieved around the world with different crop species, the period of most rapid improvement being in the 1960s and 1970s when the incorporation of dwarfing genes into cereal crops together with increased mechanization and the widespread use of nitrogen fertilizers, herbicides and pesticides led to the so-called 'Green Revolution'. The dwarfing genes concerned actually affected the synthesis of a plant hormone, gibberellin, although it was not known at the time. Their incorporation reduced the amount of resources that cereal plants put into their inedible parts, making more available to go into the seed, and at the same time made the plants less susceptible to damage under damp and/or windy conditions. One of the pioneers of their use was Norman Borlaug, who not only used the technology himself in wheat breeding but also persuaded wheat breeders in Asia to do the same. Borlaug's actions are widely believed to have averted critical food shortages in Asia; indeed, it has been suggested that he is responsible for saving more lives than any other individual in history. No doubt Louis Pasteur and others would have supporters in a debate on that point, but Borlaug's success is something that all plant scientists can be proud of.

The seemingly inexorable rise in crop yield might be taken to indicate that the improvement will continue in perpetuity. However, improvement brought about by the recombination of existing genotypes is limited by the genetic variation that is present. Yield depends on many factors and is affected by many different genes, but eventually the possible combinations of genotypes will be exhausted. Furthermore, some targets for plant breeders have been much less amenable to breeding. If a trait, whether it be for resistance to a disease, tolerance of a herbicide, the ability to survive and yield highly in a particular environment, or whatever the target might be, does not exist in any of the genotypes within a species then a breeder cannot simply invent it.

In the mid-20th century, plant breeders began to use two new methods to increase the genetic variation available in their breeding lines. The first was 'wide crossing', the creation of hybrids between crop plants and exotic relatives or even species with which they would not normally cross in nature. The second was to induce mutations by treatment with either ionizing radiation (neutrons, gamma rays, X-rays or UV radiation) or a chemical mutagen.

Wide crosses usually require rescue of the embryo to prevent abortion; the embryo is removed from a developing seed under sterile conditions and cultured in a nutrient medium until it germinates. If the cross is made between two different species then the hybrid is usually sterile. This is because the members of each pair of chromosomes have to come together at the beginning of the process of meiosis by which sperm and egg cells are formed. In a hybrid cell with one set of chromosomes from each parent species, either the chromosomes do not pair at all or they mispair; the result is that the sperm and egg cells that are formed have too many, too few or the wrong combination of chromosomes and are not viable. This can be overcome by inducing chromosome doubling, usually by treatment of anthers, immature inflorescences or cultured cells with a chemical called colchicine. The hybrid cells then have a pair of chromosomes originating from each parent and are said to be polyploid (having more than one genome).

The best known example of a crop plant produced in this way is triticale, a hybrid between wheat and rye. The hybrid is usually made between durum wheat, already a tetraploid (two genomes), and rye (a diploid) to produce a hexaploid triticale (three genomes), although it is also possible to cross hexaploid wheat with rye to produce an octoploid triticale (four genomes). The name triticale was first used in 1935 by Tschermak but it was not until 1969, after considerable improvement through breeding, that the first commercial varieties of triticale were released. Triticale is now grown on more than 2.4 million hectares worldwide, producing more than 6 million tonnes of grain per year. It combines the yield potential of wheat with the acid soil-, damp- and extreme temperature-tolerance of rye and is used mostly for animal feed.

Experiments with mutagenesis of crop plants began in the 1920s. The radiation or chemical treatment, usually of seeds, damages the DNA, resulting in changes in the DNA sequence and hence genetic variation. The process has the disadvantage of being entirely random, and therefore mutagenesis programs usually involve very large populations of at least 10 000 individuals to ensure that a useful mutant is produced. Nevertheless, it has proved successful; the first commercial varieties arising from mutation breeding programs were released in the 1950s and the technique was used widely in the 1960s and 1970s, and continues to be used today.

Mutagenesis played an important role in the improvement of oil quality of oilseed rape, the first variety produced in this way being Regina II which was released in 1953. Oilseed rape was first grown in the UK during World War II to provide oil for industrial uses, and some varieties are still grown for that purpose. Its oil was regarded as unfit for human consumption because it contained high levels of erucic acid and glucosinolates.

Both of these compounds are very poisonous and glucosinolates have a bitter flavor. Their levels were gradually reduced by breeders using mutagenesis and crossing, and oilseed rape was finally passed for human consumption and animal feed in the 1980s. The edible varieties were given the name canola in North America and this name is now used for all varieties in that part of the world. Mutagenesis has also played an important role in the improvement of pasta wheats, rice, white bean and barley.

Genetic Modification

In 1977, 4 years after the first recombinant plasmid DNA molecule had been produced, Nester, Gordon and Chilton showed that bacterial DNA was inserted into the DNA of host plant cells during infection by a bacterium called *Agrobacterium tumefaciens* (Chilton *et al.*, 1977). This bacterium causes crown gall disease, characterized by the formation of large swellings (galls) just above soil level. The piece of DNA that is inserted into the plant genome is called the transfer DNA (T-DNA) and is carried on a plasmid called the tumor-inducing or Ti plasmid. Besides causing the host cell to proliferate to form the gall, it also induces the production and secretion of unusual sugar and amino acid derivatives that are called opines, on which the *Agrobacterium* feeds. There are several types of opines, including nopaline and octopine, produced after infection with different strains of the bacterium.

The cells of the gall are not differentiated; in other words they do not develop into the specialized cells of a normal plant. They can be removed from the plant and cultured as long as they are supplied with light and nutrients and are protected from fungal and bacterial infection. A clump of these undifferentiated cells is called a callus and callus formation can be induced in the laboratory by infecting explants (e.g., leaf pieces, stem sections or tuber discs) with *A. tumefaciens*. All the cells in the callus contain the T-DNA that originated from the bacterium.

This discovery caused great excitement because it represented a means by which the genetic make-up of a plant cell could be transformed (the process is often referred to as transformation). In 1983, groups led by Schell and Van Montagu (Ghent), Schilperoort (Leiden), Chilton and Bevan (St. Louis and Cambridge) and Fraley, Rogers and Horsch (St. Louis) showed that bacterial antibiotic resistance genes could be inserted into the   T-DNA carried on a Ti plasmid and transferred into plant cells (Bevan *et al.*, 1983; Fraley *et al.*, 1983; Herrera-Estrella *et al.*, 1983; Hoekema *et al.*, 1983). Michael Bevan in Cambridge developed the so-called binary vectors, plasmids that would replicate in both *E. coli*, in which it could be manipulated and bulked up, and *A. tumefaciens* (Bevan, 1984). Binary vectors contain the left and right T-DNA borders but none of the genes present in 'wild type' T-DNA. They are unable to induce transfer of the T-DNA into a plant cell on their own because they lack genes called virulence (*VIR*) genes that are required to do so. However, when present in *A. tumefaciens* together with another plasmid containing the *VIR* genes, the region of DNA between the T-DNA borders is transferred, carrying any genes that have been placed there in the laboratory.

Calli have to be kept under sterile conditions to prevent bacterial or fungal infection. They can be induced to form a shoot by treatment with a plant hormone; once a shoot with a stem is formed, the hormone is withdrawn and hormones produced by the shoot itself then induce root formation and a complete plantlet is formed. The plantlet can be transferred to the soil and treated like any other plant. All the cells of the plant will contain the T-DNA integrated into its own DNA, and the T-DNA and all the genes in it will be inherited in the same way as the other genes of the plant. In 1983, Tim Hall used this method to produce a sunflower plant carrying a seed protein gene from French bean (Murai *et al.*, 1983). Not only was the gene present in every cell of the plant, but also it was inherited stably and was active. The era of plant transformation had begun.

Plants that have been altered genetically in this way are referred to as transformed, transgenic, genetically engineered (GE) or GM. The term transgenic is favored by scientists but GM has been adopted most widely by non-specialists. All plant breeding, of course, involves the alteration (or modification) of plant genes, whether it is through the selection of a naturally occurring mutant, the crossing of different varieties or even related species or the artificial induction of random mutations through chemical or radiation mutagenesis. Nevertheless, the term 'genetically modified' is now used specifically to describe plants produced by the artificial insertion of a single gene or small group of genes into its DNA. Genetic transformation mediated by *A. tumefaciens* is now not the only method available to scientists; other methods, including the latest advances, will be described in Chapter 2.1.

Genetic modification has been an extremely valuable tool in plant genetic research. It has been applied, amongst other things, to the analysis of gene promoter activity, the functional characterization of regulatory elements within gene promoters, the determination of gene function, studies on metabolic pathways, elucidation of the mechanisms by which plants respond to light, disease, grazing, drought, nutrition and other stimuli, and analyses of protein structure, function and regulation. However, this book is concerned with its use in crop plant breeding.

Out of the Laboratory and into the Field; Commercial GM Crops

Genetic modification has some advantages over other techniques used in plant breeding. It allows genes to be introduced into a crop plant from any source, so technically at least the genetic resources available are huge; it is relatively precise in that single or small numbers of genes can be transferred; the safety of genes and their products can be tested extensively in the laboratory before use in a breeding program; genes can be manipulated in the laboratory before insertion into a plant to change when and where they are active, or to change the properties of the proteins that they produce. These advantages have led to genetic modification becoming established as a new tool for plant breeders to add to (not replace) those already available.

Delayed Ripening/ Increased Shelf Life

The first commercial GM plant varieties to be released were tomato varieties that had been modified to slow down the ripening process, giving them a longer shelf life, the first of which were approved for food use in the USA in 1994. A major problem in fruit production is that consumers want to buy ripe fruit but ripening is often followed quite rapidly by deterioration and decay. Fruit ripening is a complex process that brings about the softening of cell walls, sweetening and the production of compounds that impart color, flavor and aroma. The process is induced by the production of a plant hormone, ethylene. Genetic modification has been used to slow ripening or to lengthen the shelf life of ripe fruit by interfering either with ethylene production or with the processes that respond to ethylene.

The development of these varieties went hand in hand with the invention of techniques that enabled scientists to use genetic modification to reduce the activity of (or silence) a specific plant gene. The first of these techniques was the so-called antisense method first described by Don Grierson in Nottingham (reviewed by Grierson, 1996). Antisense gene silencing involves the construction of a gene in which part of the gene to be silenced is spliced in the reverse orientation downstream of a promoter sequence. The promoter may derive from the same gene, but usually it is a more powerful one. When a GM plant is produced carrying this gene, it synthesizes RNA of the reverse and complementary sequence of that produced by the target gene. This antisense RNA interferes with the accumulation of RNA from the target gene, preventing it from acting as a template for protein synthesis. The second technique for silencing target genes in plants arose from the surprising observation that one or more additional copies of all or part of a gene even in the correct orientation sometimes had the same effect as antisense gene expression when introduced into a plant by genetic modification. This method of gene silencing is called co-suppression.

Gene silencing turned out to be a natural defense mechanism employed by plants against virus infection. It involves the production of small, antisense RNAs, 25 nucleotides in length, that interfere with the processing, transport and translation of RNA molecules produced by a target gene. The third method of gene silencing by genetic modification, called RNA interference (RNAi), involves inducing the plant to synthesize a double-stranded RNA molecule derived from the target gene. This has been done by splicing part of the gene sequentially in a head-to-tail formation downstream of a promoter. Introduction of such a gene into a plant causes the production of an RNA molecule that forms a hairpin loop, which is cleaved by enzymes naturally present in plant cells into short molecules, each 23 nucleotides long.

Antisense and co-suppression were used in the first GM tomato varieties to reduce the activity of a gene encoding polygalacturonase (PG), an enzyme that contributes to cell wall softening during ripening. Two competing groups developed these varieties at approximately the same time. Calgene in the USA used an antisense technique while Zeneca in collaboration with Grierson's group used co-suppression. The Calgene product was a fresh fruit variety called 'Flavr Savr'. It was first grown on a large scale in 1996 but was not a commercial success, and was withdrawn within a year.

Zeneca chose to introduce the trait into tomatoes used for processing and this proved to be much more successful. These tomatoes have a higher solid content than conventional varieties, reducing waste and processing costs in paste production and giving a paste of thicker consistency. This product went on the market in many countries and proved very popular in the UK from its introduction in 1996 until 1999 when most retailers withdrew it in response to anti-GM hostility.

Some GM tomato varieties with delayed ripening are still on the market in the USA. They have reduced activity of the enzyme aminocyclopropane-1-carboxylic acid (ACC) synthase, which is required for ethylene synthesis. ACC has also been targeted using a gene from a bacterium, *Pseudomonas chlororaphis*, that encodes an enzyme called ACC deaminase, which breaks down ACC. A similar strategy has been adopted to break down another of the precursors of ethylene, *S*-adenosyl methionine (SAM), using a gene encoding an enzyme called SAM hydrolase. Genetic modification to delay ripening and improve post-harvest shelf life is also being used in papaya, mango, pineapple and other fruits but there are no commercial varieties available yet.

Herbicide Tolerance

Tomato is an important fruit crop but its production is dwarfed by that of the major agricultural crops; and it was the release and success of GM varieties of two of these, soybean and maize (corn), that really established genetic modification as an important tool in plant breeding. These varieties were first grown on a large scale in the USA in 1996. The traits that they carried as a result of genetic modification were herbicide tolerance (soybean) and insect resistance (maize). These traits have now been introduced into other crops and combined (stacked) in some varieties.

Herbicide-tolerant GM crops were produced to simplify and cheapen weed control using herbicides. Of course, herbicides have been used since long before the advent of genetic modification, the first modern herbicide, 2,4-dichlorophenoxyacetic acid (2,4-D), was synthesized in 1941 and released in 1946. They are now an essential part of weed control for farmers in developed countries. However, besides the obvious considerations of equipment and labor costs, as well as the cost of the chemicals themselves, herbicides pose a number of problems for farmers. Most are selective in the types of plants they kill, and a farmer has to use a particular herbicide or combination of herbicides that is tolerated by the crop being grown but kill the problem weeds. Some of these herbicides have to be applied at different times during the season, including some that have to go into the ground before planting, some that pose a health risk to farm workers and some that are persistent in the soil, making crop rotation difficult.

The most successful herbicide tolerance trait to be introduced so far enables plants to grow in the presence of a broad-range herbicide, glyphosate. The soybean variety known as RoundUp Ready, marketed by Monsanto, was the first to carry this trait (Padgette *et al.*, 1995). Glyphosate is relatively safe to use, does not persist long in the soil because it is broken down by microorganisms and is taken up through the foliage of a plant, so it is effective after the weeds have established. It is also relatively cheap. Its target is an enzyme called 5-enolpyruvoylshikimate 3-phosphate synthase (EPSPS). EPSPS catalyzes the formation of 5-enolpyruvoylshikimate 3-phosphate (EPSP) from phosphoenol-pyruvate (PEP) and shikimate 3-phosphate (S3P). This reaction is the penultimate step in the shikimate pathway (Figure 1.1.2), which results in the formation of chorismate, which in turn is required for the synthesis of many aromatic plant metabolites including the amino acids phenylalanine, tyrosine and tryptophan. The shikimate pathway is not present in animals, which have to acquire phenylalanine, tyrosine and tryptophan (referred to as essential amino acids) in their diet; this is the reason for glyphosate's low toxicity in animals. The gene that confers tolerance of the herbicide is from the soil

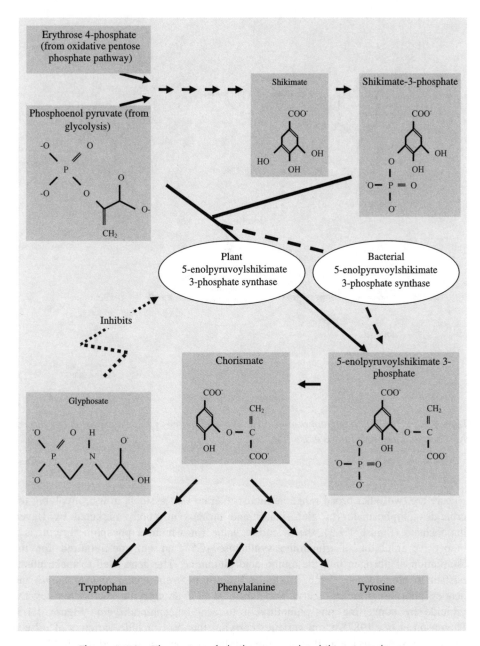

Figure 1.1.2 *The action of glyphosate on the shikimate pathway.*

bacterium *A. tumefaciens* and makes an EPSPS that is not affected by glyphosate. It has been introduced into commercial varieties of soybean, maize, cotton and oilseed rape, while glyphosate-tolerant varieties of many other crops, from wheat and sugar beet to onion, have been produced but not released yet.

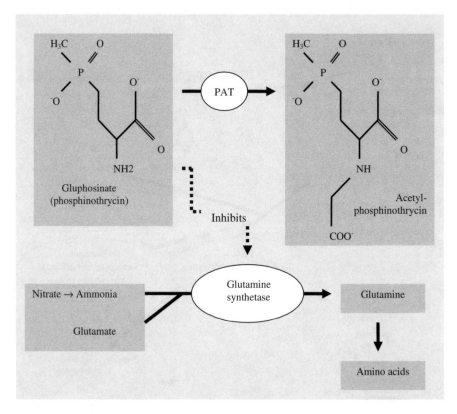

Figure 1.1.3   *The action of gluphosinate on amino acid synthesis and the detoxifying action of phosphinothricine acetyl transferase (PAT).*

There are two other broad-range herbicide-tolerant GM systems in use, involving the herbicides gluphosinate (or glufosinate) and bromoxynil, both marketed by Bayer. Gluphosinate (Figure 1.1.3), the scientific name for which is phosphinothricin, is a competitive inhibitor of glutamine synthetase (GS), an enzyme required for the assimilation of nitrogen into the amino acid glutamine. The gene used to make plants resistant to gluphosinate comes from the bacterium *Streptomyces hygroscopicus* and encodes phosphinothricine acetyl transferase (PAT), an enzyme that detoxifies the herbicide by converting phosphinothrycin to acetylphosphinothrycin (Figure 1.1.3) (Thompson *et al.*, 1987). Crop varieties carrying this trait include varieties of oilseed rape, maize, soybeans and cotton, and the trait has also been introduced into fodder beet and rice. The oilseed rape variety has been particularly successful in Canada.

The primary mode of action for bromoxynil (3,5-dibromo-4-hydroxybenzonitrile) is to inhibit photosynthesis by binding to the photosystem II complex of chloroplast membranes and blocking electron transport; tolerance is conferred by a gene isolated from the bacterium *Klebsiella pneumoniae ozanae*. This gene encodes for an enzyme called nitrilase, which converts bromoxynil into 3,5-dibromo-4-hydroxybenzoic acid, a

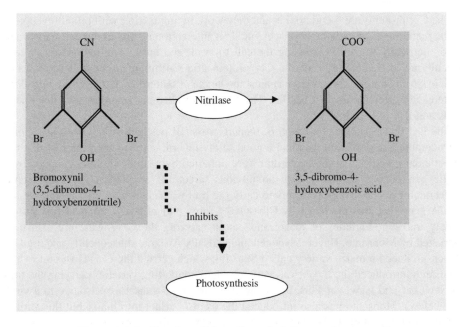

Figure 1.1.4   *The detoxifying effect of nitrilase on bromoxynil.*

non-toxic compound (Figure 1.1.4). So far this has only been used commercially in Canadian oilseed rape.

Interestingly there is a fourth broad-range herbicide tolerance trait available in commercial oilseed rape varieties in Canada. The herbicide in this case is imidazolinone and the varieties were produced by Pioneer Hi-Bred, now part of DuPont. However, the trait was produced by mutagenesis, not genetic modification.

Herbicide tolerance has now been engineered into many crop species and is undoubtedly the most successful GM trait to be used so far. In the USA in 2003, 81 % of the soybean crop, 59 % of the upland cotton and 15 % of the maize were herbicide tolerant (Benbrook, 2003). Herbicide-tolerant soybeans have been adopted even more enthusiastically in Argentina and now account for 95 % of the market, while herbicide-tolerant oilseed rape has taken 66 % of the market in Canada. This success is due to the factors such as simplified and safer weed control, reduced costs and more flexibility in crop rotation.

Insect Resistance

Organic and salad farmers have been using a pesticide based on a soil bacterium, *Bacillus thuringiensis*, for several decades. The bacterium produces a protein called the Cry (crystal) protein (often referred to now as the Bt protein); different strains of the bacterium produce different versions of the protein and these can be assigned to family groups, Cry1-40 (and counting), based on their similarity with each other. These families are further divided into subfamilies, Cry1A, B, C etc.

The Cry proteins are δ-endotoxins and they work by interacting with protein receptors in the membranes of cells in the insect gut. This interaction results in the cell membrane becoming leaky to cations, causing the cell to swell and burst. The interaction is very specific and different forms of the Cry protein affect different types of insects. Cry1 proteins, for example, are effective against the larvae of butterflies and moths, while Cry3 proteins are effective against beetles. The toxicity of all the Cry proteins to mammals, birds and fish is very low.

The fact that pesticides based on *B. thuringiensis* (Bt pesticides) had been used for a considerable length of time and had a good safety record, coupled with the fact that the insecticidal properties of the bacterium were imparted by a single protein, encoded by a single gene, made the Bt system an obvious target for adaptation for use in crop biotechnology. The first crop variety to carry the trait was a maize variety containing the *Cry1A* gene that was produced by Ciba-Geigy (now part of Syngenta) and first grown widely in 1996. Varieties of maize and cotton carrying the *Cry1* gene are also now marketed by Monsanto, Bayer, Mycogen and DeKalb. Aventis, subsequently acquired by Bayer, produced a maize variety called StarLink which carried the *Cry9C* variant, while Monsanto introduced the *Cry3A* variant into potato, marketing varieties carrying the trait as NewLeaf and NewLeaf Plus, the latter also carrying a gene for resistance to a virus (see below). Monsanto has also introduced the *Cry3B* variant into maize but this variety is not yet on the market. All these varieties are commonly referred to as Bt varieties.

The Cry1A and Cry9C proteins are effective against the European corn borer, a major pest of maize in some areas, while Cry1A is also effective against tobacco budworm, cotton bollworm and pink bollworm, three major pests of cotton. The Cry3A protein that was introduced into potato is effective against the Colorado beetle and the Cry3B protein against corn rootworm.

The benefits of using Bt varieties depend on many factors, most obviously the nature of the major insect pests in the area (not all are controlled by Bt) and the insect pressure in a given season. Bt varieties have been successful in many parts of the USA (in 2003, 29 % of the maize and 41 % of the upland cotton crop was Bt) and Bt cotton in particular is gaining ground in Australia, China, India and the Philippines. Farmers who use Bt varieties cite reduced insecticide use and/or increased yields as the major benefits (Gianessi *et al.*, 2002). A further, unexpected benefit of Bt maize varieties is that the Bt grain contains lower amounts of fungal toxins (mycotoxins) such as aflatoxin and fumicosin (Dowd, 2000).

Not all Bt varieties have been successful. NewLeaf and NewLeaf Plus potato were withdrawn in the USA due to reluctance to use them in the highly lucrative fast food industry. Farmers have adopted broad-range insecticides instead to combat the Colorado beetle. StarLink maize was an even more costly failure; it was not approved for human consumption because of doubts over the allergenicity of the Cry9C protein but, inexplicably given that maize is an outbreeding crop, the Environmental Protection Agency approved it for commercial cultivation for animal feed in 1998. Inevitably, cross-pollination occurred between StarLink and maize varieties destined for human consumption and StarLink had to be withdrawn.

Other approaches to engineering insect resistance into plants by genetic modification are being developed and tested but none have yet been used in a commercial crop variety. Many of the genes that are being used in these studies include those that encode inhibitors of digestive enzymes, including trypsin, other proteases and α-amylase, and originate from a variety of plant sources. Although they occur naturally in many crop species, some are potentially toxic or allergenic to humans and their use in crop biotechnology may not be practical.

Similar reservations are held over another group of proteins that have insecticidal properties, the plant lectins. These proteins occur naturally in many kinds of beans, but most are toxic to animals, causing the clumping of erythrocytes, reduced growth, diarrhea, interference with nutrient absorption, pathological lesions and hemorrhages in the digestive tract, amongst other symptoms. However, not all lectins are toxic to animals and one such that retained its insecticidal properties would have potential in biotechnology.

Another group of proteins that are being investigated for their use in imparting insect resistance are the chitinases, enzymes that degrade chitin. Chitin is a polysaccharide present in fungal cell walls and chitinases are believed to have evolved as a defense against fungal attack. However, chitin is also present in the exoskeleton of insects, and although naturally occurring chitinases are not present in sufficient quantities to kill a grazing insect, it might be possible to increase their level by genetic modification to the point where they would cause lesions in the midgut membrane.

A concern with any strategy for engineering insect resistance into plants is the emergence of resistant insects. In the case of Bt this would not only nullify the advantage of using Bt crops but would also render spray-on Bt pesticides useless. Indeed, concern over resistance to Bt pre-dates the development of GM crops, but the rapid increase in the use of Bt corn and cotton in the USA from 1996 onwards necessitated action. The Environmental Protection Agency devised a solution in which farmers using Bt crops would have to plant a proportion of non-GM crop as well. This provides a refuge in which insects that have developed resistance to the effects of the Bt protein do not have a selective advantage over insects that have not (in fact they have a selective disadvantage). The proportion of non-GM crop that has to be grown varies according to what other insect-resistant GM crops are being grown in a particular area, and to prevent gene flow of the trait into wild species, Bt varieties cannot be grown where wild relatives occur (in the USA this affects cotton rather than maize). So far the refuge strategy appears to have been very successful in the USA but there is doubt as to whether every country that is growing or might grow Bt crops could enforce such a policy.

Another concern over the use of insect-resistant crops is their potential effect on non-target organisms. The obvious response to such concerns is that they are likely to have a beneficial effect by reducing the use of spray-on pesticides. There is plenty of anecdotal evidence, particularly from American cotton farmers, regarding this case, but it is difficult and expensive to undertake meaningful scientific experiments to confirm or contradict this. The largest field study on the effects of GM crops on biodiversity to be conducted so far was the United Kingdom's farm-scale evaluations program, but this concerned herbicide-tolerant not insect-resistant crops. Laboratory-based experiments are much less satisfactory and can give misleading results. One example of this was a study conducted by John Losey and his team at Cornell University and published in 'Nature' in 1999 (Losey *et al.*, 1999). Losey found that caterpillars of the monarch butterfly that were forced to eat large quantities of pollen from Bt maize suffered higher mortality levels than caterpillars that were not fed the pollen. In the wild, monarch butterfly larvae eat milkweed, not maize pollen; in the experiment, pollen was spooned onto milkweed leaves so that the larvae had no choice but to eat it. Field-based studies subsequently showed that the larvae would never be exposed to such levels of maize pollen in the wild.

Similar laboratory-based experiments have shown that the survival rate of predator species such as lacewings and ladybirds can be reduced if they are fed exclusively on prey species that feed on GM insect-resistant plants. None of these results have been replicated in the field.

Virus Resistance

Virus resistance has been achieved using two methods; the first of these arose from studies on the phenomenon of cross protection, in which infection by a mild strain of a virus induces resistance to subsequent infection by a more virulent strain. Modifying a plant with a gene that encodes the viral coat protein has been found to mimic the phenomenon.

An example of the commercialization of this technology comes from the papaya industry in the Puna district of Hawaii (Ferreira *et al.*, 2002; Gonsalves, 1998). After an epidemic of papaya ringspot virus (PRSV) in the 1990s almost destroyed the industry, growers switched to a virus-resistant GM variety containing a gene that encodes a PRSV coat protein.

The second method used to impart virus resistance is to use antisense or co-suppression techniques to block the activity of viral genes when the virus infects a plant. The NewLeaf Plus potato variety discussed above, for example, carried a replicase gene from potato leaf role virus (PLRV) in combination with the Bt insect-resistance trait. This technology is being applied to many other plant virus diseases and just one example of resistance being achieved, at least under trial conditions, is with potato tuber necrotic ringspot disease (Racman *et al.*, 2001). It has tremendous potential for developing countries where losses to viral diseases are the greatest and have the most severe consequences.

Modified Oil Content

The principle components of plant oils are fatty acids and the various properties of oils from different plants are determined by their differing fatty acid contents. Many hundreds of different fatty acids have been identified in plants, with diverse food and non-food uses. Lauric acid, for example, is used in cosmetics and detergents. Palmitic acid, stearic acid and oleic acid are used in foods, while γ-linolenic acid is used in health products. Erucic acid is poisonous but is used in the manufacture of plastics and lubricating oils.

GM crop varieties with modified oil content are already on the market in the USA. Calgene, subsequently taken over by Monsanto, genetically modified an oilseed rape variety to produce high levels of lauric acid in its oil. This variety was introduced onto the market in 1995. It contains a gene from the Californian Bay plant that encodes an enzyme that causes premature termination of growing fatty acid chains. The result is an accumulation of the 12-carbon chain lauric acid to approximately 40 % of the total oil content, compared with 0.1 % in unmodified oilseed rape. Lauric acid is a detergent traditionally derived from coconut or palm oil.

The other major crop that has been modified to increase the value of its oil is soybean. The GM variety was produced by PBI, a subsidiary of DuPont; it accumulates oleic acid, an 18-carbon chain fatty acid with a single unsaturated bond (a monounsaturate) to approximately 80 % of its total oil content, compared with approximately 20 % in non-GM varieties. In conventional soybean, relatively little oleic acid accumulates because it is converted to linoleic acid, an 18-carbon chain fatty acid with two double bonds (a polyunsaturate), by an enzyme called a Δ^{12}-desaturase. Some of the linoleic acid is further desaturated to linolenic acid, a polyunsaturate with three double bonds. In the GM variety, the activity of the gene producing this enzyme is reduced so that oleic acid levels are increased while linoleic and linolenic acid levels are decreased.

Oleic acid is very stable during frying and cooking, and is less prone to oxidation than polyunsaturated fats, making it less likely to form compounds that affect flavor. The traditional method of preventing polyunsaturated fat oxidation involves hydrogenation and this runs the risk of creating *trans*-fatty acids. *Trans*-fatty acids contain double bonds in a different orientation to the *cis*-fatty acids present in plant oils. They behave like saturated fat in raising blood cholesterol, contributing to blockage of arteries. The oil produced by high-oleic acid GM soybean requires less hydrogenation and there is less risk of *trans*-fatty acid formation.

Relatively small amounts of these GM oilseed rape and soybean varieties are grown on contract, but those farmers who can get into this business benefit from a premium price for their crop.

Current Status of GM Crops

Table 1.1.1 shows the global cultivation of GM varieties of the four major crops, soybean, maize, cotton and oilseed rape, for which GM varieties have been developed and commercialized. In 2003, the International Service for the Acquisition of Agri-biotech Applications (ISAAA) (www.isaaa.org) reported that GM crops were being grown commercially in 18 countries: Argentina, Australia, Brazil, Bulgaria, Canada, China, Colombia, Germany, Honduras, India, Indonesia, Mexico, Philippines, Romania, South Africa, Spain, Uruguay and the USA. Of these, Argentina, Brazil, Canada, China and the USA dominate in terms of total area (James, 2003).

Table 1.1.1 *Global cultivation in 2003 of the four major crops for which GM varieties have been commercialized.*

Crop	Global cultivation (million hectares)		Proportion of global crop GM (%)
	All varieties	GM varieties	
Soybean	76	41.40	54
Maize	140	15.50	11
Cotton	34	7.20	21
Oilseed rape	22	3.60	16
Total	275	67.70	25

Source: Food Standards Agency.

A remarkable feature of the global status of GM crops at present is the rapid and enthusiastic uptake of GM varieties in some countries and the lack of uptake of and resistance to GM crops in other countries, notably in Europe. The only significant use of GM crops in Europe at present is the cultivation of Bt maize in Spain. At the heart of the 'problem' for plant biotechnology in Europe is the hostile attitude of European consumers. This has led legislators at the European Union and national government level to introduce legislation to control the development and marketing of GM crops and foods, apparently in the hope that strict controls would reassure consumers. These controls are discussed in detail in Chapter 3.3. Briefly, any GM crop or food derived from it has to be approved for use within the European Union by the European Commission, and approval is extremely difficult to obtain. Furthermore, any food containing GM crop material above a threshold of 0.9 % has to be labeled, while novel foods produced in any other way need not. Unfortunately this legislation has undoubtedly deterred seed companies from developing GM crops for the European market but has so far failed to reassure consumers at all.

Exactly why European consumers have been so much more fearful of GM crops than other consumers is not clear. A recent poll showed that 66 % of consumers in China, Thailand and the Philippines believed that they would benefit personally from food biotechnology during the next 5 years. A different poll in the USA found that 71 % of US consumers would be likely to choose produce that had been enhanced through biotechnology to require fewer pesticide applications. Polls in the UK and Europe continue to show much less favorable attitudes amongst consumers.

Part of the answer lies in the reluctance of Europeans to trust their governments or scientific experts. GM foods were launched in Europe shortly after the epidemic of bovine spongiform encephalopathy (BSE) in the UK cattle herd had led to one of the biggest food scares in UK history. Rightly or wrongly, consumers felt that they had been given the wrong advice by scientists and government ministers on the safety of beef. However, food 'scares' are not unique to the UK and Europe.

Another reason for consumer antipathy towards GM crops in Europe is that the debate has been dominated by anti-GM pressure groups. European consumers have been bombarded with inaccurate information, half-truths and wild 'scare' stories. Even if they do not believe the more hysterical of these stories, why should they take the risk of buying GM food products?

The first imports of GM crop products into Europe began just before Christmas in 1996, with American soybean and maize, which at that time were approximately 2 % GM. American producers refused to segregate the GM from the non-GM and there was a flurry of media activity on the issue. This died down but Greenpeace, Friends of the Earth and other campaign groups had promoted the GM issue to the top of their list of campaign priorities (the title of a Greenpeace briefing pack in February 1997 was 'The end of the world as we know it') and it was only a matter of time before it returned to the top of the news agenda. It did, thanks to two scare stories originating from the legitimate scientific literature: the work of Dr Arpad Pusztai on feeding lectin-containing GM potatoes to rats (Ewen and Pusztai, 1999) and of John Losey on the effects of feeding monarch butterfly larvae on GM corn pollen (Losey *et al.*, 1999). Pusztai's paper was subsequently debunked by the Royal Society, while monarch butterfly larvae were found never to be exposed to the levels of maize pollen used in Losey's self-described 'preliminary' study. Incidentally, the monarch butterfly prospered after the introduction of GM insect-resistant corn and cotton into large areas of the USA in 1996, although it is now threatened by habitat destruction in its Mexican wintering sites. Despite this, I have been assured several times by different people in the UK that it is extinct as a result of the introduction of GM crops.

The GM crop debate has now become entangled with campaigns against capitalism, globalization and multinational companies, and spiraled out of the control of scientists to become a potent political issue. The only factor preventing the technology being lost to Europe now is the fact that GM crops are being used widely elsewhere in the world.

Conclusions

In this chapter genetic modification is described in context of a long history of plant breeding, which had become science-based long before genetic modification was invented. Genetic modification is now an established technique in plant breeding in many parts of the world. While not being a panacea, it does hold the promise of enabling plant breeders to improve crop plants in ways that they would not be able to through other methods. GM crops now represent approximately 6 % of world agriculture, and are being used in developed and developing countries. Farmers who use them report one or more of greater convenience, greater flexibility, simpler crop rotation, reduced spending on agrochemicals, greater yields or higher prices and increased profitability at the farm gate as the benefits.

The delay in allowing plant biotechnology to develop in Europe has already damaged the European plant biotechnology industry significantly and is putting European agriculture at an increasing competitive disadvantage. Europe desperately needs politicians and the food industry to show leadership on the issue, but there is little indication that they will. Powerful, multinational pressure groups continue to call the shots on GM crops and food in Europe, and these groups remain implacably opposed to the use of the technology. Despite this, it seems inconceivable that agricultural biotechnology will not continue and develop, at least outside Europe, given the success of GM crops and their popularity with farmers in those countries where farmers are allowed to use them.

References

Avery, T., Macleod, C.M. and McCarty, M. (1944) Studies on the chemical nature of the substance inducing transformation of pneumococcal types. *Journal of Experimental Medicine* **79**, 137–1583.

Beadle, G. and Tatum, E. (1941) Genetic control of biochemical reactions in *Neurospora*. *Proceedings of the National Academy of Sciences USA* **27**, 499–506.

Benbrook, C.M. (2003) Impacts of Genetically Engineered Crops on Pesticide Use in the United States: The First Eight Years. BioTech InfoNet Technical Paper Number 6 (www.biotech-info.net/Technical_paper_6.pdf).

Bevan, M. (1984) Binary *Agrobacterium* vectors for plant transformations. *Nucleic Acids Research* **12**, 8711–8721.

Bevan, M.W., Flavell, R.B. and Chilton, M.D. (1983) A chimaeric antibiotic-resistance gene as a selectable marker for plant-cell transformation. *Nature* **304**, 184–187.

Chilton, M.D., Drummond, M.H., Merio, D.J., Sciaky, D., Montoya, A.L., Gordon, M.P. and Nester, E.W. (1977) Stable incorporation of plasmid DNA into higher plant cells: the molecular basis of crown gall tumorigenesis. *Cell* **11**, 263–271.

Cohen, S.N., Chang, A.C., Boyer, H.W. and Helling, R.B. (1973) Construction of biologically functional bacterial plasmids *in vitro*. *Proceedings of the National Academy of Sciences USA* **70**, 3240–3244.

Dowd, P.F. (2000) Indirect reduction of ear molds and associated mycotoxins in *Bacillus thuringiensis* corn under controlled and open field conditions: utility and limitations. *Journal of Economic Entomology* **93**, 1669–1679.

Ewen, S. and Pusztai, A. (1999) Effect of diets containing genetically modified potatoes expressing *Galanthus nivalis* lectin on rat small intestine. *The Lancet* **354**, 1353–1354.

Ferreira, S.A., Pitz, K.Y., Manshardt, R., Zee, F., Fitch, M. and Gonsalves D. (2002) Virus coat protein transgenic papaya provides practical control of papaya ringspot virus in Hawaii. *Plant Disease* **86**, 101–105.

Fraley, R.T., Rogers, S.G., Horsch, R.B., Sanders, P.R., Flick, J.S., Adams, S.P., Bittner, M.L., Brand, L.A., Fink, C.L., Fry, J.S., Galluppi, G.R., Goldberg, S.B., Hoffmann, N.L. and Woo, S.C. (1983) Expression of bacterial genes in plant cells. *Proceedings of the National Academy of Sciences USA* **80**, 4803–4807.

Gianessi, L.P., Silvers, C.S., Sankula, S. and Carpenter, J.E. (2002) *Plant Biotechnology: Current and Potential Impact for Improving Pest Management in US Agriculture: An Analysis of 40 Case Studies*. Washington: National Center for Food and Agricultural Policy.

Gonsalves, D. (1998) Control of papaya ringspot virus in papaya: a case study. *Annual Reviews of Phytopathology* **36**, 415–437.

Grierson, D. (1996) Silent genes and everlasting fruit and vegetables. *Nature Biotechnology* **14**, 828–829.

Herrera-Estrella, L., Depicker, A., Van Montagu, M. and Schell, J. (1983) Expression of chimaeric genes transferred into plant cells using a Ti-plasmid-derived vector. *Nature* **303**, 209–213.

Hoekema A., Hirsch P.R., Hooykaas P.J.J. and Schilperoort R.A. (1983) A binary plant vector strategy based on separation of Vir-region and T-region of the *Agrobacterium tumefaciens* Ti-plasmid. *Nature* **303**, 179–180.

Jackson, D.A., Symons, R.H. and Berg, P. (1972) Biochemical method for inserting new genetic information into DNA of Simian Virus 40: circular SV40 DNA molecules containing lambda phage genes and the galactose operon of *Escherichia coli*. *Proceedings of the National Academy of Sciences USA* **69**, 2904–2909.

James, C. (2003) *Preview: Global Status of Commercialized Transgenic Crops: 2003*. ISAAA Briefs No. 30. ISAAA: Ithaca, NY.

Lehman, I.R., Bessman, M.J., Simms, E.S. and Kornberg, A. (1958) Enzymatic synthesis of deoxyribonucleic acid. I. Preparation of substrates and partial purification of an enzyme from *Escherichia coli*. *Journal of Biological Chemistry* **233**, 163–170.

Losey, J.E., Raynor, L.S. and Carter, M.E. (1999) Transgenic pollen harms monarch larvae. *Nature* **399**, 214.

Malthus, T.R. (1798) An Essay on the Principle of Population as it Affects the Future Improvement of Society, with Remarks on the Speculations of Mr Godwin, M. Condorcet, and Other Writers. London: J. Johnson.

Maxam, A.M. and Gilbert, W. (1977) New method for sequencing DNA. *Proceedings of the National Academy of Sciences USA* **74**, 560–564.

Mullis, K.B. and Faloona, F.A. (1987) Specific synthesis of DNA *in vitro* via a polymerase-catalyzed chain-reaction. *Methods in Enzymology* **155**, 335–350.

Murai, N., Sutton, D.W., Murray, M.G., Slightom, J.L., Merlo, D.J., Reichert, N.A., Sengupta-Gopalan, C., Stock, C.A., Barker, R.F., Kemp, J.D. and Hall, T.C. (1983) Phaseolin gene from bean is expressed after transfer to sunflower *via* tumor-inducing plasmid vectors. *Science* **222**, 476–482.

Padgette, S.R., Kolacz, K.H., Delannay, X., Re, D.B., Lavallee, B.J., Tinius, C.N., Rhodes, W.K., Otero, Y.I., Barry, G.F., Eichholtz, D.A., Peschke, V.M., Nida, D.L., Taylor, N.B. and Kishore,

G.M. (1995) Development, identification and characterization of a glyphosate-tolerant soybean line. *Crop Science* **35**, 1451–1461.

Racman, D.S., McGeachy, K., Reavy, B., Strukelj, B., Zel, J. and Barker, H. (2001) Strong resistance to potato tuber necrotic ringspot disease in potato induced by transformation with coat protein gene sequences from an NTN isolate of Potato virus Y. *Annals of Applied Biology* **139**, 269–275.

Sanger, F., Nicklen, S. and Coulson, A.R. (1977) DNA sequencing with chain-terminating inhibitors. *Proceedings of the National Academy of Sciences USA* **74**, 5463–5467.

Smith, H.O. and Wilcox, K.W. (1970) A restriction enzyme from *Hemophilus influenzae*. 1. Purification and general properties. *Journal of Molecular Biology* **51**, 379–391.

Thompson, C.J., Movva, N.R., Tizard, R., Crameri, R., Davies, J.E., Lauwereys, M. and Botterman, J. (1987) Characterization of the herbicide-resistance gene *bar* from *Streptomyces hygroscopicus*. *EMBO Journal* **6**, 2519–2523.

Watson, J. and Crick, F. (1953) A structure for deoxyribose nucleic acid. *Nature* **171**, 737.

Weiss, B. and Richardson, C.C. (1967) Enzymatic breakage and joining of deoxyribonucleic acid I. Repair of single-strand breaks in DNA by an enzyme system from *Escherichia coli* infected with T4 bacteriophage. *Proceedings of the National Academy of Sciences USA* **57**, 1021–1028.

1.2

Crop Biotechnology in the United States: Experiences and Impacts

Sujatha Sankula

National Center for Food and Agricultural Policy, Washington, DC, USA

Introduction

First available for commercial planting in 1996, agricultural biotechnology applications have transformed the landscape of American agriculture by providing novel approaches to pest management. By inserting genetic material from outside a plant's normal genome, crop varieties have been developed to resist an array of pest problems. As a result, these crops have been grown without using certain pesticides necessary on conventional crops (e.g. insect-resistant or Bt crops). In some cases, the biotechnology-derived crop provides effective control of a plant pest that is not otherwise well controlled (e.g. Bt crops and virus-resistant crops). Other biotechnology-derived crops are resistant to certain herbicides that injure conventional crop varieties. Planting the biotechnology-derived herbicide-resistant crop has made it possible to use the associated herbicide, which often provides more effective and less expensive weed control.

Globally, biotechnology-derived crops were planted on 168 million acres in 2003 (James, 2003). Countries that adopted these crops in 2003 include Argentina, Australia, Brazil, Bulgaria, Canada, China, Colombia, Germany, Honduras, India, Indonesia, Mexico, Philippines, Romania, South Africa, Spain, United States and Uruguay. About 63 % of the total global commercial value of biotechnology-derived crops came from the United States alone in 2003–2004 (Runge and Ryan, 2004).

The United States has been a world leader in the field of agricultural biotechnology. American growers have planted 106 million acres or 63 % of the global total to

Plant Biotechnology. Edited by Nigel Halford.
© 2006 John Wiley & Sons, Ltd.

Table 1.2.1   *Adoption of biotechnology-derived crops.*

	Adoption (million acres)		US adoption as a percent of global total
Year	Global	US	%
1996	5	5	54
1997	25	20	63
1998	64	49	73
1999	99	72	72
2000	109	74	69
2001	131	87	68
2002	146	96	66
2003	168	106	63
2004	200	118	59

Source: James, multiple years.

biotechnology-derived crops in 2003 (James, 2003; Table 1.2.1). Recent estimates indicate that 2004 adoption of these crops increased by an additional 10 % (James, 2004). As evidenced by giant leaps in planted acreage each year, biotechnology-derived crops have been adopted with an unprecedented fervor in the United States since their first commercialization in 1996 (Table 1.2.1).

Agricultural biotechnology and its applications have triggered intense discussion in the last several years. At the heart of the debate are questions related to economic, agronomic and environmental impacts, safety and relevance of the technology. The confines of the biotechnology debate have been dynamic and ever-changing as new crops and more acres are planted to these varieties in more countries each year.

With a new technology that is planted on vast areas of the United States and one that is advancing at such a rapid pace as this, it is critical to analyze and understand the reasons driving the adoption and the impacts that stemmed from the adoption of biotechnology-derived crops. The objective of this chapter, therefore, is to examine the reasons inspiring the overwhelming adoption of biotechnology-derived crops in the United States. Also reviewed in this chapter are actual benefits realized by American growers since the adoption of these crops, which may help determine the future course of these crops. In a nutshell, the current chapter attempts to provide answers to some of the key questions that underlie the crop biotechnology debate to establish the basis to understand why American farmers have embraced biotechnology and are likely to continue to do so.

Adoption of Biotechnology-Derived Crops in the United States

Planted acreage of biotechnology-derived crops in the United States in 2003 encompassed three applications: insect resistance, virus resistance, and herbicide resistance; and six crops—canola, corn, cotton, papaya, soya bean and squash (Table 1.2.2). An overwhelming majority of these acres, about 99.98 % to be exact, consisted of large-acreage field crops (corn, cotton and soya bean) alone.

Table 1.2.2 Adoption of biotechnology-derived crops in the United States in 2003.

Crop	Trait	Adoption as % of total planted acres
Corn[a]	Insect resistant	31
Cotton[a]	Insect resistant	46
Papaya	Virus resistant	46
Squash	Virus resistant	3
Canola[a]	Herbicide resistant	75
Corn[a]	Herbicide resistant	14
Cotton[a]	Herbicide resistant	74
Soya bean	Herbicide resistant	82

[a]All planted applications included.
Source: Sankula and Blumenthal, 2004.

A total of 11 biotechnology-derived crop cultivars were planted in 2003. They included insect-resistant corn (three applications) and cotton (two applications); virus-resistant papaya and squash; and herbicide-resistant canola, corn, cotton and soya bean.

Herbicide-resistant crops were the most widely planted among all the applications (Table 1.2.2). Adoption of herbicide-resistant soya bean was the highest at 82 % followed by herbicide-resistant canola (75 %) and herbicide-resistant cotton (74 %). Since insect and disease pressure vary each year based on environmental factors, adoption of insect-resistant and virus-resistant crops varied based on the anticipated infestation level of target pests. Among insect-resistant crops, adoption was highest for cotton. Adoption of insect-resistant crops, corn in particular, is predicted to increase in future, as new varieties were commercialized in 2003 to combat important pest problems. The opening of the EU markets since 2004 to biotechnology-derived corn imports from the United States may further enhance the adoption of Bt corn in the United States.

The following discussion is focused on the pest management challenges encountered by growers in conventional crops, and how biotechnology-derived crops offer solutions to address these challenges. Also presented in the discussion below are the reasons for the adoption of individual crops along with their agronomic, economic and environmental impacts on US agriculture.

Insect-Resistant Crops

Insect-resistant crops or Bt crops were one of the first crops developed through biotechnology methods in the United States. These crops were developed to contain a gene from a soil bacterium called *Bacillus thuringiensis*, and hence the name Bt. The gene codes for protein crystals (referred to as Cry proteins) that are toxic to insect larvae of lepidoptera, diptera or coleoptera (Perlak *et al.*, 1990; Swadener, 1994). When larvae of the above insect species feed on the Bt plant, they ingest the Cry protein. Digestive enzymes specific to those insects dissolve the protein and activate a toxic component called delta-endotoxin. The endotoxin binds to certain receptors on the intestinal linings of these insects leading to the formation of pores in the membrane of the intestine. The proliferation of pores disrupts the ion balance of the intestine and causes the insect larvae to stop feeding, starve and eventually die.

Although highly toxic to certain insects, Bt is relatively harmless to humans as digestive enzymes that dissolve Cry protein crystals into their active form are absent in humans. Cry proteins from Bt have become an integral part of organic crop production in the United States for more than 40 years in view of their safety and effectiveness in controlling target insect pests.

Four insect-resistant crops were approved for commercial planting in the United States: field corn, cotton, potato and sweet corn. Though Bt potato and sweet corn were available for planting since 1996 and 1998, respectively, marketing concerns limited the adoption of these two crops. The following discussion on impacts of Bt crops, therefore, will focus on corn and cotton only.

Bt Corn

Three applications of insect-resistant corn were under commercial cultivation in the United States in 2003. They include Bt corn resistant to corn borer (trade names: YieldGard Corn Borer and Herculex I), Bt corn resistant to black cutworm and fall armyworm (trade name: Herculex I) and Bt corn resistant to rootworm (trade name: YieldGard Rootworm). YieldGard Corn Borer has been on the market since 1996 while 2003 was the first year of commercialization for Herculex I and YieldGard Rootworm.

Two genetic transformation events, Bt11 and Mon 810, each with the same endotoxin, are marketed as YieldGard Corn Borer for resistance against corn borers (Walker *et al.*, 2000). The Bt genes in YieldGard Corn Borer express Cry1A(b) and Cry1A(c) proteins that provide protection against European corn borer (ECB), south-western corn borer, fall armyworm, corn earworm and stalk borer. However, ECB is the main target pest for YieldGard Corn Borer in the United States. Acreage planted to YieldGard Corn Borer increased steadily from 8 % of the total planted acreage in 1997 to 26 % in 1999. However, adoption fell to 19 % in 2000 and 2001, before climbing up to 30 % in 2003 (Table 1.2.3). The adoption of Bt crops tends to vary as a function of predicted levels of insect infestations. Thus, adoption was lower in 2000 and 2001 due to forecasted lower insect pressure. Another reason for the drop in adoption in both years was low corn prices.

Herculex I corn expresses the Cry1F insecticidal protein, a protein different from the one expressed by the YieldGard Corn Borer corn (Cry1A) (Dow AgroSciences, 2002). The Herculex I corn offers similar protection against corn borer (European and

Table 1.2.3 *Adoption of currently planted insect-resistant Bt crops in the United States.*

	% of total acreage							
Crop	1996	1997	1998	1999	2000	2001	2002	2003
Bt corn–YieldGard Corn Borer[a]	1	8	18	26	19	19	24	30
Bt corn–Herculex I[b]	—	—	—	—	—	—	—	0.6
Bt corn–YieldGard Rootworm[b]	—	—	—	—	—	—	—	0.5
Bt cotton–Bollgard I[c]	12	15	19	25	28	37	35	46
Bt cotton–Bollgard II[b]	—	—	—	—	—	—	—	0.2

Sources: [a]USDA-NASS, multiple years; [b]Sankula and Blumenthal 2004; [c]USDA-AMS, multiple years.

southwestern) and corn earworm, and also expands protection to include black cutworm and fall armyworm (Babcock and Bing, 2001). Herculex I corn accounted for less than 1 % of the total planted corn acreage and 2 % of the total Bt corn acres planted for corn borer protection in 2003. Overall, about 98 % of Bt corn acreage in the United States in 2003 was planted to YieldGard Corn Borer varieties (Bt11 and MON 810 events together).

Biotechnology-derived rootworm-resistant/YieldGard Rootworm corn (event MON863) produces a Cry3Bb1 protein, which specifically targets the midgut lining of larval corn rootworms (Baum *et al.*, 2004). YieldGard Rootworm corn was planted on about 0.5 % of the total planted corn acreage in 2003. Seed supply was limited in 2003, due to it being an introductory year. Adoption is expected to increase rapidly in the next few years, as more seed becomes available to growers. Planting data from 2004, in fact, indicates a 10-fold increase in the acres planted to YieldGard Rootworm (Sankula and Blumenthal, 2004).

Insect Pest Problems in Corn

The most important insect pest problems in corn production in the United States are corn borer, rootworm, armyworm and cutworm. Corn borer, ECB in particular, and corn rootworm are two economically important insect pests of corn, costing growers billions of dollars each year in insecticides and lost crop yields (Mason *et al.*, 1996; Monsanto, 2003). In fact, both ECB and corn rootworm are nicknamed 'billion dollar bug problems' due to crop losses of at least 1 billion dollars each year from each of these insect pests.

The ECB damages corn in a slew of different ways with more than one generation each year. ECB larvae feed on all above-ground tissues of the corn plant and produce holes and cavities. These cavities interfere with the translocation of water and nutrients and reduce the strength of the stalk and ear shank, thereby pre-disposing the corn plants to stalk breakage and ear drop. The feeding of ECB larvae also results in reduced ear and kernel size leading to seed yield loss and/or reduced quality (Mason *et al.*, 1996). Furthermore, ECB larvae carry spores of pathogens such as ear rot fungi from the leaves of the crop to the kernels, thus causing secondary infections (Christensen and Schneider, 1950; Sobek and Munkvold, 1999).

Despite serious losses due to ECB, many growers are reluctant to use currently recommended integrated pest management approaches to control this pest. The reluc-tance stems from three main reasons. The first reason is that infestation of ECB is hard to predict and varies from year to year. Growers are usually uninterested in incurring costs on scouting to determine the profitability of insecticide applications since ECB is so unpredictable. The second reason is the intricate feeding and survival techniques of the insect itself. The corn borer larvae feed in leaf whorls after they hatch and then move into the stalks to pupate inside the stem burrows, thereby avoiding the insecticide applica-tions. In order to maintain any type of control, insecticides must be applied during the 2-3-day period between egg hatching and burrowing in the stems. The third reason why growers are unwilling to control ECB through chemical methods is due to inadequate control provided by insecticides. For example, insecticides, even if timed well, typically provide control from 60 to 95 % of first generation larvae and 40–80 % of second generation larvae.

Rootworms are the larval stage of Northern, Southern or Western corn rootworm beetles. The larval stage develops in the soil and feeds on the roots of corn. Feeding by rootworm larvae on the corn root system results in restriction of water and nutrient movement leading to yield losses and lodging of plants (Levine and Oloumi-Sadeghi, 1991; Monsanto, 2003).

Use of soil insecticides to control larval stages and insecticide sprays to control adult beetles is the most common approach to manage rootworm problems. Total expenditure for corn rootworm-targeted insecticides topped $171 million in 2000 (Alston *et al.*, 2002). However, excellent rootworm-control products have fallen by the wayside as rootworm has developed resistance to various insecticides (Levine and Oloumi-Sadeghi, 1991). In addition to insecticide use, crop rotation is another most widely used cultural method to manage corn rootworms. Since a variant of the corn rootworm became the first pest ever to develop a way of foiling crop rotations, corn growers have been seeking a breakthrough in corn rootworm management. Biotechnology was deemed to offer exciting new possibilities and was expected to mark a new era for corn rootworm management in the United States.

Cutworm is among the major soil insect problems of field corn, similar to rootworm. Many species of cutworm injure corn throughout the US, but the black cutworm is the most widespread and causes the maximum damage. Black cutworm larvae are pests of seedling corn (Minnesota Department of Agriculture's Black Cutworm Fact Sheet). On younger, small-stemmed corn plants, larvae cut the plant off at or near soil level thus reducing plant stands and necessitating replanting. If not replanted, yield losses from cutworm can be as high as 25 % (Pike, 1995).

Fall armyworm infestations are most common in the Southern corn-growing regions of the US, because of the insect's inability to overwinter in areas where the ground freezes (Sparks, 1979). Late-planted fields and later-maturing hybrids are more susceptible to damage by fall armyworm. In general, fall armyworm is not a problem on field corn, but in an outbreak year it may cause significant damage of about 10 % due to leaf and ear feeding (Anonymous, multiple years). Similar to ECB, fall armyworm control is also difficult due to its hidden feeding habit.

Corn growers employ both cultural and chemical methods for the control of cutworm and armyworm. Cultural control methods for cutworm include delaying planting 7 days or longer after seedbed preparation to lessen the numbers of cutworm surviving at crop emergence, and controlling weeds 6 weeks before planting to eliminate possible weedy hosts. Conversely, early planting of corn is the most widely recommended and effective cultural practice for lowering fall armyworm pressure in corn. Although some growers apply soil insecticides to prevent infestation of armyworm and cutworm, this practice is usually not economically justified in the United States.

Bt Cotton

Two applications of Bt cotton, Bollgard I and Bollgard II, are currently in commercial production in the United States. Bollgard I has been available to American cotton growers since 1996 (Gianessi *et al.*, 2002), while growers gained access to Bollgard II for the first time in 2003 (Mills and Shappley, 2004).

Bollgard I cotton expresses the Cry1Ac delta endotoxin. The target pests for Bollgard I cotton are tobacco budworm and pink bollworm. It also provides suppression of cotton bollworm, looper, armyworm and other minor lepidopteran cotton pests.

Bollgard II is the second generation of insect-protected cotton that offers enhanced protection against cotton bollworm, fall armyworm, beet armyworm and soya bean looper while maintaining control of tobacco budworm and pink bollworm (similar to that provided by Bollgard I). Bollgard II contains two Bt genes, *Cry1Ac* and *Cry2Ab*, compared to the single gene in its predecessor, Bollgard I. The presence of two genes in Bollgard II provides cotton growers with a broader spectrum of insect control, enhanced control of certain pests and increased defense against the development of insect resistance. Presence of the *Cry2Ab* gene in addition to the *Cry1Ac* in Bollgard II cotton provides a second, independent high insecticide dose against the key cotton pests. Therefore, Bollgard II is viewed as an important new element in the resistance management of cotton insect pests.

Since its introduction, Bollgard I acreage has increased steadily in the United States and was planted on 46 % of the total cotton acreage in 2003 (Table 1.2.3). On the other hand, Bollgard II cotton was planted on a limited basis in the introductory year of 2003. Adoption across the country represented only 0.2 % of the total planted cotton acreage. Bollgard I cotton will be phased out of commercial production in future in the United States once Bollgard II seed supply is abundant to meet the growers planting needs.

Cotton Insect Pests and Their Management Issues

Cotton is a major market for pesticide use in the United States (Gianessi and Marcelli, 1997). More than 90 % of the entire cotton acreage in the United States is treated with insecticides. The most damaging cotton pests are those that attack squares and bolls such as the cotton bollworm, tobacco budworm, pink bollworm, boll weevil and lygus bugs.

Larvae of cotton bollworm and tobacco budworm (often referred to as bollworm/budworm complex due to difficulty in identifying them in their early stages) feed on young cotton plants by devouring their apical portions thereby delaying plant growth. Feeding on mature plants leads to abnormal pollen in open flowers, squares and non-productive bolls. Damaged bolls are lost to boll rot even if not eaten completely. Yield losses due to bollworm/budworm complex are typically higher in bloom stage. Without effective control, cotton bollworm and tobacco budworm can cause yield losses of 67 % (Schwartz, 1983).

Pink bollworm is a major cotton pest in certain regions of California, Arizona, New Mexico and Texas. Pink bollworm larvae feed on the developing flowers and bolls. Larvae feed on squares in the early season without economic damage to the crop but, once bolls are present, they become the preferred food supply. Damage is caused late in the season, as developing larvae tunnel through the boll wall and then lint fiber. The burrowing activity of the larvae stains lint, destroys fibers and reduces seed weight, vitality and oil content. Pink bollworm cuts holes in boll walls as it leaves for pupation, leaving the bolls susceptible to infections from boll-rotting fungi.

Insecticides and cultural practices (such as manipulating the dates for planting, irrigation water cutoff and stalk destruction) are commonly used in the management of cotton pests. Chemical control costs for cotton bollworm, tobacco budworm and pink bollworm usually average around 60–70 % of the total pesticide costs to American cotton growers (Gianessi *et al.*, 2002). Cotton insects, bollworm/budworm complex in particular, have developed resistance to insecticides belonging to the classes of organophosphates, pyrethroids and carbamates posing serious problems in cotton pest management. Biotechnology-derived insect-resistant cotton was deemed to fill the holes left by conventional cotton pest management programs.

Impacts of Insect-Resistant Crops

Direct impacts. The most substantial impact of insect-resistant crops has been improvement in crop yields. Unlike conventional insecticides, Bt crops offered in-built, season-long and enhanced pest protection which has translated to gained yields. Another significant impact of insect-resistant crops has been the reduction in insecticide use targeted for key pest control because Bt crops eliminate the need for insecticide applications. Reduction in overall insecticide use and the number of insecticide sprays has led to a reduction in overall input costs for the adopters of Bt crops.

Indirect impacts. An indirect impact of Bt crops is the influence they exert on local target insect populations, leading to an overall reduction of insects in a field. This effect is termed 'halo effect' and has been noted in Bt corn and cotton. Volunteer crop plants have been reduced in the following season in Bt corn and cotton, as dropped ears and bolls were significantly reduced (Alstad *et al.*, 1997).

By targeting specific insects through the naturally occurring protein in the plant, Bt crops reduced the need for and use of chemical insecticides. By eliminating chemical sprays, the beneficial insects that naturally inhabit agricultural fields are maintained and even provided a secondary level of pest control. Beneficial insect-feeding bird populations were reported to be higher in numbers in Bt cotton fields compared to conventional fields (Edge *et al.*, 2001).

Another indirect benefit from insect-resistant crops relate to environment due to reduction in insecticide use. Energy use and atmospheric CO_2 are projected to decline, as insecticides require fossil fuel in their production, transportation, and application. Below are specific impacts that resulted from the planting of Bt corn and cotton.

Impacts of Corn Borer-Resistant Bt Corn

Bt corn varieties (YieldGard Corn Borer and Herculex I) provided high levels of protection against corn borer, which is equal to, if not greater than, the previously used conventional pest management options. Bt corn protection against the previously uncontrolled corn borers aided in preventing yield losses as a result of which gains have been noted in corn yields. Overall yield advantage from Bt corn ranged from 4% to 8 %, depending on ECB pressure in a particular year, in spite of planting 20 % of the fields to conventional corn refuge to prevent resistance development in insects (Marra *et al.*, 2002; Sloderbeck *et al.*, 2000).

Similar to crop yields, economic impacts due to Bt corn varied based on the level of insect infestation. When insect infestation levels were higher, Bt-corn delivered clear economic benefits (Alstad *et al.*, 1997). On the other hand, economic benefits were lower in low infestation years (Marra *et al.*, 1998). Overall, net economic returns have been higher from Bt corn compared to conventional varieties (Fernandez-Cornejo and McBride, 2000).

In addition to yield improvements, Bt corn has also led to reduction in insecticide use. Unlike other Bt crops, such as cotton, reductions in insecticide use from Bt corn were only moderate. This is due to the fact that only a minor acreage gets treated for ECB control in the United States each year (Gianessi *et al.*, 2002; Phipps and Park, 2002). Moreover, the insecticides used to control ECB are also effective against other insect pests to which Bt corn does not provide any protection. Nevertheless, surveys of corn growers in various mid-western states of the United States unanimously indicated that insecticide use decreased significantly since the planting of Bt corn hybrids (Rice, 2004).

An indirect benefit noted with Bt corn was reduction in the outburst of podworm (referred to as earworm in corn) infestations in rotational crops such as soya bean and fall vegetables. Research in the mid-Atlantic region of the United States consistently showed that corn earworm suppression in YieldGard Corn Borer corn (especially event Bt 11) was significantly better than the Herculex I corn (Dively, G., University of Maryland, personal communication, 2004). In the mid-Atlantic area, use of Bt corn hybrids reduced the recruitment of earworm moths from corn by 90 % or more and delayed emergence by 2 weeks. Thus, the risks of podworm outbreaks in soya bean and several vegetable crops during the fall were significantly reduced. This has resulted in substantial indirect savings to farmers.

Adoption of Bt corn has also led to reduced incidence of ear rot and stalk rot diseases in the United States. The wounds made on the plant by corn borer act as open infection sites for fungi and, in some cases, corn borer larvae themselves act as vectors of pathogenic fungi such as *Fusarium* species by carrying the fungal spores directly into the wounds. The primary importance of the above diseases is their association with mycotoxins, particularly the fumonisins. Fumonisins are a group of mycotoxins that can be fatal to livestock and are probable human carcinogens (Munkvold and Desjardin, 1997). The importance of fumonisins in human health is still a subject of debate, but there is evidence that they have some impact on cancer incidence in some parts of the world (Marasas, 1995). Multi-year studies showed that kernel feeding by insects, extent of ear rot infestation and fumonisin levels in Bt corn were significantly lower than in conventional corn (Munkvold *et al.*, 1999).

Impacts of Rootworm-Resistant Bt Corn

Excellent root protection was noted in university trials with Bt hybrids. The consistency of Bt corn hybrids was 100 %, whereas insecticide use was only 63 % consistent in protecting roots against economic damage (Rice, 2004). However, information is sparse on yield response of rootworm-resistant corn hybrids as 2003 was the first field year. Most of the field research with Bt corn hybrids in 2003 has focused on root injury. However, available information indicates that Bt hybrids yielded 1.5 %–4.5 % higher relative to a soil insecticide treatment (Lauer, 2004; Rice, 2004).

Unlike Bt corn resistant to corn borer, pesticide use reductions due to rootworm-resistant corn would be higher as predictability of insect infestations, overall insecticide use and total number of acres treated for rootworm control are higher. Less than 5 % of corn acres get treated for ECB control in US each year, while corn growers treat more than 10 % of the acres with insecticides for rootworm control (Alston *et al.*, 2002). A 75 % reduction in insecticide use has been predicted with the adoption of rootworm-resistant corn in the United States (Rice, 2004). With planted acres of only 0.5 % of the total, rootworm-resistant Bt corn hybrids reduced insecticide use by 0.7 lb active ingredient per acre (ai/A) or 225 000 pounds and $4.4 million in insecticide costs in 2003 (Sankula and Blumenthal, 2004). An ex-ante analysis of the impacts of rootworm-resistant corn based on acres treated in 2000 reported $58 million savings due to reduced use of insecticides (Alston *et al.*, 2002).

Similar to corn borer-resistant corn, rootworm-resistant corn will also decrease the incidence of stalk rot in corn due to reduced feeding of rootworm larvae on corn roots. Other intangible benefits associated with the use of rootworm-resistant corn would be safety of reduced handling of insecticides, better and consistent pest control, time, equipment and labor savings (Rice, 2004).

Impacts of Bt Corn Resistant to Cutworm and Armyworm

Due to full season and full plant expression of Cry1F protein, the larvae of both cutworm and armyworm are exposed to Bt toxin at all stages in their life cycle. Consequently, yield losses have been significantly reduced in Bt corn. Based on corn acreage treated for cutworm control with insecticides in 2003 and planted Bt corn acreage of less than 1 %, it was estimated that net economic impact of planting insect-resistant varieties was $9.6 million due to reduced yield losses and insecticide use (Sankula and Blumenthal, 2004). Since 2003 was the first year of commercial production of Herculex I corn and since fall armyworm is a sporadic pest, impact information is sparse.

Impacts of Bt Cotton

Bt cotton provided the best arsenal against the key lepidopteran pest problems. It served as a valuable alternative pest management tool in regions where budworms had become resistant to conventional pyrethroid insecticides. Since Bt cotton productions fit well with boll weevil eradication programs in cotton, the adoption of Bt cotton has been high in areas where boll weevil eradication programs are used. Bt cotton functioned as an insurance against unchecked budworm and bollworm populations in these areas because of the lack of natural predators (Gianessi *et al.*, 2002).

Overall, Bt cotton plantings have led to highest per acre grower benefits and largest reduction in insecticide use among all the insect-resistant crops. Number of acres treated, applications and lost production have all declined significantly. Numerous studies have confirmed that, in general, Bt cotton conferred a significant economic advantage relative to conventional technologies, due to improved yields and reduced insecticide use (Bryant *et al.*, 2000; Cooke *et al.*, 2001; Gianessi and Carpenter, 1999; Mullins and Mills, 1999). Yield advantage for Bt cotton generally ranged from 7 % to 12 % (Bryant *et al.*, 1999; Stark, 1997).

Cotton growers applied fewer insecticide treatments, two to eight treatments fewer in certain states (Fernandez-Cornejo and McBride, 2000). Averaged across various cotton-growing states, insecticide applications were reduced by at least two in 2003, which translated to time, labor and energy savings for cotton growers (Mullins and Hudson, 2004; Sankula and Blumenthal, 2004). A 1999 estimate by EPA (2001) indicated a reduction of 1.6 million pounds of insecticide use due to Bt cotton. By 2003, insecticide use was further reduced by another 50 % (Sankula and Blumenthal, 2004).

Higher yields coupled with lowered insecticide use in cotton production have led to improved grower returns, in spite of associated technology fees. Grower benefits were reported to be 175 % higher in 1999 compared to 1996 due to Bt cotton (EPA, 2001). In 2003, Bt cotton (Bollgard I) delivered net economic benefits worth $190 million in the United States (Sankula and Blumenthal, 2004).

Local ecosystems were impacted favorably since the planting of biotechnology-derived Bt cotton. Research has shown that beneficial insect-eating bird populations flock more to Bt cotton fields as opposed to conventional fields (Edge *et al.*, 2001).

The need for supplemental remedial insecticide applications to fully control pests such as cotton bollworm has been a minor drawback for Bollgard I cotton. Bollgard I cotton has been consistently efficacious on tobacco budworm and pink bollworm. However, Bollgard I provides only suppression of cotton bollworm, looper, armyworm and other minor lepidopteran cotton pests. As a result, growers may have to spray for these pest problems under certain circumstances, especially during bloom stage.

In 2003, about 74 % of the US cotton crop was infested with the bollworm/budworm pest complex of which 86 % were bollworms (Williams, 2003). Approximately 52 % of the Bt cotton acreage (Bollgard I) was sprayed with insecticide applications to control bollworms in 2003 (Williams 2003). Number of insecticide applications for bollworm control in Bt cotton averaged 0.54 per acre in 2003.

Evidence indicates that Bollgard II cotton enhanced insecticidal activity against pests on which Bollgard I was weakest. The enhanced control with Bollgard II of the principal cotton bollworm/budworm complex and control of secondary lepidopteran insect pests (such as the armyworm and looper) has resulted in further yield increases and reductions in insecticide use in the United States. Multi-state trials in 2003 indicated that lint yields were improved by 26 %, returns were 37 % higher and insecticide treatments were 83 % lower with Bollgard II cotton compared to Bollgard I (Mullins and Hudson 2004). In comparison to the conventional non-Bt cotton, Bollgard II cotton averaged 3.6 fewer insecticide applications, $17 less insecticide costs, 74 pounds more lint yields and $40 higher economic returns in 2003 (Mullins and Hudson, 2004).

Insect-Resistant Crops and Refuge Requirements

To slow the evolution of insect adaptation to Bt toxin, which may render an otherwise valuable technology useless, the Environmental Protection Agency mandated in 2000 that Bt corn and cotton growers set aside some acres where non-Bt crop will be grown to serve as a 'refuge'. The refuge fields will support populations of insects not exposed to the Bt toxin and will help prevent resistance development when they cross-breed with insects in the Bt fields.

Mandatory refuge requirements for corn obligate that growers plant at least 20 % of the area with conventional varieties. This means that the maximum amount of Bt corn on any farm would be 80 % of the corn acreage planted. On the other hand, three distinct refuge designs were available for cotton growers since 2001: a 20 % sprayed refuge option, 5 % unsprayed refuge option outside Bt corn and 5 % embedded refuge option. The probability of resistance development is not a major issue in cotton at present as Bollgard II cotton is available commercially.

Companies that developed Bt crops (Dow AgroSciences; Pioneer Hi-Bred International, Inc; Monsanto Company; and Syngenta Seeds, Inc) are engaged in an aggressive and broad-based awareness campaign aimed at ensuring that growers understand resistance management obligations in Bt crops. Some of these efforts include informative collateral material, a web-based training module, on-farm visits and other education- and compliance-based activities. The Compliance Assurance Program (CAP), introduced by the seed industry in 2002, has increased the awareness of the growers of insect resistance management (IRM) strategies in Bt crops. Under the CAP, growers who do not meet their IRM refuge requirements in two consecutive years can be denied access to Bt varieties in the third year by their Bt corn seed provider.

Consequently, grower compliance to refuge establishment and management has been high in the United States. The National Corn Growers Association reported that 92 % of the nation's growers met the IRM requirements for Bt corn in 2003, which was 87 % higher compliance, than in 2000 (National Corn Growers Association's Press Release, 2004). A survey of 550 Bt corn growers in the Corn Belt and Cotton Belt during the 2004 growing season by Agricultural Biotechnology Stewardship Technical Committee highlighted that 91 % met regulatory requirements for refuge size while 96 % met refuge distance requirements. An internal survey by the Environmental Protection Agency indicated that 77 % of cotton farmers were in compliance with the refuge requirements for Bt cotton in 2002.

Virus-Resistant Crops

Biotechnology-derived virus-resistant crops were developed to express genes derived from the pathogenic virus itself. Plants that express these genes interfere with the basic life functions of the virus. Use of coat protein genes is the most common application of pathogen-derived resistance. The specific mechanism for coat protein-mediated resistance is not clearly elucidated, but evidence indicates interruption of critical processes like replication, post-transcriptional gene expression, virion coating/uncoating and intercellular transport (Beachy *et al.*, 1990; Kaniewski and Lawson, 1998). The expression of the coat protein gene gives protection against infection of the virus from which the gene is derived, and possibly other viruses as well (Di *et al.*, 1996).

Adoption of Virus-Resistant Crops

Two virus-resistant crops, papaya and squash, have been planted in the United States since 1998. Of the two, biotechnology-derived papaya is a dramatic illustration of biotechnology success in the United States. Virus-resistant papaya acreage increased steadily since its first commercial planting and was planted on 46 % of the total acreage by 2003 (Table 1.2.4). Since biotechnology-derived papaya is not approved for human consumption in export markets such as Japan and the European Union, American papaya

Table 1.2.4 Adoption of biotechnology-derived virus-resistant papaya in the United States.

Year	Planted papaya acreage	VR papaya acreage as a % of total planted acres	VR papaya acres
	Acres	%	Acres
1999	3205	37	1186
2000	2775	42	1166
2001	2720	37	1060
2002	2145	44	944
2003	2380	46	1095

Source: Hawaii Agricultural Statistics Service, multiple years.

growers are forced to plant conventional varieties to meet the trade requirements. Adoption of virus-resistant papaya, however, will grow significantly once export markets approve the shipments of biotechnology-derived varieties.

Biotechnology-derived transgenic squash production in the United States is concentrated mostly in Georgia followed by Florida. Planted acreage of virus-resistant squash in these two states accounted for 3 % of the total US acreage in 2003. The adoption of virus-resistant squash has been low and stagnant for several reasons. Biotechnology-derived squash does not provide protection against papaya ringspot virus, a virus of significance in squash production. Lack of availability of the virus-resistance trait in the myriad squash varieties that are currently under cultivation in the United States is a second factor that limited the widespread adoption of biotechnology-derived varieties. In the last few years, several traditionally bred varieties with tolerance to key virus problems have been introduced. As a result, these varieties are being used on more acres than the biotechnology-derived varieties. The high seed costs of biotechnology-derived varieties further hindered the adoption of transgenic squash. Seed costs of biotechnology-derived squash varieties are two to four times higher than susceptible conventional varieties. In contrast, traditionally bred varieties that have some virus tolerance are only 50 % more costly than the susceptible ones.

Virus Problems in Conventional Papaya and Squash

Papaya ring spot virus is the most important disease that affects papaya. Papaya production in the United States, concentrated mainly in Hawaii, was declining in the 1990s due to epidemics of papaya ringspot virus. Hawaiian farmers relied on surveying and rouging the infected trees to keep the virus from spreading to other fields. This process of identification and destroying of infected trees turned out to be expensive and ineffective, and led to a collapse of papaya industry in Hawaii.

Four viruses affect summer squash production in the United States. They are the zucchini mosaic virus, watermelon mosaic virus 2, cucumber mosaic virus and the papaya ringspot virus. These viral diseases can cause devastating losses to squash due to leaf mottling and yellowing, stunted plant growth and deformed fruit. Growers often use foliar applications of petroleum oil to create a barrier between the aphids that transmit the virus and the plant to prevent the attachment of the virus when aphids probe infected plants with their stylets. To be effective, oil applications must be made before aphids appear and thus are applied in the absence of the virus in some years.

The most conventional way to manage viruses is to limit their transmission by controlling insect vectors and by planting resistant varieties developed through conventional tactics. Control of viruses through insect vectors is rather difficult for two reasons. The first is that virus transmission through insects is almost immediate, which makes insecticide applications futile; the other is that the secondary hosts that harbor the viruses do not show any symptoms of the virus. Natural virus resistance, on the other hand, is not available in all crops and the protection offered is highly variable. While use of conventionally developed virus-resistant squash varieties has yielded some success, resistant varieties do not exist for papaya.

Impacts of Virus-Resistant Crops

Papaya. The papaya industry owes its continued existence in Hawaii to biotechnology. Virus-resistant papaya has facilitated strategic planting of conventional varieties in areas that were previously infested with the ringspot virus, and also planting of conventional and biotechnology-derived varieties in close proximity to each other (Gonsalves *et al.*, 2004).

Papaya production, which had fallen 45 % from the early 1990s to 1998, rebounded by 44 % by 2003 (Hawaii Agricultural Statistics Service, 2004). Experts credit this increase in papaya production to planting of virus-resistant varieties. Biotechnology-derived papaya, overall, has restored the economic viability of an industry that was on the verge of extinction.

Squash. American growers have planted biotechnology-derived squash varieties as an insurance against yield losses from fall plantings, during which time infestations are more prevalent. In-built virus protection in squash has led to an increase in the number of harvests, higher yield per harvest and higher quality fruit (Fuchs *et al.*, 1998; Schultheis and Walters, 1998). Virus-resistant squash did not reduce insecticide use because the chemicals that control aphids also control white flies. Insecticide applications need to be made to biotechnology-derived squash to prevent whitefly infestations.

Herbicide-Resistant Crops

Herbicide-resistant crops that were planted on a commercial scale in 2003 in the United States include bromoxynil-resistant cotton (BXN), glufosinate-resistant corn and canola, and glyphosate-resistant canola, corn, cotton and soya bean. BXN was introduced in 1995 while glufosinate-resistant corn and canola were commercialized in 1997 and 1999, respectively. Glyphosate-resistant canola, corn, cotton and soybean have been available in the United States since 1999, 1998, 1997 and 1996, respectively. Glyphosate-resistant sugar beet has been available for commercial planting since 1999; however, adoption has been non-existent due to marketing issues.

Herbicide-resistant crops have experienced the most widely used application of agricultural biotechnology in the United States. Adoption has increased steadily since

Table 1.2.5 *Adoption of herbicide-resistant crops in the United States.*

Crop	Resistance to	1995	1996	1997	1998	1999	2000	2001	2002	2003
Canola[a]	Glufosinate/ Glyphosate	—	—	—	—	31	47	~50	~70	75
Corn[b]	Glufosinate/ Glyphosate	—	—	—	9	8	6	7	11	14
Cotton[b,c]	Bromoxynil	<1	<1	1	6	8	7	4	2	2
Cotton[b,c]	Glyphosate	—	—	4	21	37	54	55	61	72
Soya bean[d,b]	Glyphosate	—	2	13	37	47	54	68	75	82

Sources: [a]Coleman, B., North Dakota Canola Growers' Association, personal communication, 2005; [b]USDA-NASS, multiple years; [c]USDA-AMS, multiple years; [d]Marshall, K., Monsanto, personal communication, 2000.

they were first commercialized. While soya bean has been the most predominantly planted herbicide-resistant crop, corn has been adopted at a slightly slower pace. In 2003, herbicide-resistant canola, corn, cotton and soybean were planted on 75%, 14%, 74% and 82% of the total planted acreage respectively (Table 1.2.5).

Of the 75% of the canola acres planted to biotechnology-derived varieties in 2003, 55% were planted to glyphosate-resistant varieties while the rest comprised glufosinate-resistant varieties. Acreage planted to glufosinate-resistant canola increased significantly in 2002 and 2003 due to awareness of and increased knowledge about the trait, availability of the trait in high-yielding varieties and also due to a greater choice of varieties.

Both glyphosate- and glufosinate-resistant corn varieties were planted in 2003 in the United States. However, adoption of glufosinate-resistant corn has been low in several states and insignificant in some states compared to glyphosate-resistant corn. Competitive pricing of glyphosate, good seed distribution systems and effectiveness of glyphosate in controlling weeds were the major driving forces behind the rapid increase in the adoption of glyphosate-resistant corn compared to glufosinate-resistant corn.

The lack of approval for biotechnology-derived corn imports into the European Union and the lack of availability of the herbicide-resistance trait in the varieties adapted for corn-growing regions have weakened the adoption of herbicide-resistant corn. However, with the end of the 5-year moratorium and the approval of imports of herbicide-resistant corn into the European Union in 2004, herbicide-resistant corn adoption is projected to increase significantly in the next few years.

About 97% of the herbicide-resistant cotton acreage in the United States in 2003 was planted to glyphosate-resistant varieties. Adoption of BXN was only 2% in the US in 2003. Deficiencies associated with the BXN system, such as the inability of bromoxynil to control certain key broadleaf weeds (e.g., sicklepod) and its lack of activity on grass weeds, were the main contributing factors for the poor and declining adoption of BXN cotton. Restrictions placed by the Environmental Protection Agency on bromoxynil and lack of availability of stacked varieties (herbicide- and insect-resistance together) further limited its adoption. A weed management system that was available to growers for the first time in 2004 was glufosinate-resistant cotton. Planted acreage of glufosinate-resistant cotton has been limited in the introductory year of 2004.

The adoption of herbicide-resistant soya bean is the most rapid case of technology diffusion in the history of agriculture. Available for planting since 1996, adoption increased briskly each year and reached 85 % in 2004 (Fernandez-Cornejo, 2004). This is the highest adoption rate for any biotechnology-derived crop in the world.

Weed Management Deficiencies in Conventional Crops

Common weed management programs in conventional crops incorporate the use of many herbicides, targeted at a specific weed or groups of weeds. Herbicides are usually applied either as pre-plant incorporated (PPI) treatments prior to planting, pre-emergence (PRE) applications at planting or before crop emergence, post-emergence (POST) applications after the crop has emerged or a combination of PRE followed by POST applications.

Several factors limit the success of PRE and PPI herbicide applications. Both PRE and PPI applications involve guesswork since herbicides are applied anticipating the weed species that may emerge. In addition, soil-applied PRE herbicides rely greatly upon rainfall resulting in poor weed control under extremely low or high rainfall. As a result of the unpredictable nature of PPI and PRE treatments, there is an increasing trend toward total POST programs. In a POST program, herbicides are chosen based on weed species that are in the field, taking into consideration the limits of crop and weed growth (Carpenter *et al.*, 2002). Herbicide-resistant crops facilitate the use of POST herbicides such as glyphosate and glufosinate.

In conventional crops, control of annual and perennial broadleaf and grass weeds cannot be achieved with one herbicide application. In most cases, a tank-mix partner is needed for complete weed control. Conventional crops such as cotton, on average, receive three herbicide applications consisting of three active ingredients along with one to three cultivations. Herbicide-resistant crops, on the other hand, eliminate the need for multiple herbicides and applications as complete weed control is achieved with a single application of one herbicide alone. This simplicity in weed management is a major reason why growers have embraced herbicide-resistant crops (Carpenter *et al.*, 2002).

The flexibility associated with herbicide-resistant crops in managing weeds is another reason for their widespread adoption. Weed management programs in herbicide-resistant crops are less restricted by growth stage and size of crop and weed species. Herbicides used in conjunction with herbicide-resistant crops can be applied at later crop growth stages as opposed to conventional herbicides. For example, glyphosate can be used up to the five-leaf stage on cotton, six-leaf stage on canola and up to flowering on the soya bean (Carpenter *et al.*, 2002). These application windows are much wider compared to herbicides used in conventional crops. The maximum height up to which glyphosate applications can be made to control two corn weeds, foxtail and fall panicum, is 6 inches, as opposed to 3 inches with the premix of conventional herbicides atrazine/rimsulfuron/ nicosulfuron (trade name: Basis Gold) (Curran *et al.*, 1999).

Prior to herbicide-resistant crops, the potential for crop injury was a huge issue with herbicides used in conventional crops, especially in soya bean and cotton. Herbicide-resistant crops allayed the fear of crop injury, as damage to crop is non-existent with herbicides used in these crops. Moreover, unlike conventional crops, herbicide carryover is not a concern in herbicide-resistant crops. Herbicides used in biotechnology-derived crops such as glyphosate and glufosinate have no residual activity, and therefore crops can be planted without waiting for the herbicide residues to break down.

Impacts of Herbicide-Resistant Crops

Herbicide-resistant crops have been adopted very enthusiastically in the United States as these crops have simplified weed management, thereby increasing the overall crop production efficiency of growers and reducing reliance on intense herbicide use. Impacts that resulted from the adoption of individual crops are detailed below.

Canola. Weed management is extremely critical in canola, more so than in other crops, for three reasons: its slow initial growth leading to its poor ability to compete with weeds; the narrow row plantings which prevent the use of cultivations thereby increasing dependence on the use of herbicides; quality concerns due to weed seed contamination leading to severe price cuts and oftentimes market rejection. Though conventional herbicides provide effective control of target weeds, weed management is often challenging in canola due to crop injury from herbicides, fewer and expensive options for perennial weed management and weed resistance to conventional herbicides.

Growers have embraced herbicide-resistant canola varieties due to increased ease in controlling problem weeds such as wild mustard, kochia and Canada thistle (Jenks, B., North Dakota State University, personal communication). Control of these weeds is costly with the available conventional options and necessitates the use of numerous herbicides. Both glyphosate- and glufosinate-resistant canola varieties provided weed control equivalent to that achieved with conventional herbicides but with the use of one or two herbicides and one or two applications only, and at a reduced rate and cheaper cost. The one-pass chemical operation, characteristic of herbicide-resistant canola systems, increased the simplicity in managing weeds and also eliminated the cost of additional machine operations over the field. Overall, estimated grower savings due to herbicide-resistant canola include a reduction in herbicide use of 0.7 pounds and weed management costs of $15 per acre (Johnson *et al.*, 2000). A most recent impact estimate indicated that herbicide use was reduced by 0.4 lb ai/A and grower cost savings improved by $11/A due to decreased herbicide use and applications (Sankula and Blumenthal, 2004).

Another benefit to using herbicide-resistant canola is the preservation of moisture on the soil surface due to the elimination of mechanical cultivations for herbicide incorporation and supplemental weed control. Moisture creates a good seedbed that is difficult to achieve with conventional weed management strategies.

Corn. The niche for herbicide-resistant corn was in the control of specific difficult-to-control problem weeds such as Johnsongrass, Bermudagrass, crabgrass, burcucumber, bindweed and herbicide-resistant weeds such as kochia and pigweed for which conventional weed control programs have limitations. Besides being cost-effective, weed management programs in herbicide-resistant corn enhanced flexibility in timing herbicide applications because glyphosate and glufosinate can be applied at later growth stages without injuring the crop (Brunoehler, 1999).

Herbicide-resistant corn proved to be an excellent choice in a dryland production system, where crop competes poorly with weeds and weed control from soil-applied herbicides is dependent upon timely rainfall events that are needed for herbicide incorporation. Use of glyphosate or glufosinate in herbicide-resistant corn eliminated the need for herbicide activation as these herbicides are applied POST and do not need activation to be effective. Furthermore, herbicide-resistant corn alleviated carryover concerns in rotational crops such as alfalfa and vegetables.

The weed management program in conventional corn is typically based on PRE soil applications of atrazine, a chemical that was used on two-thirds of the United States corn acreage in 1998 (USDA-NASS, 1999). Herbicide programs in biotechnology-derived corn replaced the previously used herbicide programs in conventional corn in two ways: by facilitating the use of reduced rates of soil-applied PRE herbicides followed by a POST application of glyphosate or glufosinate for problem weed management, or substitution of the conventional herbicides with a total POST program with glyphosate or glufosinate. This switch led to overall reduction in herbicide use of 0.96 lb/A and weed control costs of $10/A in 2003 (Sankula and Blumenthal, 2004). Earlier estimates suggested a 30 % or 0.7 lb ai/A reduction in herbicide use in herbicide-resistant corn (Phipps and Park, 2002). Overall number of herbicide applications has however remained the same due to substitutions of herbicides.

Cotton. Weed management in conventional cotton is often complicated due to its slow early growth and sensitivity to herbicides, resulting in limited options when compared with other row crops. As a result, conventional cotton requires a combination of mechanical, manual and chemical control methods.

Biotechnology-derived herbicide-resistant varieties have led to a new era in cotton weed management. Weed management has become simpler since the introduction of herbicide-resistant cotton as one herbicide and few herbicide applications replaced a multitude of control methods and herbicide applications. Another major advantage of herbicide-resistant cotton was the increased ease in applying the POST over the top herbicides with excellent crop safety.

Crop safety will further be enhanced with a second-generation glyphosate-resistant cotton called Roundup Ready Flex cotton that is due for commercial release in the next few years. The first generation of glyphosate-resistant cotton provided very good vegetative tolerance but marginal reproductive resistance. Thus, any glyphosate applications beyond the five-leaf stage caused crop loss if the application was not directed. The use of Roundup Ready Flex cotton will extend the window of application for glyphosate and allow the use of its POST applications beyond the five-leaf stage, with the additional benefit of higher use rates. This will provide growers additional flexibility when timely herbicide application is delayed by environmental conditions. The second generation of glyphosate-resistant cotton may further increase grower efficiency as herbicide applications can be combined with other applications of insecticide, plant growth regulators and other topical applications. Herbicide-resistant cotton acreage is expected to further increase when Roundup Ready Flex cotton is commercially available.

Herbicide-use patterns, herbicide rates and number of applications have changed due to herbicide-resistant cotton (glyphosate-resistant mainly) in the United States. Glyphosate use increased from 8 % of the acres treated in 1994 to 69 % in 2003 (USDA-NASS, 2004). Conversely, there was a decrease in the use of herbicides used in conventional cotton such as MSMA, trifluralin and fluometuron. Prior to herbicide-resistant cotton, growers typically used 2.7 herbicide-active ingredients for weed control (USDA-ERS, 1997), while number of active ingredients has come down by more than 50 % in glyphosate-resistant cotton. Average herbicide use in cotton, based on NASS surveys, has decreased by 5 % in 2003 compared to 1994. Altogether, cotton growers reduced herbicide use by 9.7 million pounds in 2003 (Sankula and Blumenthal, 2004).

Cotton growers have adopted herbicide-resistant varieties as a way to reduce production costs also. Production costs have decreased as growers made fewer trips across fields applying herbicides, made fewer cultivation trips and performed fewer handweeding operations. The number of herbicide applications declined by 1.4, tillage operations by 1.7 and handweeding hours by 2.8 per acre in 2003 due to planting of herbicide-resistant cotton varieties (Sankula and Blumenthal, 2004). In general, reduction in production costs delivered net returns of $221 million to cotton growers in 2003.

Soya Bean. Weed management in soya bean production has changed radically since the widespread adoption of glyphosate-resistant soya bean. It has become simpler, more flexible and less costly with the use of herbicide-resistant varieties. Simplicity in weed management has resulted from the replacement of multiple treatments of conventional herbicides with one to two treatments of a single broad-spectrum herbicide. Furthermore, crop injury, a common occurrence with conventional herbicides, ceased to be an issue in glyphosate-resistant soya bean. The effectiveness of glyphosate in controlling weeds such as waterhemp that has developed resistance to many conventional soya bean herbicides has been another driving factor for the rapid adoption of glyphosate-resistant soya bean (Johnson and Smeda, 2001).

A major impact of glyphosate-resistant soya bean has been a significant change in herbicide use patterns in the United States. Since the introduction of glyphosate-resistant soya bean, the use of most conventional herbicides has decreased, while the use of glyphosate has increased. Glyphosate was used on 20 % of soya bean acreage in 1995, as a burndown herbicide prior to planting or as a spot treatment during the growing season. By 2002, glyphosate-treated acreage had increased to 78 % of soya bean acreage (USDA-NASS, 2003). Imazethapyr was the most commonly used herbicide in soya bean, applied to 44 % of acreage in 1995. Since the commercialization of glyphosate-resistant soya bean in 1996, however, imazethapyr usage steadily decreased to just 9 % of acreage by 2002. Similar trends were also noted with the other soya bean herbicides such as pendimethalin (9 %), trifluralin (7 %) and chlorimuron (6 %).

A second major impact of glyphosate-resistant soya bean has been the reduction in the amount of herbicide applications per acre. Evidence indicates that the number of herbicide applications in herbicide-resistant soya bean decreased by 22 million or 13 %, in spite of a 19 % increase in soya bean acreage in 2000 (Carpenter and Gianessi, 2002). The reduction in the number of trips across the field led to energy and fuel savings, increased the efficiency of the farming operations and reduced soil compaction problems.

Herbicide use information on NASS database indicates that herbicide application rates in soya bean increased slightly from 1995 to 2002 (USDA-NASS, 2003). Substitution with glyphosate of low-use rate herbicides such as imazethapyr increased soya bean acreage, and increased no-till production practices might have influenced this slight increase in herbicide application rates.

Another reason for the rapid expansion of herbicide-resistant soya bean acreage in the US is the lower cost associated with the weed management programs in glyphosate-resistant soya bean. Since glyphosate, the herbicide associated with herbicide-resistant soya bean, is competitively priced and necessitates fewer applications, soya bean weed management has become cheaper than the conventional alternatives. Based on technology charges and herbicide application costs in 2003, per acre weed management costs in glyphosate-resistant soya bean were \$20 lower than conventional soya bean (Sankula and Blumenthal, 2004).

Several herbicides commonly used in conventional soya bean production have ground water advisories, including alachlor, metolachlor and metribuzin. Although the amount of herbicide that runs off a field is normally small ($<2\%$ in terms of total amount applied), the yearly flow-weighted average herbicide concentrations frequently exceed drinking water standards (Shipitalo and Malone, 2000). As a result, there is growing pressure to reduce the use of PRE herbicides in soya bean. Herbicides used POST such as glyphosate are less subject to transport in runoff since they are foliar-applied. The use of ground water-polluting herbicides was reduced by 60% or 17 million pounds since the introduction of herbicide-resistant soya bean (Krueger, 2001).

Herbicide-Resistant Crops and Crop Yields

Weed management programs in conventional crops are usually intensive, involving multiple applications of several herbicides in combination with two or more cultivations to obtain good weed control needed to prevent yield losses. As a result, no differences were reported in crop yields between conventional and herbicide-resistant crops (Fernandez-Cornejo, 2004).

Yield is a complex phenomenon governed by environment and genetics. Herbicide-resistant crops differ from conventional crops in the introduced trait alone, and thus they have no inherent ability to increase crop yields. However, the herbicide-resistant trait enables crop plants to withstand non-selective herbicide applications, which are more effective than individual conventional herbicides in controlling weeds. More effective weed control means decreased yield losses and consequent yield increase.

Impact of Herbicide-Resistant Crops on Conservation Tillage Practices

In addition to positive agronomic and economic impacts, the adoption of biotechnology-derived herbicide-resistant crops has led to significant environmental impacts. Conservation tillage practices, no-till in particular, have increased significantly since the adoption of herbicide-resistant crops. Grower surveys and expert polls strongly indicate that the adoption of herbicide-resistant crops correlated positively with increase in no-till acreage since 1996, the year when herbicide-resistant crops were first planted.

Weed control is a major concern in no-till fields as poor weather conditions hamper the effectiveness of burndown herbicides used in no-till systems. Herbicide-resistant crops increased growers' confidence in their ability to control weeds without relying on tillage because herbicides used in biotechnology-derived crops are more effective than those used before. With that increased confidence, American growers planted 45%, 14% and

Table 1.2.6   *Impact of herbicide-resistant cotton on no-till acreage in the United States.*

Year	No-till acreage (million acres)	No-till acreage as a % of total	% Increase in no-till acreage based on 1996
1996	0.51	3.4	—
1997	0.53	3.7	4
1998	0.67	4.9	31
2000	1.35	8	166
2002	2.03	14	300

Source: Conservation Technology Information Center, multiple years.

300 % more acres to no-till in soya bean, corn and cotton, respectively, in 2003, compared with years before their introduction. A survey by Doane Marketing Research (2002) also revealed that herbicide-resistant crops have enabled growers to successfully incorporate no-till production practices into their farming operations.

The increase in no-till acreage has been higher in cotton than any other crop (Table 1.2.6). Several reasons have been cited for the dramatic increase in no-till cotton acreage. These include adoption of herbicide-resistant crops which enable the over-the-top herbicide applications, enhanced awareness in growers of the benefits of conservation tillage practices, increase in fuel prices, access to better no-till equipment and availability of better herbicides to control weeds in no-till fields. However, biotechnology-derived cotton is by far the leading reason for this increase in no-till production practices in cotton.

The Conservation Technology Information Center reported in 2002 that increased use of no-tillage reduced soil erosion by 90 % or nearly 1 billion tons and saved $3.5 billion in sedimentation treatment costs (Fawcett and Towery, 2002). Other benefits from no-tillage included significant fuel savings (3.9 gallons of fuel per acre), reduced machinery wear and tear, reduced pesticide runoff (70 %) and water runoff (69 %), reduced greenhouse gas emissions due to improved carbon sequestration and improved habitat for birds and animals.

Some experts have credited herbicide-resistant crops for transforming American agriculture from a carbon-intensive operation to a potential carbon sink. By providing more assured weed control, biotechnology-derived herbicide-resistant crops have facilitated the increase in no-till production practices and the associated environmental and economic benefits.

Conclusion

American experience from almost a decade-long use of biotechnology-derived crops indicate that these crops have revolutionized crop production and provided best hope to growers by helping to meet one of the key goals of production agriculture: improving yields with the use of minimal inputs. Evidence thus far indicates that crop biotechnology is critical in a world where natural resources are finite. Continuing improvements in productivity facilitated by biotechnology-derived crops will enable growers in the United States and worldwide to increase food security without having to bring more forestland into agricultural use.

American growers have increased planting of biotechnology-derived crops from 5 million acres in 1996 to 106 million acres in 2003. The fact that adoption of biotechnology-derived crops has continued to grow each year since their first introduction is a testimony to the ability of these products to deliver tangible positive impacts and to the optimistic future they hold.

Adoption increased at a phenomenal pace in the United States due to the positive impacts derived in the form of increased yields, improved insurance against pest problems, reduced pest management costs and pesticide use and overall increase in grower returns. Biotechnology-derived crops becoming such a dominant feature of the American landscape also indicate the confidence of American farmers in these crops. While control of key insect pests that resulted in increased yields and reduced insecticide use were the reasons for the success of Bt crops, increased ease and flexibility of weed management afforded by herbicide-resistant crops enhanced their adoption.

In spite of proven potential and documented positive impacts, opponents continue to argue about issues such as the impacts of these crops on pest resistance and human health, while many researchers have concluded that biotechnology-derived crops are as safe as, if not safer than, their conventional counterparts. Concerns such as pest-resistance and gene flow not only are akin to biotechnology-derived crops, but also relate to conventional pest management practices as well. Therefore, it is important to weigh risks against benefits to judge the value of the technology.

Biotechnology-derived crops in production to date in the United States have modified crop protection characteristics only. The second generation of biotechnology-derived crops is already underway and includes traits that may solve production challenges such as cold tolerance, drought tolerance and increased nitrogen efficiency, and output traits such as better flavor and appearance, greater shelf life and improved nutritive value. With a pipeline that is packed with crops that may further improve yields and deliver health and safety benefits to consumers, public approval for these crops will only continue to increase in the near future.

References

Alstad, D.N., Witkowsi, J.F., Wedberg, J.L., Steffey, K.L., Sloderbeck, P.E., Siegfried, B.D., Rice, M.E., Pilcher, C.D., Onstad, D.W., Mason, C.E., Lewis, L.C., Landis, D.A., Keaster, A.J., Huang, F., Higgins, R.A., Haas, M.J., Gray, M.E., Giles, K.L., Foster, J.E., Davis, P.M., Calvin, D.D., Buschman, L.L., Bolin, P.C., Barry, B.D., Andow, D.A. and Alstad, D.N. (1997) In *Bt-Corn & European Corn Borer: Long-Term Success Through Resistance Management* (eds K.R. Ostlie, W.D. Hutchison and R.L. Hellmich). Available at www.extension.umn.edu/distribution/cropsystems/DC7055.html.

Alston, J.M., Hyde, J., Marra, M. and Mitchell, P.D. (2002) An ex ante analysis of the benefits from the adoption of corn rootworm resistant transgenic corn technology. *AgBio Forum* 5, 71–84.

Anonymous (multiple years) Summary of losses from insect damage and costs of control in Georgia. University of Georgia.

Babcock, J.M. and Bing, J.W. (2001) Genetically enhanced Cry1F corn: broad-spectrum lepidopteran resistance. *Down to Earth* 56, 10–15.

Baum, J.A., Chu, C-R., Rupar, M., Brown, G.R., Donovan, W.P., Huesing, J.E., Ilagan, O., Malvar, T.M., Pleau, M., Walters, M. and Vaughn, T. (2004) Binary toxins from *Bacillus thuringiensis* active against the western corn *rootworm, Diabrotica virgifera virgifera LeConte. Applied and Environmental Microbiology* **70**, 4889–4898.

Beachy, R.N., Loesch-Fries, S. and Tumer, N.E. (1990) Coat protein-mediated resistance against virus infection. *Annual Review of Phytopathology* **28**, 451–474.

Brunoehler, R. (1999) Farmers, researchers rate Roundup Ready corn. *Soybean Digest*, January edition.

Bryant, K.J., Allen, C.T., Bourland, F.M., Smith, K.L. and Earnest, L.D. (2000) Cost and return comparisons of transgenic cotton systems in Arkansas. *Proceedings of the Beltwide Cotton Production Conference* **1**, 343–345.

Bryant, K.J., Roberston, W.C. and Lorenz, G.M. (1999) Economic evaluation of Bollgard cotton in Arkansas. *Proceedings of the Beltwide Cotton Conference* **1**, 349–359.

Carpenter, J., Felsot, A., Goode, T., Hammig, M., Onstad, D. and Sankula, S. (2002) *Comparative Environmental, Impacts of Biotechnology-Derived and Traditional Soybean, Corn, and Cotton Crops*, Council for Agricultural Science and Technology, Ames, Iowa.

Carpenter, J.E. and Gianessi, L.P. (2002) Trends in pesticide use since the introduction of genetically engineered crops in *Economic and Environmental Impacts of Agbiotechnology: A Global Perspective* (ed Nicholas Kalaitzandonakes), Kluwer-Plenum, NY.

Christensen, J.J. and Schneider, C.L. (1950) European corn borer (*Pyrausta nubilalis* HBN.) in relation to shank, stalk, and ear rots of corn. *Phytopathology* **40**, 284–291.

Cooke, F.T., Scott, W.P., Martin, S.W. and Parvin, D.W. (2001) The Economics of Bt Cotton in the Mississippi Delta 1997–2000. *Proceedings of the Beltwide Cotton Conference* **1**, 175–177.

Curran, W.S., Lingenfelter, D.D., Calvin, D.D., Ayers, J.E. and Dewolf, E.D. (1999) Corn pest management. Part 2, Section 2. Available at www.agguide.agronomy.psu.edu/pm/default.html.

Conservation Technology Information Center (multiple years) Crop residue management. Available at www.ctic.purdue.edu/Core4/Core4Main.html

Di, R., Purcell, V., Collins, G.B. and Ghabrial, S.A. (1996) Production of transgenic soybean lines expressing the bean pod mottle virus coat protein precursor gene. *Plant Cell Reports* **15**, 746–750.

Dively, G. (2004) Personal Communication. University of Maryland.

Doane Marketing Research (2002) Conservation Tillage Study prepared for the Cotton Foundation. Available at www.cotton.org/tech/biotech/presentation/doanecontillfinalreport.ppt.

Dow Agrosciences News Center (2002) Herculex I earns Japanese approval, corn growers gain new option. Available at www.dowagro.com/herculex/news/japan_app.htm.

Edge, J.M., Benedict, J.H., Carroll, J.P. and Reding, H.K. (2001) Contemporary issues— Bollgard® cotton: an assessment of global economic, environmental, and social benefits. *Journal of Cotton Science* **5**, 121–136.

EPA (2001) Revised risks and benefits sections *Bacillus thuringiensis* plant-pesticides. Biopesticides Registration Action Document. Environmental Protection Agency, pp. 1–306.

Fawcett, R. and Towery, D. (2002) Conservation tillage and plant biotechnology: how new technologies can improve the environment by reducing the need to plow. Available at www.ctic.purdue.edu/CTIC/BiotechPaper.

Fernandez-Cornejo, J. (2004) Agricultural biotechnology: adoption of biotechnology and its production impacts. Available www.ers.usda.gov/Briefing/biotechnology/chapter1.htm.

Fernandez-Cornejo, J. and McBride, W.D. (2000) Genetically engineered crops for pest management in US agriculture: farm level impacts. Economic Research Service/USDA—Agricultural Economic Report—786.

Fuchs, M., Tricoli, D.M., Carney, K.J., Schesser, M., McFerson, J.R. and Gonsalves, D. (1998) Comparative virus resistance and fruit yield of transgenic squash with single and multiple coat protein genes. *Plant Disease* **82**, 1350–1356.

Gianessi, L. and Carpenter, J. (1999) *Agricultural Biotechnology Insect Control Benefits*, National Center for Food and Agricultural Policy, Washington, DC, pp. 1–78.

Gianessi, L.P. and Marcelli, M.B. (1997) *Pesticide Use in US Crop Production: 1997*, National Center for Food and Agricultural Policy, Washington, DC. Available at www.ncfap.org.

Gianessi, L.P., Silvers, C.S., Sankula, S. and Carpenter, J.E. (2002) The potential for biotechnology to improve crop pest management in the US: 40 case study. Available at www.ncfap.org.

Gonsalves, D., Gonsalves, C., Ferreira, S., Pitz, K., Fitch, M., Manshardt, R. and Slightom, J. (2004) Transgenic virus-resistant papaya: from hope to reality for controlling papaya ringspot virus in Hawaii. APSnet Feature, American Phytopathological Society. Available at www.apsnet.org/online/feature/ringspot.

Hawaii Agricultural Statistics Service (multiple years) Papaya acreage information from Online Publication Archive. Available at www.nass.usda.gov/hi.

James, C. (2003) Global status of commercialized transgenic crops: 2003. Number 30. ISAAA. Ithaca, New York.

James, C. (2004) Global status of commercialized biotech/GM crops: 2004. Number 32. ISAAA. Ithaca, New York.

James, C. (multiple years) Global status of commercialized transgenic crops. Available at www.isaaa.org.

Johnson, B., Zollinger, R., Hanson, B., Erikson, E., Riveland, N., Henson, R. and Jenks B. (2000) Herbicide-tolerant and conventional canola production systems comparisons, 2000 North Dakota Weed Control Research, pp. 8–10.

Johnson, W. and Smeda, R. (2001) Weed management issues in Roundup Ready soybeans in *Integrated Pest and Crop Management Newsletter,* Volume 11 (1), University of Missouri.

Kaniewski, W. and Lawson, C. (1998) Coat protein and replicase-mediated resistance to plant viruses, in *Plant Virus Disease Control* (eds A. Hadidi, R.K. Khetarpal and H. Koganezawa), APS Press, St. Paul, MN, pp. 65–78.

Kreuger, R.W. (2001) The public debate on agrobiotechnology: a biotech company's perspective. *AgBioForum* **4**, 209–220.

Lauer, J. (2004) 2003 performance of Bt-CRW in university trials. *Wisconsin Crop Manager* **11**, 14–15.

Levine, E. and Oloumi-Sadeghi, H. (1991) Management of Diabroticite rootworms in corn. *Annual Review of Entomology* **36**, 229–255.

Marasas, W.F.O. (1995) Fumonisins: their implications for human and animal health. *Natural Toxins* **3**, 193–198.

Marra, M., Carlson, G. and Hubbell, B. (1998) Economic impacts of the first crop biotechnologies. Available at www.ag-econ.ncsu.edu/faculty/marra.firstcrop/img001.gif

Marra, M., Pardey, P. and Alston, J. (2002) The payoffs of agricultural biotechnology—an assessment of the evidence. *International Food Policy Research Institute (IFPRI) Environment and Production Technology Division discussion paper* **87**, 1–57.

Mason, C.E., Rice, M.E., Calvin, D.D., Van Duyn, J.W., Showers, W.B., Hutchison, W.D., Witkowski, J.F., Higgins, R.A., Onstad, D.W. and Dively, G.P. (1996) *European Corn Borer: Ecology and Management*. North Central Regional Extension Publication 327, Iowa State University, Ames, IA.

Mills, J.A. and Shappley, Z. (2004) Performance review of Bollgard II in the midsouth. *2004 Beltwide Cotton Conferences*, p. 1617.

Monsanto (2003) Background on the corn rootworm. Available at www.monsanto.com/monsanto/content/media/backgrounders/rw_backgrounder.pdf.

Minnesota Department of Agriculture. Black Cutworm Factsheet. Available at www.mda.state.mn.us/pestsurvey/Factsheets/blkctworm.html.

Mullins, W. and Hudson, J. (2004) Bollgard II versus Bollgard sisterline economic comparisons. *2004 Beltwide Cotton Conferences*, pp. 1660–1661.

Mullins, J. and Mills, J. (1999) Economics of Bollgard versus non-bollgard cotton in 1998. *Proceedings of the Beltwide Cotton Conference* **2**, 958–961.

Munkvold, G.P. and Desjardin, A.E. (1997) Fumonisins in maize: can we reduce their occurrence? *Plant Disease* **81**, 556–565.

Munkvold, G.P., Hellmich, R.L. and Rice, L.F. (1999) Comparison of fumonisin concentrations in kernels of transgenic Bt maize hybrids and non-transgenic hybrids. *Plant Disease* **83**, 130–138.

National Corn Growers Association (2004) Press Release: IRM refuges are a must for Bt corn growers, compliance assurance program infractions may result in denied access to Bt technology in 2005. Available at www.fifra.info/x/press/2004/NCGA-20040423A.html.

Perlak, F.J., Deaton, R.W., Armstrong, T.A., Fuchs, R.L., Sims, S.R., Greenplate, J.T. and Fischoff, D.A. (1990) Insect resistant cotton plants. *Biotechnology* **8**, 939–943.

Phipps, R.H. and Park, J.R. (2002) Environmental benefits of genetically modified crops: global and European perspectives on their ability to reduce pesticide use. *Journal of Animal and Feed Sciences* **11**, 1–18.

Pike, D.R. (1995) Biologic and economic assessment of pesticide use on corn and soybeans. *USDA National Agricultural Pesticide Impact Assessment Program: Report Number 1-CA-95.*

Rice, M.E. (2004) Transgenic rootworm corn: assessing potential agronomic, economic and environmental benefits. *Plant Health Progress.* March 2004 online publication.

Runge, F. and Ryan, B. (2004) The global diffusion of plant biotechnology: international adoption and research in 2004. Available at www.apec.umn.edu/faculty/frunge/globalbiotech04.pdf.

Sankula, S. and Blumenthal, E. (2004) Impacts on US agriculture of biotechnology-derived crops planted in 2003—an update of eleven case studies. Available at www.ncfap.org.

Schultheis, J.R. and Walters, S.A. (1998) Yield and virus resistance of summer squash cultivars and breeding lines in North Carolina. *HortTechnology* **8**, 31–39.

Schwartz, P.H. (1983) Losses in yield of cotton due to insects, in *Cotton Insect Management with Special Reference to the Boll Weevil, Agricultural Handbook* (eds R.L. Ridgeway *et al.*), p. 589.

Shipitalo, M.J. and Malone, R.W. (2000) *Runoff Losses of Pre- and Post-Emergence Herbicides From Watersheds in a Corn–Soybean Rotation*, The Ohio State University, Ohio Agricultural Research and Development Center, pp. 1–4.

Sloderbeck, P., Buschman, L., Dumler, T. and Higgins, R. (2000) Economic comparison of Bt corn refuge planting strategies for south central and southwestern Kansas. Available at www.oznet.ksu.edu/ex_swao/Entomology/Bt_Folder/ btecfd.pdf.

Sobek, E.A. and Munkvold, G.P. (1999) European corn borer (Lepidoptera: Pyralidae) larvae as vectors of *Fusarium moniliforme*, causing kernel rot and symptomless infection of maize kernels. *Journal of Economic Entomology* **92**, 503–509.

Sparks, A.N. (1979) Review of the biology of the fall armyworm. *Florida Entomologist*, June edition.

Stark, C.R. (1997) Economics of transgenic cotton: some indications based on Georgia producers. *Proceedings of the Beltwide Cotton Conference* **1**, 251–253.

Swadener, C. (1994) Insecticide fact sheet: *Bacillus thuringiensis* (Bt). *Journal of Pesticide Reform* **14**, 13–20.

USDA-AMS (multiple years) Cotton varieties planted, United States. Available at www.ams.usda.gov/cotton/mncs/index.htm.

USDA-ERS (1997) Pest management in major field crops. AREI Updates. Number 19.

USDA-NASS (1999) National Agricultural Statistics Service: Agricultural Chemical Usage. Available at www.usda.gov/nass/pubs/estindx1.htm.

USDA-NASS (2003) National Agricultural Statistics Service: Agricultural Chemical Usage. Available at www.usda.gov/nass/pubs/estindx1.htm.

USDA-NASS (2004) National Agricultural Statistics Service: Agricultural Chemical Usage. Available at www.usda.gov/nass/pubs/estindx1.htm.

USDA-NASS (multiple years) National Agricultural Statistics Service: reports by commodity—acreage. Available at www.usda.gov/nass/pubs/estindx1.htm.

Walker, K.A., Hellmich, R.L. and Lewis, L.C. (2000) Late-instar European corn borer (Lepidoptera: Crambidae) tunneling and survival in transgenic corn hybrids. *Journal of Economic Entomology* **93**, 1276–1285.

Williams, M. (2003) Cotton insect loss estimates—2003. Available at www. msstate.edu/Entomology/CTNLOSS/2003/2003loss.html.

1.3

Development of Biotech Crops in China

Qingzhong Xue and Xianyin Zhang

College of Agriculture and Biotechnology, Zhejiang University, People's Republic of China

Yuhua Zhang

Crop Performance and Improvement, Rothamsted Research, Harpenden, Hertfordshire, AL5 2JQ, United Kingdom

Introduction

Since the early 1990s, plant biotechnology research has been developing rapidly in China. Transgenic technology has been used to develop new crop varieties with increased resistance to pathogen/insect attacks, increased yield and improved quality. Compared with conventional crop breeding programs, transgenic technology can reduce the time to produce a new variety significantly. It only takes 2–3 years to produce a new variety by the transgenic approach, while it normally takes 5–8 years by traditional means. The new green revolution exemplified by transgenic technology will definitely have a significant impact on future agriculture in China.

In 1993, China's national biosafety committee for genetic engineering was established. Out of 353 GM crop applications made between 1996 and 2000, it granted permission to 45 for field trial, 65 for release to the environment and 31 for commercial use (Huang *et al.*, 2002). Up to now, more than 3 million hectares of GM crops have been grown in China, and six GM crop species have been granted permission for commercial growth. These are cotton, soybean, maize, oilseed rape, tomato and pepper. The major GM crop grown in China is cotton. Except for occasional use for oil production, it is generally not

Plant Biotechnology. Edited by Nigel Halford.

for food use. In 2002, China grew the largest areas of GM cotton in the world, with about 1.5 million hectares, accounting for one third of the total grown in China.

The rapid development of agricultural biotechnology in China has been due mainly to the continuous support from the Chinese government in various funding schemes such as 'National High-Tech Research and Development Program (863 Program)', 'National Plant Transgenic Research and Commercialization Project', 'National Key Basic Research and Development Program (973 Program)' and the National Natural Science Foundation. Funding for agricultural biotechnology is to be increased by four times, reaching an annual amount of $500 million. It is also very important that China has been building up an excellent human resource in the area of plant biotechnology. According to incomplete statistics, the number of properly trained scientists in this field was increased from 740 in 1986 to 1988 in 1999 (Huang *et al.*, 2002). China will have the largest number of research scientists in the biotechnology sector among the developing countries. A new revolution of agricultural technology led by transgenesis has been developing and will accelerate the development of high-yield, high-quality and high-performance agriculture. China has an abundant bioresource and a potentially very large and demanding market. This has also been drawing the attention of the outside world.

In this chapter, we will mainly discuss the current research activities in China in the area of plant transgenesis and its application in agriculture.

Approaches to Produce Transgenic Plants

PEG-Mediated Protoplast Transformation

In the mid-1980s, technological breakthroughs and improvements enabled efficient rice protoplast culture to be achieved (Abdullah and Cocking, 1986). This provided an opportunity for plant scientists to put foreign genes directly into plant cells. Zhang and Wu (1988) established an efficient embryogenic cell suspension culture from mature embryo callus tissues of a japonica variety, 'Taibei 309'. They regenerated transgenic rice plants for the first time from suspension cell-derived protoplasts via PEG mediation. However, this method had only limited applications because of the time-consuming and genotype-dependent nature of establishing efficient protoplast culture systems. In addition, PEG is also toxic to plant cells.

Particle Bombardment

Klein *et al.* (1987) from Cornell University developed a new method to deliver foreign genes into a plant. They innovatively proposed to coat DNA molecules on the surface of gold or tungsten particles of approximately 1μm, and foreign genes were delivered into plant cells by physically bombarding these DNA-coated particles into intact plant cells or tissues. Finally they made a gene delivery instrument known as the 'gene gun'. Compared with protoplast transformation, this transformation method is simpler and much less genotype dependent. It soon became a widely-used method to transform monocot plant species.

Agrobacterium-mediated Transformation

Agrobacterium-mediated transformation is the method of choice to make transgenic plants for dicot plant species but has been difficult to apply to monocot species. Chan *et al.* (1992) first reported transgenic roots and calli generated by *Agrobacterium*-infected seedlings of an indica rice variety, 'TN1'. They demonstrated the expression of foreign genes in these transgenic tissues. Later, they were able to increase the transformation efficiency and finally regenerated transgenic plants from cultured rice immature embryos after co-cultivating with potato cell suspension and agrobacterial cells. The transgenes were shown to be inherited in the next generation (Chan *et al.*, 1993). This transformation technology for monocot species was proved and further improved by a Japanese group (Hiei *et al.*, 1994). Compared with particle bombardment, *Agrobacterium*-mediated transformation has many advantages such as relatively well-defined gene transfer (T-DNA), high percentage of single or low-copy transgene integration, potentially high-transformation efficiency, cost-effectiveness and ease of use. It has been used by more and more laboratories.

The Pollen-Tube Pathway

In 1993, an *in planta* method to produce transgenic plants was proposed by a Chinese scientist, Professor Guangyu Zhou. Based on the formation and growth of pollen tubes during pollination, she proposed the injection of a DNA solution into the plant ovary after pollination. Foreign DNA might be able to get into the fertilized egg via the growing pollen tube, and transgenic plants might be generated if the foreign DNA were to integrate into the genome of the developing zygotes. This transformation method is very unique; it uses the egg cell or the fertilized egg cell as the transformation target and transgenic plants can be generated by simply allowing the fertilized egg cells to develop into embryos. Therefore, it avoids the tissue culture process and obviously is also genotype independent. In addition, it is technically simple and can be used by conventional breeding workers. Although it is a low-efficiency method, and the nature of its transformation mechanism and the molecular evidence for it are not widely accepted, its usefulness has been proved by more and more research (Zhou *et al.*, 1993). Transgenic cotton varieties and strains with insect- or disease-resistance currently grown on a large scale in China were generated by this pollen-tube pathway.

This is an important contribution from Chinese scientists to the area of agro-biotechnology. Transgenic plants have also been produced in maize (*Zea mays* L.) by the pollen-mediated method (Wang *et al.*, 2001a). In this method, the maize pollen grains were mixed with plasmid DNA pGLII_RC_1 containing a chitinase gene expression cassette and treated by sonication; the DNA–pollen mixture was then used to pollinate a maize inbred line artificially. The delivery of foreign genes into the progeny plants was confirmed by DNA dot blotting, PCR and PCR–Southern blotting (Wang *et al.*, 2001a).

There are other methods which could be used to make transgenic plants. The choice of method depends on the nature of the target plant. Furthermore, the following things need to be considered before making transgenic plants: the expression cassette containing the gene of interest, the co-ordination of transgenes and endogenous genes, the transgene heritability and its expression stability, the removal of marker genes and the selection efficiency.

Important Factors for Transformation

Agrobacterium-mediated transformation has now become the method of choice for most crop plants and it has been used to generate transgenic plants for cereal crops such as rice, maize, wheat and barley (Cheng *et al.*, 1997; Hiei *et al.*, 1994). However, different laboratories have used different strains of *Agrobacterium* as well as different expression constructs and plant materials to make transgenic plants, with various efficiencies reported. Plant transformation efficiency has been shown to be affected by many factors, some of which are very important for a reproducibly efficient system. These include the *Agrobacterium* strain, the binary vector, the expression cassette, the plant genotype, medium composition, the *vir* gene-inducing compound (e.g. acetosyringone) and the growth status of embryogenic callus. In order to improve transformation efficiency, it is necessary to optimize all of these important factors.

Promoters

In order to define transgene expression in plants, appropriate promoter and termination sequences are added at the 5' and 3' ends of the gene of interest. Constitutive promoters such as the 35S gene promoter from cauliflower mosaic virus and the *Actin1* gene promoter from rice were widely used to make transgenic plants in the early days. While more and more genes of interest have been cloned and characterized, and in order to control transgene expression spatially and/or temporally during plant growth and development, many tissue-specific and/or inducible promoters have been used to direct transgene expression. Examples are to use seed-specific promoters to improve seed quality, to use inducible promoters to confer insect resistance, and to use anther-specific promoters to create male sterility. Table 1.3.1 shows some tissue-specific promoters used in plant transgenesis.

Xie *et al.* (2000) investigated the functional regions of the complementary sense promoter from cotton leaf curl virus by deletional analysis. They generated transgenic plants containing a GUS reporter gene fused with five fragments of the promoter region of various lengths. Analysis of GUS activity showed that the promoter lacking the negative *cis* element was stronger than the full-length one, and the average strength was 12 times that of the CaMV 35S promoter.

The Function of Introns

Many plant gene introns, such as the first intron of *Adh1* and *Sh1* genes in maize and *Act1* in rice, are known to be able to enhance gene expression. This characteristic of introns has been exploited in transgenic research, with varying results. While an intron inserted between the CaMV 35S promoter and a GUS gene was shown to increase GUS expression up to 15 times in transgenic rice, it made no difference to the expression of the gene in transgenic tobacco protoplasts. Xu *et al.* (2003) investigated the role of the first intron of a 5-enolpyruvylshikimate-3-phosphate synthase (EPSPS) gene in transgene expression. Their results showed that the presence of this intron was able to increase GUS activity by an average of threefold in transgenic tobacco plants with a six-fold increase, the highest achieved. Chen and Wang (2004) reported that GUS gene expression driven by the OsBP-73 gene promoter needed the involvement of its intron, while the intron

Table 1.3.1 *Tissue-specific promoters used in plant transgenesis.*

Promoter	Origin	Expression tissue	Transgenic plant	Transgene
2A12	Tomato	Fruit	Tomato	ipt GUS
Rch10	Rice	Flower	Tobacco	iaaL
A9	Arabidopsis	Anther	Tobacco	Barnase
TA29+35S	CaMV	Anther	Arabidopsis	GUS
RTS	Rice	Anther	Rice	GUS
Zm13	Maize	Anther	Maize	Barnase
RTS	Rice	Anther	Rice	Barnase
Patatin	Potato	Tube	Potato	Hepatitis B virus surface antigen gene
PNZ1P	Pharbitis nil(L.) Choisy	Green tissue	Tobacco	GUS
Gt1	Rice	Endosperm	Rice	GUS
4a	Rice	Endosperm	Rice, tobacco	GUS
BP	Poplar	Phloem	Tobacco	GUS
CP	*Commelina yellow mottle virus*, CoYMV	Phloem	Tobacco	GUS
SP	*C. maxina*	Phloem	Tobacco	GUS
BSP	Poplar	Phloem	Tobacco	GUS
PP2	*C. maxina*	Phloem meristem	Tobacco	GUS, modified GNA
RSP1 RSP2	Rice	Vascular tissue	Rice	GUS
GRP1.8	Jinkgo	Vascular tissue	Tobacco	GUS
profilin2	Arabidopsis	Vascular tissue	Kalanchoe	GUS
CoYMV	*Commelina yellow mottle virus*	Vascular tissue	Cotton	GUS
CoYMV	*Commelina yellow mottle virus*	Vascular tissue	Tobacco	GUS
HRGP	Carrot	Vascular tissue	Tobacco	GUS
po1,po2,po3	Banana bunchytop virus	Vascular tissue	Tobacco	GUS GFP
16S	Tobacco	Chloroplast	Tobacco	aadA
Glutellin	Rice	Seed	Tobacco	GUS
napinB	Brassica	Seed	Tobacco	GUS
BcNAI	Brassica	Seed	Tobacco	GUS
Alcohol-soluble protein gene promoter	Maize	Seed	Tobacco	GUS

alone had no promoter activities. In contrast, in an investigation of the activity of a rice sucrose synthase gene promoter, Li *et al*. (2002a) reported that the presence of the first intron did not affect the strength of the promoter significantly when driving GUS expression in transgenic rice plants.

Selection Markers

Selection marker genes together with selective agents are used to enable individual transformed plant cells to grow out of a non-transformed cell mass, facilitating the generation of transgenic plants. The most widely used selection marker genes/selective agents are the neomycin phosphotransferase gene (*NPTII*) together with the selective agent kanamycin, the hygromycin phosphotransferase gene (*HPT*) together with hygromycin and the phosphinothricin acetyltransferase (*PAT*) or *BAR* genes with phosphinothricin (PPT).

An Accelerated, High-Efficiency *Agrobacterium*-mediated Transformation System for Rice

By systematic optimization of various factors based on a published protocol (Hiei *et al.*, 1994), an accelerated high-efficiency *Agrobacterium*-mediated rice transformation system has been established in Professor Qingzhong Xue's laboratory in Zhejiang University. A binary vector containing both a Bt insecticidal gene, *CryIA(c)*, and a cowpea trypsin inhibitor gene, *CpTI*, was constructed and transformed into a japonica rice variety, 'Zheda 19', via the *Agrobacterium*-mediated method. Approximately 2000 callus tissues derived from scutella were treated with *Agrobacterium* cells and this resulted in about 1300 hygromycin-resistant calli, from which 1500 plants were regenerated. Seventy plants regenerated from different resistant calli were screened for transgenes by PCR and PCR–Southern blotting; this showed that 62 plants (88.6 %) were positive for both the *CryIA(c)* and *CpTI* genes. Furthermore, high toxicity to the striped stem borer, *Chilo supperssalis* (Walker), was observed in the T1 generation of three independent transgenic lines (Li *et al.*, 2002b).

Using 80–90 % Maturity Fresh Embryos to Induce Callus Culture

Since the availability of inflorescences and immature embryos is dependent on season, mature embryos are normally used as starting materials to induce callus culture for *Agrobacterium*-mediated transformation in rice. The effects of seed maturity and storage condition on callus induction and growth have been investigated. It was found that calli derived from fresh seeds of 80–90 % maturity with a greenish seed coat were the best in terms of uniform induction, rapid growth and high transformation efficiency.

Using Primary Calli for *Agrobacterium* Infection

Subcultured calli (about 3 weeks) were normally used for co-cultivation in the published protocols (Hiei *et al.*, 1994). However, a transient GUS assay showed that only 20 % of calli subcultured for 3 weeks developed blue foci, whereas this frequency was increased to 30–80 % when primary calli, either fresh or subcultured for 2–4 days on fresh medium, were used.

Using Normal MS Medium (sucrose 3 %) to Re-suspend *Agrobacterium* Cells

Before co-cultivation, *Agrobacterium* cells harvested either from solid or liquid medium were generally re-suspended in AAM medium (Hiei *et al.*, 1994) with a high sugar concentration (sucrose at 6.85 % and glucose at 3.6 %). We compared the effects of different re-suspension media on transformation efficiency and found that the normal MS medium plus 3 % sucrose was actually better than the AAM medium. This suggests that *Agrobacterium* cells do not need high osmotic treatment to transform rice callus cells.

Pre-culture Duration

Transformation efficiency is also affected by pre-culture duration before co-cultivation as indicated by transgene transient expression. While 1-day pre-culture only resulted in a very low level of GUS expression, an extension of pre-culture up to 3 days gave rise to 80 % of calli with GUS expression.

Reducing the Level of 2,4-dichlorophenoxyacetic Acid (2,4-D) in the Medium

We also investigated the effects on transformation efficiency of plant growth regulators in the pre-culture medium. This showed that a combination of benzyl amino purine (BAP) and 1-naphthalenacetic acid (NAA) was much better than 2,4-D. When a medium containing 2,4-D (N6-2D) was used for pre-culture (2.5–3 days), 30–80 % of the calli showed GUS expression but less than 20 % developed a large GUS staining area. In contrast, when 2,4-D was replaced by BAP (0.5 ppm) and NAA (0.5 ppm) (N6-BA), the proportion of calli with GUS expression was consistently above 80 %, and more than 50 % developed a large GUS staining area.

The level of 2,4-D in the co-cultivation medium seemed to have no significant effects on the percentage of calli with transient GUS expression, but it did affect the speed of GUS staining. While it took more than 6 h to develop blue foci resulting from GUS expression on the callus cells when co-cultivation was carried out on 2N6 medium with 2,4-D, it only took 4 h to get the blue color when the same medium without 2,4-D was used. This indicates that high level of 2,4-D in the callus might affect the transformation efficiency.

Thus far, we have established an accelerated, high-efficiency *Agrobacterium*-mediated rice transformation system based on the above major modifications to the established protocol. This is summarized in Figure 1.3.1, in which the black arrows show the modified steps. Compared with other published procedures, this system has advantages such as higher transformation efficiency, good repeatability and less time from callus induction to transgenic plant regeneration (2 months compared with 3–5 months).

Agricultural Applications of Transgenic Plants in China

Transgenic Cotton

The research, development and use of transgenic cotton in China started in the 1990s. Generally, Bt and *CpTI* genes were used to produce insect-resistant transgenic plants. In 2001, further development was achieved in the research and field application of transgenic insect-resistant cotton. The Biotechnology Research Institute of the Chinese Academy of Agricultural Sciences was granted a patent for an expression vector

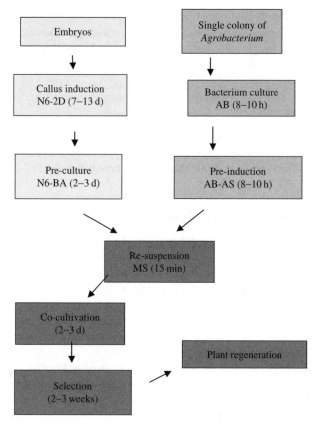

Figure 1.3.1   *The accelerated, high-efficiency Agrobacterium-mediated transformation of rice.*

containing a fusion gene coding for an insecticidal protein. Till then, 14 transgenic insect-resistant varieties had been approved in China, with 11 containing a single insect-resistant gene and three containing two transgenes. All these varieties were highly resistant to cotton bollworm (*Helicoverpa armigera* (Hubner)), of good quality and high yield.

In 2001, more than 3.5 million Chinese farmers grew transgenic insect-resistant cotton varieties and the growth area of transgenic insect-resistant cotton varieties was 31 % of the total. Chinese-made transgenic insect-resistant cotton varieties were grown in 17 provinces, including Hebei, Henan, Shanxi, Shandong, Hunan, Hubei, Jiangsu, Anhui, Xingjiang and Liaoning, and the growth area reached 0.6 million hectares. This accounted for 43.3 % of the total transgenic cotton grown in China.

Regimes were developed to ensure that the transgenic Bt cotton was used properly. These include maintenance of genetic purity, seed multiplication, plant cultivation, an insect-resistance assay, a Bt toxin assay, management of resistance to BT in the cotton bollworm, integrated insect control and biosafety assessment (Jia, 2001).

Although transgenic insect-resistant cotton has been grown widely in China, problems such as the development of resistance in insects, the relatively weak insect-resistance of the BT varieties, dynamic changes in insect resistance, the limited target insect spectrum of Bt and biosafety management still exist. This makes it necessary to search for large spectrum insect-resistance genes, use tissue-specific, development-specific or inducible promoters, produce multigene transgenic cotton varieties and strengthen field management regimes (Wang, 2003).

Transgenic research of insect-resistant cotton has currently been switching from using a single resistance gene to using two. Compared with transgenic cotton expressing a single insect-resistance gene, transgenic cotton expressing both *CryIAc* and *CpTI* genes was found to be more resistant to insect feeding in the tissue of flower bud, sepal, petal, anther and cotton boll, although not in leaf tissue. This was particularly obvious for old larvae. After feeding on plant tissues expressing both transgenes, the larvae grew and developed much more slowly, with a significant loss of weight and lower rate of pupation and maturation. The larvae also had a relatively reduced rate of feeding and digestion and a relatively high rate of metabolism. Even the development of the pupa and the adult was affected significantly, with a significant loss of weight and prolonged pupa growth, as well as a shortened life expectancy of surviving adults (Cui *et al.*, 2002).

Rui *et al.* (2002) investigated the spatial and temporal changes of the insecticidal activity of different transgenic insect-resistant cotton varieties. They found that insecticidal activity was generally higher in early rather than late developmental stages. In the early stage the higher activity was found in leaf tissue, whereas higher activity existed in the boll and bud in mid and late stages. Moreover, during mid to late developmental stages, plants expressing two insecticidal genes (Bt and *CpTI*) had significantly higher and more stable insecticidal activity than plants expressing only a single Bt gene.

While the growth of transgenic Bt cotton continues to increase, the population of most non-target insects in the field tends to increase as well. Compared with a conventional cotton field in which an integrated insect-controlling management was used, Deng *et al.* (2003) found that the occurrence of aphids was increased by 37.9 % in a Bt cotton field in which agrochemicals were used to control insects, and 71.4 % in a Bt cotton field in which the insects were controlled only by their natural enemies. This was increased to 92.5 % and 134.9 %, respectively, in the second year. In addition, they also observed dynamic changes in the population of different insect natural enemies due to the expanding growth of Bt cotton.

Transgenesis has also been used to improve restorer lines in cotton. Wang and Li (2002) transformed a cytoplasmic male sterility restorer line with a glutathione-*S*-transferase gene (*GST*). They found a strong restorer named 'Zhedaqianhui' from a transgenic progeny population. This restorer was 25.8 % stronger than the donor plant control 'DES HAF2 77'; it improved the F1 boll formation by more than 3.6, reduced the occurrence of sterile seeds by 10.1 %, and increased the cotton yield by 10.6 %. Southern and Northern blotting using *GST* as probe showed that the new restorer had the transgene integrated in its genome and that the gene was highly expressed.

Transgenic Rice

Rice transgenic research in China started from international collaborations in the 1990s. The collaboration between Qingzhong Xue's laboratory and Ray Wu's laboratory resulted in the production for the first time of insect-resistant and herbicide-resistant transgenic rice strains selected from transgenic plants containing *CpTI* and *PINII* genes (Duan *et al.*, 1996; Xu, *et al.*, 1996). In collaboration with its Canadian partner, Qingyao Shu's laboratory of Zhejiang University produced transgenic Bt rice plants and created novel rice germplasm, KMD1 and KMD2, which were highly resistant to rice stem borer (Cheng *et al.*, 1998; Shu *et al.*, 2000; Ye *et al.*, 2001).

Zhen Zhu's laboratory in the Genetics Institute of the Chinese Academy of Sciences expressed a modified *CpTI* gene in transgenic rice plants. Activity assays showed that the modified version of *CpTI* had two- to four-fold higher activity than the original one. This corresponded with a very high resistance to the rice stem borer and other lepidopteran insects.

Wang *et al.* (2001b) examined the genetics of transgene expression in crossing populations between or within rice subspecies, namely Bt japonica × wild-type indica and Bt japonica × wild-type japonica. Using GUS assays they found transgene segregation ratios of 1:1 and 3:1, respectively, in the BC_1 and BCF_2 generations of a Bt japonica × wild-type indica cross. This suggested that the transgene was inherited as a dominant single gene. Western dot blotting showed that Bt toxin expression in F1, F2 and BC1 was higher than that of the transgenic parent, indicating a role played by heterosis. In addition, no significant difference was found in plant height, length of spike, number of tillers and 1000 grain weight in the progenies of both crosses.

In order to increase the expression of insect-resistance genes in transgenic rice, Li *et al.* (2002b) investigated the strength of rice sucrose synthase gene promoters, *RSP1* and *RSP2*, in transgenic rice. They detected high expression of a GUS gene driven by these two promoters only in roots, stem, leaf and grain husk; no expression was detected in embryo or endosperm. This suggested that the *RSP1* and *RSP2* promoters could be used to direct the expression of transgenes and would cause less food safety problems because the transgene would not be active in the edible parts of the grain.

The heritability and stability of transgene expression were also examined in progenies of crosses between a photoperiod-sensitive male-sterile line, 'ZAU11S', and three transgenic strains (Cheng and Xue, 2003). Using PCR, Basta painting and an *in vitro* trypsin inhibiting activity assay, it was found that transgenes *BAR* and *PINII* were tightly linked and inherited in Mendelian fashion in the F2 generations. The wound-inducible expression of the *PINII* gene was found to have clear spatial and temporal regulation. The inducing signal could be transduced upward as well as downward in the plants. In addition, the inducibility of the *PINII* gene expression was slightly different among the three transgenic strains and partial *BAR* silencing was detected in some plants.

Sun *et al.* (2001) generated transgenic plants expressing a snowdrop lectin gene, *GNA*, in japonica rice by particle bombardment. PCR and Southern analyses showed that 79 % of the regenerated plants contained transgenes. Western blot indicated that the GNA protein made up 0.5 % of the total soluble protein in high-expressing plants. The high-expressing lines were resistant to brown rice plant hopper.

Transformation has also been used to improve rice disease resistance and stress tolerance. Zhao *et al.* (2000) first transformed a male-sterile rice variety, 'Pei-ai 64S', with gene *Xa21* and the transgenic plant was reported to be highly resistant to rice bacterial blight disease. Later, Chen *et al.* (2002) made F1 plants by crossing the progeny of *Xa21* transgenic Pei-ai 64S with its non-transgenic parent. PCR analysis showed that all the F1 plants contained the *Xa21* gene. Most of them were resistant to blight. In the F2 population, resistant and sensitive plants exhibited a 3:1 ratio, suggesting that the *Xa21* gene was inherited as a single dominant gene. Under long-day and high-temperature conditions, the F1 plants were all fully fertile, whereas sterility occurred at a rate of 1 out of 28 in the F2 population.

The *Xa21* gene was also transformed into five major rice varieties in China (Zhai *et al.*, 2000). In total, 110 independent transgenic lines were produced. The integration of the transgene in the plant genome was demonstrated by Southern blotting. The *Xa21* gene was shown to be stably inherited and the plants with a single-copy transgene showed the expected 3:1 ratio of transgene segregation in the T1 generation. Disease inoculation experiments showed that the transgenic T0 plants as well as the PCR-positive T1 plants were highly resistant to rice blight disease.

Feng *et al.* (2001) reported the generation of 49 independent transgenic rice plants containing multiple fungal-resistance chitinase genes (1–4). Chitinase activity of every transgenic plant was shown to be higher than that in its non-transgenic control. It was also shown that chitinase activity was higher in plants with multiple transgenes than in plants with a single transgene.

Wang *et al.* (2000) transformed rice with bacterial genes *MtlD*, coding for mannitol-1-phosphate dehydrogenase, and *GutD*, coding for sorbitol-6-phosphate dehydrogenase. Gas chromatography detected the accumulation of mannitol and sorbitol in the transgenic plants. Salt tolerance experiments showed that the transgenic plants were much more tolerant to salt treatment than the non-transgenic controls.

In order to modify rice seed storage protein composition, Zhang and Xue (2001) cloned the rice glutelin gene, *Gt1*, promoter and used it to drive the expression of soybean glycinin subunit genes, *A1a* and *B1b*. Transgenic plants were generated via the *Agrobacterium*-mediated method. The transgene was confirmed to be integrated in the rice genome and stably transmitted to the next generation.

Insect-resistant transgenic rice has recently been used in conventional rice breeding programs. This has been developing rapidly, resulting in high yields as well as good control of insect damage. Due to government regulations on biosafety issues, however, as yet no transgenic rice strains are allowed to be released into the field.

Transgenic Maize

Bt toxin Cry1Ie1 is highly toxic to the Asian corn borer. Although it is not expressed in *Bacillus thuringiensis*, the gene encoding this toxin was cloned by the scientists in China Agricultural University. Both *Cry1Ie1* and another Bt toxin gene, *Cry1Ac,* with modified codon usage, were subcloned in binary vectors for expression in plants. Liu and Wang (2003) generated transgenic plants containing the *Cry1A* gene and showed that the transgene was inherited as a single dominant genetic locus in transgenic plants. The expression of the Bt toxin was significantly different among independent transgenic plants. Transgene expression also changed in different plant tissues, with significantly higher expression in green tissues than in non-green ones. Among three independent transgenic lines, the expression level of *Cry1A* was not significantly different among R2, R3, and R4 generations.

Using the pollen-tube method, Wang *et al.* (2002) attempted to transform plants from nine maize inbred lines with a Bt toxin gene expression vector. Of the 800 plants germinated, 30 % survived the initial PPT screening. Of the 60 plants analyzed further, three were confirmed as transgenic by PCR. Zhang *et al.* (2003a) used particle bombardment of immature embryos to transform two maize inbred lines, 501 and C111, with a Bt gene. Putative transgenic plants were regenerated from Bialaphos-resistant calli and confirmed by PCR amplification of the Bt gene. Liu *et al.* (2003) investigated the insect toxicity of four inbred lines (i.e. Mo17Bt, 478Bt, Zhong 3Bt and 871Bt) derived from backcrossing of transgenic lines, expressing the Bt gene. A field bioassay showed that all four Bt inbred lines performed better than the wild-type control in terms of insect resistance, although they were less resistant than the positive control.

Li *et al.* (2002c) were able to induce embryogenesis and the formation of shoot clusters from a shoot meristem culture of an elite inbred line. They established a rapid and efficient maize shoot cluster culture system as a convenient source of target tissues for transformation. Using particle bombardment with the herbicide-resistance gene, acetolactate synthase (*ALS*), isolated from Arabidopsis, they produced regenerated plants from chlorsulfuron-resistant shoot cluster cultures.

Zhang *et al.* (2003b) reported for the first time the expression of a rabbit defensin gene, *NP-1*, in maize in order to produce disease-resistant plants. They used particle bombardment to make the transgenic plants. Effective resistance was observed in the T1 progeny of transgenic lines when challenged by the physiological race 0 of the corn leaf blight disease. In addition, transgenic plants containing the structural protein P1 of foot and mouth disease virus (FMDV) were also produced (Yu *et al.*, 2003).

Transgenic Oil Crops

Chen *et al.* (1999) established an *Agrobacterium*-mediated transformation system for oilseed rape. They expressed an antisense version of the *PEP* gene in the current elite varieties 'Zheyou 758' and 'Zheyouyou 1' and an increase in oil content was observed in the transgenic plants.

Zhou *et al.* (2001) reported the generation of transgenic soybean with a Bt toxin gene, *Cry1A*, by *Agrobacterium* infection of the cotyledon tissue. Using the pollen-tube pathway, Cui *et al.* (2003) tried to transform 14 soybean varieties with a chitinase gene. In order to elucidate the mechanism for the low transformation efficiency of this technique, they investigated the process of pollen-tube germination, elongation and penetration in the ovary using fluorescence microscopy, and proposed possible ways to improve this transformation method.

Transgenic Tobacco

Tobacco was the first plant species to be used as a model plant for molecular biological research and genetic engineering. Consequently, efficient technical procedures involved in these researches have been established and used for many years. These include genetic characterization, the building of expression constructs, gene transfer, cell culture and plant regeneration, and the breeding of strains of interest. Lots of disease-resistant and insect-resistant transgenic lines and varieties have already been made available. In China, transgenic tobacco with resistance to tobacco mosaic virus (TMV), cucumber mosaic virus (CMV) and pepper mottle virus (PMV) has been produced. However, tobacco transgenic research is still to be carried out in the laboratory or in a strictly controlled environment due to government regulations (Zhao, 2000).

References

Abdullah, R. and Cocking, E. (1986) Efficient plant regeneration from rice protoplasts through somatic embryogenesis. *Biotechnology* **4**, 1087–1090.

Chan, M.T., Lee, T.M. and Chang, H.H. (1992) Transformation of indica rice (*Oryza sativa* L.) mediated by *Agrobacterium tumefaciens*. *Plant Cell Physiology* **33**, 577–583.

Chan, M.T., Chang, H.H., Ho, S.L., Tong, W.F. and Yu, S.M. (1993) *Agrobacterium*-mediated production of transgenic rice plants expressing a chimeric alpha-amylase promoter/beta-glucuronidase gene. *Plant Molecular Biology* **22**, 491–506.

Chen, J.Q., Huang, R.Z., Lang, C.X., Hu, Z.H. and Liu, Z.H. (1999) Molecular cloning and sequencing of the *PEP* gene from *Brassica napus* and the construction of the antisense *PEP* gene. *Journal of Zhejiang University (Agricultural and Life Sciences)* **25**, 365–367.

Chen, J. and Wang, Z.Y. (2004) Expression of *OsBP-73* gene requires involvement of its intron. *Rice Journal of Plant Physiology and Molecular Biology* **30**, 81–86.

Chen, X.R., Qian, H.F. and Xue, Q.Z. (2002) Resistance to bacterial blight and fertility of the progenies from *Xa21* transgenic rice Pei'ai 64S. *Chinese Journal of Rice Science* **16**, 270–272.

Cheng, M., Fry, J.E., Pang, S., Zhou, H., Hironaka, C.M., Duncan, D.R., Conner, T.W. and Wan, Y. (1997) Genetic transformation of wheat mediated by *Agrobacterium tumefaciens*. *Plant Physiology* **115**, 971–980.

Cheng, Z.Y. and Xue, Q.Z. (2003) Inheritance and expression of potato proteinase inhibitor gene II (*pinII*) in transgenic rice. *Scientia Agricultura Sinica* **2**, 728–735.

Cheng, X.Y., Sardana, R., Kaplan, H. and Altosaar, I. (1998) *Agrobacterium*-transformed rice expressing synthetic *cry1Ab* and *cry1Ac* genes are highly toxic to striped stem borer and yellow stem borer. *Proceedings of the National Academy of Sciences USA* **95**, 2767–2772.

Cui, J.J., Xia, J.Y., Ma, L.H. and Wang, C.Y. (2002) Studies on the efficacy dynamic of the transgenic *CryIAc* plus *CpTI* cotton (ZGK9712) to cotton bollworm. *Acta Gossypii Sinica* **14**, 323–329.

Cui, Y., Yang, Q.K., Zhou, S.J. and Kong, F.J. (2003) An investigation of improving soybean transformation via pollen-tube pathway. *Soybean Science* **22**, 75–77.

Deng, S.D., Xu, J., Zhang, Q.W., Zhou, S.W. and Xu, G.J. (2003) Effect of transgenic Bt cotton on population dynamics of non-target pests and natural enemies of pests. *Acta Entomologica Sinica* **46**, 1–5.

Duan, X., Li, X., Xue, Q., Abo-el-Saad, M., Xu, D. and Wu, R. (1996) Transgenic rice plants harbouring an introduced potato proteinase inhibitor II gene are insect resistant. *Nature Biotechnology* **14**, 494–498.

Feng, D.R., Xu, X.P., Fan, Q., Li, B.J., Lei, C.L. and Ling, Z.Z. (2001) Rice plants of multiple transgenes for resistance to rice blast and sheath blight diseases. *Acta Agronomica Sinica* **27**, 293–301

Hiei, Y., Ohta, S., Komari, T. and Kumashiro, T. (1994) Efficient transformation of rice (*Oryza sativa* L.) mediated by *Agrobacterium* and sequence analysis of the boundaries of the T-DNA. *The Plant Journal* **6**, 271–282.

Huang, J., Rozelle, S., Pray, C. and Wang, Q. (2002) Plant biotechnology in China. *Science* **295**, 674–676.

Jia, S.R. (2001) *Transgenic Cotton*. China Science Press, Beijing.

Klein, T.M., Wolf, E.D., Wu, R. and Sanford, J.C. (1987) High-velocity microprojectiles deliver nucleic acids into living cells. *Nature* **327**, 70–73.

Li, Y.C., Zhang, X.Y. and Xue, Q.Z. (2002a) Cloning and specific expression of the rice sucrose synthase gene promoter in transgenic rice plants. *Acta Agronomica Sinica* **28**, 586–590.

Li, Y.C., Zhang, X.Y. and Xue, Q.Z. (2002b) The production of large number of transgenic rice plants with two insect-resistant genes by *Agrobacterium*-mediated transformation. *Journal of Agricultural Biotechnology* **10**, 60–63.

Li, G.S., Zhang, Q.W., Zhang, J.R., Bi, Y.P. and Shen, L. (2002c) Establishment of multiple shoot clumps from maize (*Zea mays* L.) and regeneration of herbicide-resistant transgenic plantlets. *Science in China, Ser.C* **1**, 40–49.

Liu, Y.J. and Wang, G.Y. (2003) The inheritance and expression of *cry1A* gene in transgenic maize. *Acta Botanica Sinica* **45**, 253–256.

Liu, Z.H., Tang, J.Z., Li, B.J., Fan, Y.L., Li, G.L., Ji, L.Y. and Chen, W.C. (2003) Primary study on identification of corn-borer resistance in corn inbred-lines transformed with Bt gene. *Acta Agronomica Sinica* **29**, 621–625.

Rui, C.H., Fan, X.L., Dong, F.S. and Guo, S.D. (2002) Temporal and spatial dynamics of the resistance of transgenic cotton cultivars to *Helicoverpa armigera* (Hübner). *Acta Entomologica Sinica* **45**, 567–570.

Shu, Q.Y., Ye, G.Y., Cui, H.R., Cheng, X.Y., Xiang, Y.B., Wu, D.X., Gao, M.W., Xia, Y.W., Hu, C., Sardana, R. and Altosaar, I. (2000) Transgenic rice plants with a synthetic *cry1Ab* gene from *Bacillus thuringiensis* were highly resistant to eight lepidopteran rice pest species. *Molecular Breeding* **6**, 433–439.

Sun, X.F., Tang, K.X., Wan, B.L., Qi, H.R. and Lu, X.G. (2001) Pure transgenic rice plants expressing the snowdrop lectin GNA are resistant to brown rice plant hopper. *Chinese Science Bulletin* **46**, 1108–1113.

Wang, G., Zhang, Y.Z., Wei, S.H., Hu, Y.G., Wu, Y. and Ji, J. (2002) Transferring insecticidal protein from *Bacillus thuringiensis* (Bt) into excellent maize inbred line. *Journal of Jilin Agricultural University* **24**, 40–44.

Wang, H.Z., Huang, D.N., Lu, R.F., Liu J.J., Qian, Q. and Pen, X.Y. (2000) Salt tolerance of transgenic rice by transformation of *mtlD/gutD* gene. *Chinese Science Bulletin* **45**, 724–729.

Wang, J.X., Sun, Y., Cui, G.M. and Hu, J.J. (2001a) Transgenic maize plants obtained by pollen-mediated transformation. *Acta Botanica Sinica* **43**, 275–279.

Wang, Z.H., Wu, G., Cui, H.R., Altosaar, I., Xia, Y.W. and Shu, Q.Y. (2001b) Genetic analysis of *cry1Ab* gene of Bt rice. *Acta Genetica Sinica* **28**, 846–851.

Wang, R.X. (2003) The development and application of transgenic insect-resistant cotton in China. *Cotton Science* **15**, 180–184.

Wang, X.D. and Li, Y.Y. (2002) Development of transgenic restorer of cytoplasmic male sterility in upland cotton. *Scientia Agricultura Sinica* **35**, 7–141.

Xie, Y.Q., Liu, Y.L. and Zhu, Z. (2000) Deletional analysis of functional regions of complementary sense promoter from cotton leaf curl virus. *Science in China (Series C)* **5**, 498–506.

Xu, J.W., Chang, T.J., Feng, D.J., Zhou, Z. and Li, X.G. (2003) Cloning of genomic DNA of rice 5-enolpyruvylshikimate 3-phosphate synthase gene and chromosomal localization of the gene. *Science in China (Series C)* **33**, 224–230.

Xu, D., Xue, Q., McElroy, D., Mawal, Y., Hilder, V.A. and Wu, R. (1996) Constitutive expression of a cowpea trypsin inhibitor gene, *CpTi*, in transgenic rice plants confers resistance to two major rice insect pests. *Molecular Breeding* **2**, 167–173.

Xue, Q., Duan, X., Xu, D. and Wu, R. (1996) Production and testing of insect-resistant transgenic rice plants, in *Rice Genetics III, Proceedings of the Third International Rice Genetics Symposium*, pp. 239–245.

Ye, G.Y., Shu, Q.Y., Yao, H.W., Cui, H.R, Cheng, X.Y., Hu, C., Xia, Y.W., Gao, M.W. and Altosaar, I. (2001) Field evaluation of resistance of transgenic rice containing a synthetic *cry1Ab* gene from *Bacillus thuringiensis* Berliner to two stem borers. *Journal of Economic Entomology* **94**, 271–276.

Yu, Y.Z., Wang, G., Jin, N.Y., Du, J. and Ji, J. (2003) Transformation of type II calli of maize inbred lines mediated via *Agrobacterium tumefaciens*. *Maize Science* **11**, 28–33.

Zhai, W.X., Li, X.B., Tian, W.Z., Zhou, Y.L., Pian, X.B., Cao, S.Y., Zhao, X.F., Zhao, B., Zhang, Q. and Zhu, L.H. (2000) Introduction of a blight resistance gene, *Xa21*, into Chinese rice varieties through an *Agrobacterium*-mediated system. *Science in China (Series C)* **43**, 361–368.

Zhang, H.M., Jia, L.M., Wang, G.Y. and Tan, Z.B. (2003a) Transfer of Bt gene into maize elite inbred line 501 and C111 by microprojectile bombardment. *Maize Science* **11**, 7–9.

Zhang, W.H., Zhao, Q., Yu, J.J., Zhu, D.Y. and Ao, G.M. (2003b) To obtain transgenic maize plants with rabbit defensin (*NP-1*) gene and analyze their disease resistance. *Journal of Agricultural Biotechnology* **11**, 342–346.

Zhang, X.Y. and Xue, Q.Z. (2001) Introduction of soybean glycinin gene into rice (*Oryza sativa* L.) with *Agrobacterium*-mediated transformation. *Journal of Zhejiang Agricultural University (Agricultural and Life Sciences)* **27**, 495–499.

Zhang, W. and Wu, R. (1988) Efficient regeneration of transgenic plants from rice protoplasts and correctly regulated expression of the foreign gene in the plants. *Theoretical and Applied Genetics* **76**, 835–840.

Zhao, B.D. (2000) Biotechnology and tobacco production in China. *Acta Tabacaria Sinica* **6**, 37–42.

Zhao, B., Wang, W.M., Zheng, X.W., Wang, C.L., Ma, B.J., Xue, Q.Z., Zhu, L.H. and Zhai, W.X. (2000) Introduction of wide spectrum rice bacterial blight resistance gene *Xa21* into two-line genic male sterile rice variety Pei'ai 64S. *Chinese Journal of Biotechnology* **16**, 137–141.

Zhou, G.Y., Chen, S.B. and Hang, J.Q. (1993) *Research Advances in Crop Molecular Breeding.* Chinese Agricultural Science and Technology Press, Beijing.

Zhou, S.J., Li, X.C., Liu, S.J., Liu, L.Y. and Yang, Q.K. (2001) Optimization of the *Agrobacterium*-mediated transformation system of soybean. *Journal of Northeast Agricultural University* **32**, 325–331.

PART II
NEW DEVELOPMENTS

2.1

Advances in Transformation Technologies

Huw Jones

Crop Performance and Improvement, Rothamsted Research,
Harpenden, Hertfordshire, AL5 2JQ, United Kingdom

Introduction

The genetic modification of plants depends on two key processes: the ability to transfer and integrate DNA into the genome of a host cell and the ability to regenerate adult, fertile plants from that transformed cell. The development of these processes, from proof of concept experiments in the early 1980s to the robust platform technologies that they represent today, have followed separate paths in monocotyledonous and dicotyledonous crop plants characterized by different modes of DNA delivery and different tissue culture regimes.

The first reports of successful dicot transformation utilized oncogenic and disarmed strains of *Agrobacterium tumefaciens* and a range of explant types (see reviews by Gasser and Fraley, 1989; Klee *et al.*, 1987; Wordragen and Dons, 1992). Early work in tobacco targeted regenerable protoplasts or cells derived from them as plant hosts (Horsch *et al.*, 1984; Krens *et al.*, 1982). However, regeneration for protoplasts proved more difficult for many crop species and alternative, more easily regenerable explants were developed. For example, leaf discs were successful in tomato (McCormick *et al.*, 1986) while in soybean, cotyledonary nodes (Hinchee *et al.*, 1988) and immature cotyledons (Parrott *et al.*, 1989) were preferred. Stem sections were widely utilized for *Brassica* species (Fry *et al.*, 1987; Pua *et al.*, 1987) and potato (Ooms *et al.*, 1987). Direct gene transfer (DGT) methods have also been successfully demonstrated in dicot species; the first report

Plant Biotechnology. Edited by Nigel Halford.
© 2006 John Wiley & Sons, Ltd.

was for tobacco transformation (Paszkowski *et al.*, 1984). Although a few specialized examples of DGT are used routinely for dicot crops, such as PEG-mediated transformation of sugarbeet guard cell protoplasts (Hall *et al.*, 1996), in general *Agrobacterium*-mediated transformation of cell cultures or explants is the current method of choice and protocols are now available to genetically modify a wide range of dicot crops (Curtis, 2004) (Table 2.1.1).

The first fertile transgenic monocots were regenerated from protoplasts transformed with naked DNA using electroporation or polyethylene glycol (PEG), including rice (Datta *et al.*, 1990; Shimamoto *et al.*, 1989; Toriyama *et al.*, 1988; Zhang *et al.*, 1988; Zhang and Wu, 1988) and maize (Rhodes *et al.*, 1988). For other cereals, difficulties in regenerating from protoplasts led to the targeting of alternative cell types such as regenerable cell cultures, callus or immature embryo explants. This, combined with improved DNA delivery methods via particle bombardment led to a second phase of technology development with breakthroughs in the transformation of maize (Fromm *et al.*, 1990; Gordon-Kamm *et al.*, 1990), sugarcane (Bower and Birch, 1992), wheat (Vasil *et al.*, 1992), tritordium (Barcelo *et al.*, 1994) and rye (Castillo *et al.*, 1994). In addition, particle bombardment removed the germplasm dependency of other species, enabling the transformation of elite varieties of indica and japonica rice (Christou *et al.*, 1991) and cotton (McCabe and Martinell, 1993).

The robust tissue culture methods from regenerable explants such as immature zygotic embryos were then adapted for T-DNA delivery and enabled the development of *Agrobacterium*-mediated transformation of many monocots (Table 2.1.2). Hiei *et al.* (1994) reported transformation of japonica rice with super-binary vectors containing additional *VIR* genes. Subsequently, high-efficiency transformation with little genotype dependency has been reported for japonica, indica and javanica varieties (reviewed by Tyagi and Mohanty, 2000; see Chapter 1.3). In maize, barley and wheat, *Agrobacterium* transformation has been achieved mainly in genotypes selected for their high regeneration capacity such as the maize inbred line A188 (Ishida *et al.*, 1996), the barley cultivars Golden Promise (Tingay *et al.*, 1997), Dissa (Wu *et al.*, 1998) and Schooner (Wang *et al.*, 2001), and the ill-defined 'Bobwhite' wheat line (Cheng *et al.*, 1997). Modifications in the tissue culture regimes indicate that it may be possible to broaden the application of some protocols to include previously untransformed elite cereal varieties (Gordon-Kamm *et al.*, 2002; Murray *et al.*, 2004; Wu *et al.*, 2003). The regeneration of stably transformed sorghum (Zhao *et al.*, 2000) and forage grass (Bettany *et al.*, 2003) via *Agrobacterium* co-cultivation has also been reported.

Agrobacterium and Agrovectors

For most research and biotechnology applications, *Agrobacterium* has advantages over other forms of DNA delivery methods, including biolistics, because it can introduce larger segments of DNA with minimal rearrangement and with fewer copies of inserted transgenes at higher efficiencies and at lower cost (see reviews by Gelvin, 2003; Gheysen *et al.*, 1998; Hansen and Wright, 1999; Kohli *et al.*, 2003). These advantages have been major drivers in the move away from DGT to *Agrobacterium* transformation in monocots. In addition to the wide range of plant types transformed, *Agrobacterium*-mediated T-DNA

Table 2.1.1 *Selected reports detailing the explant and* A. tumefaciens *strain/plasmid used to transform a wide range of dicotyledonous food and horticultural crops.*

Species	Variety/ cultivar	Explant	Agrobacterium strain	Plasmid	Reference
Almond (*Prunus dulcis* Mill.)	Boa Casta	Wounded leaves	LBA4404/ EHA105	p35SGUSint/ pFAJ3003	(Miguel and Oliveira, 1999)
Apple (*Malus* × *domestica* Borkh.)	Rootstock M26	Shoots/leaves	A281/C58/ A348	pCGP257/ pBI121	(Maheswaran et al., 1992)
Apple	Jonagold	Wounded leaves	EHA101	pFAJ3003 pFAJ3027	(DeBondt et al., 1996)
Apple (*Malus* × *pumila* Mill.)	Queen Cox	Shoots/leaves	EHA101	pSCV1.6	(Wilson and James, 2003)
Apricot (*Prunus armeniaca*)	Kecskemeter	Immature cotyledons	LBA4404	pBinGUSint/ pBinPPVm	(Machado et al., 1992)
Brassica rapa (*Brassica campestris* ssp. *pekinensis*)	Spring Flavor	Immature cotyledons	LBA4404	pTOK/BKS1	(Jun et al., 1995)
Brassica napus (*Brassica napus* L.)	Westar and Sabine	Floral bud dipping	C58ClRif[R]	pGV3101/ pNOV264	(Wang, Menon and Hansen, 2003)
Brassica oleracea Brassica oleracea	Hercules, Cape spitz and others	Immature cotyledons	A281	pKK6/pGA472	(Pius and Achar, 2000)
Buckwheat (*Fagopyrum esculentum* Moench.)	Darja	Cotyledon fragments	A281	pGA427/ pTiBo542	(Miljusdjukic et al., 1992)
Cotton (*Gossypium hirsutum* L.)	CUBQHRPIS	Shoot apex	LBA 4404	pBI121	(Zapata et al., 1999)
Cotton (*Gossypium hirsutum* L.)	Coker201	Cotyledon fragments	LBA4404/ 15955	pH575	(Firoozabady et al., 1987)
Cyclamen (*Cyclamen persicum* Mill.)	Anneke	Petiole segments	AGL0/ LBA4404	pIG121Hm	(Aida et al., 1999)
Hot chilli (*Capsicum annuum* L.)	Pusa jawala	Shoot buds	EHA 105	pBI 121	(Manoharan, Vidya and Sita, 1998)
Orange (*Citrus sinensis*) hybrid	Sweet orange	Epicotyl segments	EHA 105	p35SGUSint	(Pena et al., 2004)
Pear (*Pyrus communis* L.)	Conference	Wounded leaves	EHA101	pFAJ3000	(Mourgues et al., 1996)
Peppermint (*Mentha* × *piperita* L.)	Black Mitcham 38	Leaf disks	EHA105 LBA4404	pMOG410	(Diemer et al., 1998)
Pine (*Pinus radiata*)	GF17 GF19	Cotyledon	AGL1	pGA643	(Grant, Cooper and Dale, 2004)

(*continued*)

Table 2.1.1   *(Continued)*

Species	Variety/ cultivar	Explant	*Agrobacterium* strain	Plasmid	Reference
Rhododendron (*Rhododendron sp.*)	America, Mars and others	Stem segments	LBA4404	pAL4404/ p35SGUSint	(Pavingerova, *et al.*, 1997)
Soybean (*Glycine max* L. Merr.)	Hefeng 35 and 39 and Dongnong 42	Embryonic tips	EHA 105	pCAMBIA2301	(Liu, Yeng and Wei, 2004)
Soybean (*Glycine max* L. Merr.)	Jack	Immature cotyledons	EHA 105	pHIG/Z	(Yan *et al.*, 2000)
Strawberry (*Fragaria × ananassa* Duch.)	Chandler	Leaf squares	LBA4404 (used in PDS1000)	pAL4404/ pGUSint	(de Mesa *et al.*, 2000)
Sunflower (*Helianthus annus* L.)	KBSH-1	Wounded seedling	LBA 4404	pKIWI105	(Rao and Rohini, 1999)
Sunflower (*Helianthus annus* L.)	HA300B	Wounded shoot apicies	GV2260	p35SGUSint	(Weber *et al.*, 2003)
Sugar beet (*Beta vulgaris* and *B. maritima*)	9 accessions	Shoot bases	EHA101 LBA4404	pGM221 pBI121	(Hisano *et al.*, 2004)
Sugar beet (*Beta vulgaris*)	O-type 272	Cotyledons	EHA101	p35SPMI	(Joersbo *et al.*, 1999)

transfer has also been demonstrated in yeast (Bundock *et al.*, 1995), filamentous fungi (de Groot *et al.*, 1998; dos Reis *et al.*, 2004; Park and Kim, 2004) and human HeLa cells (Kunik *et al.*, 2001).

Agrobacterium tumefaciens is a pathogenic soil bacterium that infects susceptible plants and causes neoplastic growths characteristic of crown gall disease. It does so by transferring a portion of its large Ti (tumor inducing) plasmid into a host plant cell. The transferred DNA, usually called the 'T-region' when located on the native Ti plasmid (and 'T-DNA' when part of a modified plasmid), contains genes that become integrated into the plant's genome and encode auxins and cytokinins (plant hormones that lead to tumor formation) and opines to feed the bacteria. The T-region is flanked by left and right border sequences (24 or 25 bp imperfect repeats) that define the DNA to be transferred. The Ti plasmid also contains a virulence (*vir*) region including the genes *VirA*, *VirB*, *VirC*, *VirD*, *VirE*, *VirG* and *VirH* whose products, along with those of chromosomal genes, mediate the formation, transfer and integration of the T-complex (for reviews see Gelvin, 2003; Tzfira *et al.*, 2004; Zupan *et al.*, 2000).

The development of *Agrobacterium* for plant transformation required the removal of the oncogenic functions to generate disarmed strains and modifications to allow the insertion of foreign DNA between the border sequences, and to reduce the large size of the native Ti plasmid (which in *Agrobacterium* strain C58 is 214 233 bp (Goodner *et al.*, 2001)). Two approaches have been used. One involved *cis*-integration of transgenes, via homologous recombination or co-integration, to generate a disarmed but otherwise intact

Table 2.1.2 *Selected reports detailing the explant and A. tumefaciens strain/plasmid used in A. tumefaciens transformation of rice, maize, wheat and barley. PCIE, pre-cultured immature embryos; EC, embryogenic callus cultures; IE, fresh immature embryos.*

Variety (S—spring) (W—winter)	Explant	Agrobacterium strain (vector)	Transformation Frequency (%)	Plant Selection	No. of plants reported	Refs.
Wheat						
Bobwhite (S)	PCIE EC	C58 pMON18365	1.4–4.3	G418	>100	(Cheng et al., 1997)
Fielder (S)	PCIE	AGL0 pBGX1	1.8	GFP, Timentin, Bialaphos	4	(Weir et al., 2001)
Florida (W)	IE	AGL1 pAL154/156	0.3–3.3	PPT (L-phosphinothricin)	39	(Wu et al., 2003)
Cadenza (S)	IE	AGL1 pAL154/156	2.5	PPT (L-phosphinothricin)	5	(Wu et al., 2003)
Veery-5 (S)	EC	LBA4404 pHK21	1.2–3.9	Glufosinate ammonium	17	(Khanna and Daggard, 2003)
Bobwhite (S)	PCIE; EC	C58 pMON18365	4.4	Glyphosate	3354	(Hu et al., 2003)
Bobwhite (S)	PCIE; EC	C58 pMON18365	4.8–19	Paromomycin Glyphosate	154	(Cheng et al., 2003)
Bobwhite (S)	PCIE	C58 pPTN155	0.5–1.5	G418	13	(Haliloglu and Baenziger, 2003)
Barley						
Golden Promise (S)	IE	AGL1 pDM805	4.2	Bialaphos	54	(Tingay et al., 1997)
Golden Promise (S)	IE	AGL1 pBGXYNR: pBHXYNR	6.0	Bialaphos	24	(Patel et al., 2000)
Golden Promise (S)	IE	AGL1; AGL0 pWBVec82b	2–12	Hygromycin	63	(Matthews et al., 2001)
Schooner (S)	EC	AGL0; AGL1; LBA4404 pTO134; pHPAVpol; pB2Pol	0.5–2.9	Hygromycin; Biolaphos	19	(Wang et al., 2001)

Table 2.1.2 (Continued)

Variety (S—spring) (W—winter)	Explant	Agrobacterium strain (vector)	Transformation Frequency (%)	Plant Selection	No. of plants reported	Refs.
Golden Promise (S)	PCIE	AGL0 pBU-B35S.IG	1.7–6.3	Bialaphos	42	(Trifonova, Madsen and Olesen, 2001)
Golden Promise (S)	IE	AGL0 pYF133	3.5	Hygromycin GFP	11	(Fang, Akula and Altpeter, 2002)
Golden Promise (S)	IE	H228 pNRG040	N/A	Bialaphos	42	(Horvath et al., 2003)
Golden promise; Schooner; Chebec and Sloop (all S)	IE	AGL1 pWBVec8; pVec8-GFP; pVec8-Gusl	0.6–12.0	Hygromycin; GFP; GUS	34	(Murray et al., 2004)
Maize						
Funk's G90	Shoot apices	EHA1 TiBO542; pGUS3	N/D	Kanamycin; Carbenicillin	6	(Gould et al., 1991)
A118 and others	EC, IE	LBA4404 pSB131; pTOK233	5–30	PPT; Bialaphos	>100	(Ishida et al., 1996)
A118	IE	LBA4404 pNOV117	0.7–32	Mannose	>50	(Negrotto et al., 2000)
Hi II	IE	LBA4404 pSB11 Ubi bar/35SGUS	40	Bialaphos	>500	(Zhao et al., 2002)
Hi II	IE	EHA101 pTF102	5.5	Bialaphos	50	(Frame et al., 2002)
Hi II	IE	LBA4404 p108–p111, based on pSB11	0.5–25	Bialaphos	>100	(Gordon-Kamm et al., 2002)
A118, Hi II and hybrids	IE	LBA4404 pWCO38; pWCO39	N/D	Butafenecil; mannose	>2500	(Li et al., 2003)

Rice

(*Oryza sativa* L.) Javavica cv. Rojolele	EC	EHA101 pAFT14	7–23	Hygromycin	71	(Rachmawati et al., 2004)
Javanica cv. Taipei 309 Gulfmont Jefferson	EC	LBA4404/EHA101 pTOK233/pIG121	8–13	Hygromycin	>100	(Dong et al., 1996)
Japonica cv. Nortai and Radon	IE	LBA4404 pTOK233	27	Hygromycin	346	(Aldemita and Hodges, 1966)
Indica cv. IR72 and TCS10	IE	At656 pCNL56	1–5	Hygromycin	9	(Aldemita and Hodges, 1996)
Indica cv. Basmati 370, 385 and 6129	EC	EAH101 pIG121Hm	cv 370 22 cv 385 5	Hygromycin	30	(Rashid et al., 1996)
Indica cv. Pusa Basmati 1	EC	LBA4404 pTOK233	184	Hygromycin	25	(Mohanty, Sarma and Tyagi, 1999)
Japonica cv. Tsukinohikari, Asanohikari and Koshihikari	EC IE Scutella	LBA4404/EHA101 pTOK233/pIG121Hm	13–29	Hygromycin	300	(Hiei et al., 1994)

Ti plasmid (Ruvkun and Ausubel, 1981; Vanhaute *et al.*, 1983; Zambryski *et al.*, 1983). These procedures were laborious and this approach is now used only for specialist applications. The other, which represented a major advance in the application of *Agrobacterium* to plant genetic engineering was made when it was reported that, apart from the border sequences themselves, the functions of the Ti plasmid necessary for transformation could be supplied *in trans* via a binary vector system (Bevan, 1984; Deframond *et al.*, 1983; Hoekema *et al.*, 1983). These vector systems comprise two plasmids: one with a convenient multiple cloning site flanked by T-border sequences, a selectable marker gene and an origin of replication for easy maintenance in *E. coli*; and another, disarmed Ti plasmid, lacking the tumor-inducing genes but retaining the *vir* loci whose products interact with the T-strand and facilitate DNA transfer to the plant cell.

Numerous specialist vector systems based on the binary approach have been developed for plant transformation (for review see Hellens *et al.*, 2000a). For example, the BIBAC and TAC vectors were designed for large genomic inserts (Hamilton, 1997; Liu *et al.*, 1999). A series of mini binary vectors were made by removing half the DNA from the plasmid backbone to generate more unique restriction sites in the multiple cloning site (Xiang *et al.*, 1999). The modular vector pMVTBP was designed specifically for wheat and incorporated a ribosomal attachment sequence to increase translational efficiency and FLAG and His epitope tags to allow immunodetection and purification (Peters *et al.*, 1999). In the pSoup/pGreen suite of vectors, the *RepA* replication function was removed from pGreen, which meant it could replicate in *Agrobacterium* only if another plasmid, pSoup, was co-resident in the same strain to provide this function *in trans* (Hellens *et al.*, 2000b). Specific pSoup/pGreen suites of vectors also included examples that incorporated additional copies of *VirB, VirC* and *VirG* found on the so-called Komari fragment, implicated in the super-virulence of strains haboring the plasmid pTiBo542 (Hood *et al.*, 1987; Komari, 1989; 1990). An alternative suite of GATEWAY-compatible (Invitrogen, Gaithersburg, MD, USA) destination vectors incorporating a range of scorable and selectable marker genes for *Agrobacterium*-mediated plant transformation has been developed (Karimi *et al.*, 2002). Many *A. tumefaciens* strains and vectors suitable for plant transformation can be ordered online, for example: www.cambia.org, www.pgreen. ac.uk, www.cbs.knaw.nl/index.

Other DNA-Delivery Methods

Although *Agrobacterium*-based transformation is the preferred method for most plant species, DGT methods have also been widely used, both for general transformation and for specialist applications. DGT methods were developed and utilized mainly in cereals due to the early difficulties with *Agrobacterium* transformation. The transfer of naked DNA to plant cells and successful transgene expression have been achieved with methods as diverse as: electroporation, macro- and microinjection, vortexing with silicon carbide fibers, polyethylene glycol, ultrasound- and laser-mediated uptake and particle bombardment (see review by Barcelo *et al.*, 2001). Some of these methods, such as polyethylene glycol-mediated uptake or electroporation, were most successful with protoplasts, which in cereals are laborious to isolate and difficult to regenerate from. Others were inefficient or not reproducible. However, particle or microprojectile bombardment (also called biolistics) became a widely used robust method for cereal transformation. The helium-driven particle delivery system first developed by DuPont and subsequently marketed by BioRad as the PDS1000/He was widely used for this purpose, but other devices such as the particle inflow gun (PIG) and the ACCELL^{TM} electrical discharge technology have also been used successfully. All involve the adsorption of circular or linear forms of DNA onto the surface of microcarrier particles, which are driven at high velocity into recipient plant cells using an acceleration device (Sanford, 1988; 1990). Biolistics has also been used to deliver DNA into the chloroplast and mitochondrion genomes (for review see Sanford *et al.*, 1993) and in a twist to the conventional approach, *Escherichia coli* or *Agrobacterium* cells have been used directly as microprojectiles (Kikkert *et al.*, 1999; Rasmussen *et al.*, 1994).

In vitro and *in planta* Transformation

For most biotechnology applications, the successful integration of transgenes into the genome of a target plant cell must be followed by regeneration of whole, fertile, non-chimeric plants. This is commonly achieved by micropropagation through *in vitro* tissue culture which utilizes the totipotency and plasticity of plant cells. There are two principle routes by which plants can be recovered via tissue culture: somatic embryogenesis and organogenesis. Somatic embryogenesis is an asexual propagation process where somatic cells differentiate into embryo-like structures with shoot and root meristems. With appropriate plant hormones and other culture medium additions, somatic embryos can be 'germinated' and give rise to viable adult plants. This is the route used in the regeneration of most cereal crops (for review see Barcelo *et al.*, 2001). Organogenesis refers to the ability of some plant tissues (e.g. hypocotyls, cotyledons, leaf bases or callus derived from them) to re-organize into shoot meristems, which can subsequently be rooted to generate complete plants. The transformation step can be targeted to explant tissue such as leaf disks or shoots prior to de-differentiation or to de-differentiated callus cells.

Although plant regeneration via *in vitro* tissue culture is the only currently available route to genetically engineer crops, it is a major bottleneck to high-efficiency transformation in many species; it is time-consuming, labor- and material-intensive and increases the chance of genetic instability due to unpredictable somaclonal variation. This has led to considerable interest in various *in planta* (also called germline) transformation methods, which depend on the ability to introduce naked DNA or T-DNA into gametes, pre-gametic tissue or zygotes around the time of fertilization and avoids the need for tissue culture. Well-developed 'vacuum infiltration' and 'floral dip' methods are available for *Agrobacterium* transformation of Arabidopsis (Bechtold *et al.*, 1993; Clough and Bent, 1998) that target the unfertilized ovules (Bechtold *et al.*, 2000; Desfeux *et al.*, 2000; Ye *et al.*, 1999). The infiltration of flowers with *Agrobacterium* has also been used successfully to transform the model legume *Medicago truncate* and *Brassica campestris* (pakchoi) (Liu *et al.*, 1998; Trieu *et al.*, 2000). Efficient male germline transformation by biolistics has been achieved in tobacco (Aziz and Machray, 2002) and there are reports that *in planta* transformation of soybean has been achieved in China (Hu and Wang, 1999). No *in planta* methods are currently available for monocots but, if perfected, would represent a major breakthrough for cereal transformation.

Transgene Integration and Gene Targeting

Molecular analysis of integration sites resulting from untargeted DGT suggests that transgenes insert via double-stranded illegitimate recombination, utilizing the plant's own DNA repair machinery, at a single (or sometimes more than one) locus. The locus may contain one or multiple copies of the transgene, which may have undergone rearrangements and/or may have generated short lengths of 'filler' DNA homologous to flanking plant genomic DNA (Kohli *et al.*, 1998; Pawlowski and Somers, 1998; Svitashev *et al.*, 2000; 2002). The processing, transfer and integration of T-DNA into plant chromosomal DNA have been studied, both biochemically and in *Agrobacterium* strains containing mutant forms of the *Vir, Chv, Psc* and *Att* alleles (reviewed in Christie, 1997; Ward *et al.*, 2002; Zupan *et al.*, 1998; 2000). While the precise mechanism of T-DNA integration remains unclear, it is likely that DNA in the T-complex is made double-stranded just prior to integration via non-homologous end joining. Although there appears to be no sequence specificity for T-DNA transformation, there is evidence that integrations occur preferentially into transcriptionally active regions (Feldmann, 1991; Koncz *et al.*, 1989; Lindsey *et al.*, 1993; Topping *et al.*, 1991) and can therefore potentially disrupt native genes. However, recent data suggest that these observations may be partly explained by selection bias with 30 % of PCR-identified primary T-DNA insertion sites silenced (Francis and Spiker, 2005).

The conventional way of overcoming the problem of transgene rearrangements, high copy number or unintentional disruption of native genes is to generate a large number of lines and screen for single-copy events. However, this unpredictability is a major concern for opponents of GM technologies, and gene targeting by homologous recombination offers an alternative, more precise method of genetic manipulation. It is commonly used in mice, in prokaryotes, in the nuclear genomes of simple plants such as moss and the chloroplast genome (Capecchi, 2001; Evans, 2001; Maliga, 2004; Schaefer and Zryd, 1997; Smithies, 2001) but appears to operate at low frequency in higher plants. However, two recent reports have illustrated that gene targeting might be a feasible approach to modify crop plants. In over four million Arabidopsis plants transformed to induce two simultaneous mutations, only one was identified as a true gene targeting event with no ectopic T-DNA insertions (Hanin *et al.*, 2001). A more encouraging result was reported by Terada *et al.* (2002) who obtained six rice plants in which the *waxy* gene had been disrupted by true homologous recombination with a frequency of 6.5×10^{-4}. All the plants were heterozygous at the waxy locus, with one wild-type and one mutant allele, and no random or ectopic targeting could be detected.

With further development, gene targeting technologies could provide a means by which any gene could be mutagenized and reintroduced into its original chromosomal location. This could be used to knockout individual genes or to alter the properties of proteins encoded by them. In addition, it could quell some of the fears over the random nature of genetic manipulation by targeting transgene cassettes into pre-defined, stable genomic locations.

Benign Methods of Selection and Removal of Marker Genes

Selectable marker genes are used routinely for research purposes, enabling highly efficient DNA cloning procedures in *E. coli*, yeast and plant transformation. However, perceived risks of horizontal and vertical gene transfer have made the use of such selectable genes undesirable, particularly where field trials are proposed, and efforts have been made to develop benign selection systems or to reduce the dependence on selectable marker genes in the transformation process. The most widely adopted benign selection system utilizes the *ManA* gene encoding phosphomannose isomerase (PMI) (Syngenta), which converts the predominant carbon source available to the plant, mannose-6-phosphate, to fructose-6-phosphate for respiration (Joersbo, 2001; Joersbo *et al.*, 1999). Not only are the untransformed cells deprived of a carbon source, but also the unutilized mannose-6-phosphate accumulates and has additional negative effects including inhibition of glycolysis, possibly due to phosphate starvation. PMI selection has been shown to be an effective selection system for crop species including wheat and maize (Negrotto *et al.*, 2000; Reed *et al.*, 2001; Wang *et al.*, 2000; Wright *et al.*, 2001) rice (Lucca *et al.*, 2001) and pearl millet (O'Kennedy *et al.*, 2004). Other gene/substrate combinations used for plant selection include the xylose isomerase gene and medium containing xylose as the predominant carbon source (Haldrup *et al.*, 1998; 2001), the *E. coli* threonine deaminase gene in combination with the isoleucine analog L-*O*-methylthreonine (Ebmeier *et al.*, 2004) and genes encoding enzymes that deactivate D-amino acids, which inhibit plant growth (Erikson *et al.*, 2004).

In a small proportion of lines, the co-bombardment of trait and marker genes on separate plasmids generates integrations into unlinked loci and the potential to identify progeny individuals that are null for the marker gene through segregation away from the trait gene. However, even in lines that have lost the plant selectable plasmid by segregation, the bacterial selectable marker, origin of replication and plasmid backbone linked to the trait gene remain in the transgenic plant. In an attempt to overcome this, Fu *et al.* (2000) used DNA fragments, comprising only the transgene expression cassette, to transform rice biolistically. They found that, not only was the vector backbone eliminated from the plant line but also the proportion of low-copy number, structurally simple transgenic loci were increased. Purifying clean DNA fragments containing only the sequences necessary for transgene expression is not a trivial task and *Agrobacterium* T-DNAs should be inherently cleaner because, in an ideal model, only the sequences delineated by the left and right borders are transferred. The potential for twin T-DNA systems to generate selectable marker-free plants has been assessed in barley. One third of the transformed plants contained both T-DNAs, and in approximately one quarter of those the T-DNAs segregated independently to yield marker-free transgenic plants (Matthews *et al.*, 2001). A range of elite rice cultivars transformed by *Agrobacterium* containing a vector with two T-DNA constructs recently confirmed this to be a useful system for generating marker-free rice plants (Breitler *et al.*, 2004).

Other approaches to marker gene elimination involve the use of site-specific recombinases or transposases to excise or relocate the marker gene for elimination by segregation. These methods have proved successful in model dicot species (reviewed by Ebinuma *et al.*, 2001; Hohn *et al.*, 2001) and have been successfully applied in rice (e.g. Hoa *et al.*, 2002) and maize (Zhang *et al.*, 2003). However, these methods can still leave sequence fragments in the plant genome and the trends in commercial biotechnology are toward total avoidance rather than the utilization of molecular tools to remove unwanted sequences. In a radically new approach to the removal of foreign DNA, it has been demonstrated that analogs of the *Agrobacterium* T-DNA border sequences exist in plants (so-called P-DNA) and that they can function to delimit T-strand synthesis and DNA transfer during the transformation process (Rommens *et al.*, 2004). In this study, *Agrobacterium* was used to produce marker- and backbone-free potato plants displaying reduced expression of an enzyme responsible for post-harvest discoloration. This represents a significant breakthrough in crop improvement through the modification of the plant's own genome and could be applicable to cereal species if T-DNA analogs could be found.

References

Aida, R., Hirose, Y., Kishimoto, S. and Shibata, M. (1999) *Agrobacterium tumefaciens*-mediated transformation of *Cyclamen persicum* Mill. *Plant Science* **148**, 1–7.

Aldemita, R.R. and Hodges, T.K. (1996) *Agrobacterium tumefaciens*-mediated transformation of japonica and indica rice varieties. *Planta* **199**, 612–617.

Aziz, N. and Machray, G.C. (2002) Efficient male germ line transformation for transgenic tobacco production without selection. *Plant Molecular Biology* **51**, 203–211.

Barcelo, P., Hagel, C., Becker, D., Martin, A. and Lorz, H. (1994) Transgenic cereal (Tritordeum) Plants obtained at high-efficiency by microprojectile bombardment of inflorescence tissue. *The Plant Journal* **5**, 583–592.

Barcelo, P., Rasco-Gaunt, S., Thorpe, C. and Lazzeri, P.A. (2001) Transformation and gene expression, in *Advances in Botanical Research Incorporating Advances in Plant Pathology*, Volume 34 (eds P.R. Shewry, P.A. Lazzeri, and K.J. Edwards), pp. 59–126.

Bechtold, N., Ellis, J. and Pelletier, G. (1993) *In planta Agrobacterium*-mediated gene transfer by infiltration of adult *Arabidopsis thaliana* plants. *Comptes Rendus De L Academie Des Sciences Serie Iii-Sciences De La Vie-Life Sciences* **316**, 1194–1199.

Bechtold, N., Jaudeau, B., Jolivet, S., Maba, B., Vezon, D., Voisin, R. and Pelletier, G. (2000) The maternal chromosome set is the target of the T-DNA in the *in planta* transformation of *Arabidopsis thaliana*. *Genetics* **155**, 1875–1887.

Bettany, A.J.E., Dalton, S.J., Timms, E., Manderyck, B., Dhanoa, M.S. and Morris, P. (2003) *Agrobacterium tumefaciens*-mediated transformation of *Festuca arundinacea* (Schreb.) and *Lolium multiflorum* (Lam.). *Plant Cell Reports* **21**, 437–444.

Bevan, M. (1984) Binary agrobacterium vectors for plant transformation. *Nucleic Acids Research* **12**, 8711–8721.

Bower, R. and Birch, R.G. (1992) Transgenic sugarcane plants via microprojectile bombardment. *The Plant Journal* **2**, 409–416.

Breitler, J.C., Meynard, D., Van Boxtel, J., Royer, M., Bonnot, F., Cambillau, L. and Guiderdoni, E. (2004) A novel two T-DNA binary vector allows efficient generation of marker-free transgenic plants in three elite cultivars of rice (*Oryza sativa* L.) *Transgenic Research* **13**, 271–287.

Bundock, P., Dendulkras, A., Beijersbergen, A. and Hooykaas, P.J.J. (1995) Transkingdom T-DNA transfer from *Agrobacterium tumefaciens* to *Saccharomyces cerevisiae*. *EMBO Journal* **14**, 3206–3214.

Capecchi, M.R. (2001) Generating mice with targeted mutations. *Nature Medicine* **7**, 1086–1090.

Castillo, A.M., Vasil, V. and Vasil, I.K. (1994) Rapid production of fertile transgenic plants of rye (*Secale cereale* L.). *Bio-Technology* **12**, 1366–1371.

Cheng, M., Fry, J.E., Pang, S.Z., Zhou, H.P., Hironaka, C.M., Duncan, D.R., Conner, T.W. and Wan, Y.C. (1997) Genetic transformation of wheat mediated by *Agrobacterium tumefaciens*. *Plant Physiology* **115**, 971–980.

Cheng, M., Hu, T.C., Layton, J., Liu, C.N. and Fry, J.E. (2003) Desiccation of plant tissues post-*Agrobacterium* infection enhances T-DNA delivery and increases stable transformation efficiency in wheat. *In Vitro Cellular and Developmental Biology-Plant* **39**, 595–604.

Christie, P.J. (1997) *Agrobacterium tumefaciens* T-complex transport apparatus: a paradigm for a new family of multifunctional transporters in eubacteria. *Journal of Bacteriology* **179**, 3085–3094.

Christou, P., Ford, T.L. and Kofron, M. (1991) Production of transgenic rice (*Oryza sativa* L.) plants from agronomically important indica and japonica varieties via electric discharge particle acceleration of exogenous DNA into immature zygotic embryos. *Bio-Technology* **9**, 957–962.

Clough, S.J. and Bent, A.F. (1998) Floral dip: a simplified method for *Agrobacterium*-mediated transformation of *Arabidopsis thaliana*. *The Plant Journal* **16**, 735–743.

Curtis, I.S. (2004) *Tansgenic Crops of the World—Essential Protocols*. Kluwer, Dordrecht.

Datta, S.K., Peterhans, A., Datta, K. and Potrykus, I. (1990) Genetically engineered fertile indica-rice recovered from protoplasts. *Bio-Technology* **8**, 736–740.

de Groot, M.J.A., Bundock, P., Hooykaas, P.J.J. and Beijersbergen, A.G.M. (1998) *Agrobacterium tumefaciens*-mediated transformation of filamentous fungi. *Nature Biotechnology* **16**, 839–842.

de Mesa, M.C., Jimenez-Bermudez, S., Pliego-Alfaro, F., Quesada, M.A. and Mercado, J.A. (2000) *Agrobacterium* cells as microprojectile coating: a novel approach to enhance stable transformation rates in strawberry. *Australian Journal of Plant Physiology* **27**, 1093–1100.

DeBondt, A., Eggermont, K., Penninckx, I., Goderis, I. and Broekaert, W.F. (1996) *Agrobacterium*-mediated transformation of apple (*Malus* × *domestica* Borkh): an assessment of factors affecting regeneration of transgenic plants. *Plant Cell Reports* **15**, 549–554.

Deframond, A.J., Barton, K.A. and Chilton, M.D. (1983) Mini-Ti—a new vector strategy for plant genetic-engineering. *Bio-Technology* **1**, 262–269.

Desfeux, C., Clough, S.J. and Bent, A.F. (2000) Female reproductive tissues are the primary target of *Agrobacterium*-mediated transformation by the *Arabidopsis* floral-dip method. *Plant Physiology* **123**, 895–904.

Diemer, F., Jullien, F., Faure, O., Moja, S., Colson, M., Matthys-Rochon, E. and Caissard, J.C. (1998) High efficiency transformation of peppermint (*Mentha* × *piperita* L.) with *Agrobacterium tumefaciens*. *Plant Science* **136**, 101–108.

Dong, J.J., Teng, W.M., Buchholz, W.G. and Hall, T.C. (1996) *Agrobacterium*-mediated transformation of javanica rice. *Molecular Breeding* **2**, 267–276.

dos Reis, M.C., Fungaro, M.H.P., Duarte, R.T.D., Furlaneto, L. and Furlaneto, M.C. (2004) *Agrobacterium tumefaciens*-mediated genetic transformation of the entomopathogenic fungus *Beauveria bassiana*. *Journal of Microbiological Methods* **58**, 197–202.

Ebinuma, H., Sugita, K., Matsunaga, E., Endo, S., Yamada, K. and Komamine, A. (2001) Systems for the removal of a selection marker and their combination with a positive marker. *Plant Cell Reports* **20**, 383–392.

Ebmeier, A., Allison, L., Cerutti, H. and Clemente, T. (2004) Evaluation of the *Escherichia coli* threonine deaminase gene as a selectable marker for plant transformation. *Planta* **218**, 751–758.

Erikson, O., Hertzberg, M. and Nasholm, T. (2004) A conditional marker gene allowing both positive and negative selection in plants. *Nature Biotechnology* **22**, 455–458.

Evans, M.J. (2001) The cultural mouse. *Nature Medicine* **7**, 1081–1083.

Fang, Y.D., Akula, C. and Altpeter, F. (2002) *Agrobacterium*-mediated barley (*Hordeum vulgare* L.) transformation using green fluorescent protein as a visual marker and sequence analysis of the T-DNA: barley genomic DNA junctions. *Journal of Plant Physiology* **159**, 1131–1138.

Feldmann, K.A. (1991) T-DNA insertion mutagenesis in *Arabidopsis*—mutational spectrum. *The Plant Journal* **1**, 71–82.

Firoozabady, E., Deboer, D.L., Murray, E.E., Merlo, D.J., Adang, M.J. and Halk, E.L. (1987) Transformation of cotton (*Gossypium hirsutum* L.) by *Agrobacterium tumefaciens* and regeneration of transgenic plants. *In Vitro Cellular and Developmental Biology* **23**, A67.

Frame, B.R., Shou, H.X., Chikwamba, R.K., Zhang, Z.Y., Xiang, C.B., Fonger, T.M., Pegg, S.E.K., Li, B.C., Nettleton, D.S., Pei, D.Q. and Wang, K. (2002) *Agrobacterium tumefaciens*-mediated transformation of maize embryos using a standard binary vector system. *Plant Physiology* **129**, 13–22.

Francis, K.E. and Spiker, S. (2005) Identification of *Arabidopsis thaliana* transformants without selection reveals a high occurrence of silenced T-DNA integrations. *The Plant Journal* **41**, 464–477.

Fromm, M.E., Morrish, F., Armstrong, C., Williams, R., Thomas, J. and Klein, T.M. (1990) Inheritance and expression of chimeric genes in the progeny of transgenic maize plants. *Bio-Technology* **8**, 833–839.

Fry, J., Barnason, A. and Horsch, R.B. (1987) Transformation of *Brassica napus* with *Agrobacterium tumefaciens* based vectors. *Plant Cell Reports* **6**, 321–325.

Fu, X.D., Duc, L.T., Fontana, S., Bong, B.B., Tinjuangjun, P., Sudhakar, D., Twyman, R.M., Christou, P. and Kohli, A. (2000) Linear transgene constructs lacking vector backbone sequences generate low-copy-number transgenic plants with simple integration patterns. *Transgenic Research* **9**, 11–19.

Gasser, C.S. and Fraley, R.T. (1989) Genetically engineering plants for crop improvement. *Science* **244**, 1293–1299.

Gelvin, S.B. (2003) *Agrobacterium*-mediated plant transformation: the biology behind the "gene-jockeying" tool. *Microbiology and Molecular Biology Reviews* **67**, 16–37.

Gheysen, G., Angenon, G. and Van Montague, M. (1998) *Agrobacterium*-mediated plant transformation: a scientifically intriguing story with significant application in *Transgenic Plant Research* (ed K. Lindsey), Harwood Academic Press, The Netherlands, pp. 1–33.

Goodner, B., Hinkle, G., Gattung, S., Miller, N., Blanchard, M., Qurollo, B., Goldman, B.S., Cao, Y.W., Askenazi, M., Halling, C. Mullin, L., Houmiel, K., Gordon, J., Vaudin, M., Iartchouk, O., Epp, A., Liu, F., Wollam, C., Allinger, M., Doughty, D., Scott, C., Lappas, C., Markelz, B., Flanagan, C., Crowell, C., Gurson, J., Lomo, C., Sear, C., Strub, G., Cielo, C. and Slater, S. (2001) Genome sequence of the plant pathogen and biotechnology agent *Agrobacterium tumefaciens* C58. *Science* **294**, 2323–2328.

Gordon-Kamm, W., Dilkes, B.P., Lowe, K., Hoerster, G., Sun, X.F., Ross, M., Church, L., Bunde, C., Farrell, J., Hill, P., Maddock, S., Snyder, J., Sykes, L., Li, Z., Woo, Y-M., DeBidney, D. and Larkins, B.A. (2002) Stimulation of the cell cycle and maize transformation by disruption of the plant retinoblastoma pathway. *Proceedings of the National Academy of Sciences USA* **99**, 11975–11980.

Gordon-Kamm, W.J., Spencer, T.M., Mangano, M.L., Adams, T.R., Daines, R.J., Start, W.G., Obrien, J.V., Chambers, S.A., Adams, W.R., Willetts, N.G., Rice, T.B., Mackey, C.J., Krueger, R.W., Kausch, A.P. and Lemaux, P.G. (1990) Transformation of maize cells and regeneration of fertile transgenic plants. *Plant Cell* **2**, 603–618.

Gould, J., Devey, M., Hasegawa, O., Ulian, E.C., Peterson, G. and Smith, R.H. (1991) Transformation of *Zea mays* L. using *Agrobacterium tumefaciens* and the shoot apex. *Plant Physiology* **95**, 426–434.

Grant, J.E., Cooper, P.A. and Dale, T.M. (2004) Transgenic *Pinus radiata* from *Agrobacterium tumefaciens*-mediated transformation of cotyledons. *Plant Cell Reports* **22**, 894–902.

Haldrup, A., Noerremark, M. and Okkels, F.T. (2001) Plant selection principle based on xylose isomerase. *In Vitro Cellular and Developmental Biology, Plant* **37**, 114–119.

Haldrup, A., Petersen, S.G. and Okkels, F.T. (1998) Positive selection: a plant selection principle based on xylose isomerase, an enzyme used in the food industry. *Plant Cell Reports* **18**, 76–81.

Haliloglu, K. and Baenziger, P.S. (2003) *Agrobacterium tumefaciens*-mediated wheat transformation. *Cereal Research Communications* **31**, 9–16.

Hall, R.D., RiksenBruinsma, T., Weyens, G.J., Rosquin, I.J., Denys, P.N., Evans, I.J., Lathouwers, J.E., Lefebvre, M.P., Dunwell, J.M., vanTunen, A. and Krens, F.A. (1996) A high efficiency technique for the generation of transgenic sugar beets from stomatal guard cells. *Nature Biotechnology* **14**, 1133–1138.

Hamilton, C.M. (1997) A binary-BAC system for plant transformation with high-molecular-weight DNA. *Gene* **200**, 107–116.

Hanin, M., Volrath, S., Bogucki, A., Briker, M., Ward, E. and Paszkowski, J. (2001) Gene targeting in *Arabidopsis*. *The Plant Journal* **28**, 671–677.

Hansen, G. and Wright, M.S. (1999) Recent advances in the transformation of plants. *Trends in Plant Science* **4**, 226–231.

Hellens, R., Mullineaux, P. and Klee, H. (2000a) A guide to *Agrobacterium* binary Ti vectors. *Trends in Plant Science* **5**, 446–451.

Hellens, R.P., Edwards, E.A., Leyland, N.R., Bean, S. and Mullineaux, P.M. (2000b) pGreen: a versatile and flexible binary Ti vector for *Agrobacterium*-mediated plant transformation. *Plant Molecular Biology* **42**, 819–832.

Hiei, Y., Ohta, S., Komari, T. and Kumashiro, T. (1994) Efficient transformation of rice (*Oryza sativa* L.) mediated by *Agrobacterium* and sequence analysis of the boundaries of the T-DNA. *The Plant Journal* **6**, 271–282.

Hinchee, M.A.W., Connorward, D.V., Newell, C.A., McDonnell, R.E., Sato, S.J., Gasser, C.S., Fischhoff, D.A., Re, D.B., Fraley, R.T. and Horsch, R.B. (1988) Production of transgenic soybean plants using *Agrobacterium*-mediated DNA transfer. *Bio-Technology* **6**, 915–921.

Hisano, H., Kimoto, Y., Hayakawa, H., Takeichi, J., Domae, T., Hashimoto, R., Abe, J., Asano, S., Kanazawa, A. and Shimamoto, Y. (2004) High frequency *Agrobacterium*-mediated transformation and plant regeneration via direct shoot formation from leaf explants in *Beta vulgaris* and *Beta maritima*. *Plant Cell Reports* **22**, 910–918.

Hoa, T.T.C., Bong, B.B., Huq, E. and Hodges, T.K. (2002) Cre/lox site-specific recombination controls the excision of a transgene from the rice genome. *Theoretical and Applied Genetics* **104**, 518–525.

Hoekema, A., Hirsch, P.R., Hooykaas, P.J.J. and Schilperoort, R.A. (1983) A binary plant vector strategy based on separation of Vir-region and T-region of the *Agrobacterium tumefaciens* Ti-plasmid. *Nature* **303**, 179–180.

Hohn, B., Levy, A.A. and Puchta, H. (2001) Elimination of selection markers from transgenic plants. *Current Opinion in Biotechnology* **12**, 139–143.

Hood, E.E., Fraley, R.T. and Chilton, M.D. (1987) Virulence of *Agrobacterium tumefaciens* strain-A281 on legumes. *Plant Physiology* **83**, 529–534.

Horsch, R.B., Fraley, R.T., Rogers, S.G., Sanders, P.R., Lloyd, A. and Hoffmann, N. (1984) Inheritance of functional foreign genes in plants. *Science* **223**, 496–498.

Horvath, H., Rostoks, N., Brueggeman, R., Steffenson, B., von Wettstein, D. and Kleinhofs, A. (2003) Genetically engineered stem rust resistance in barley using the *Rpg1* gene. *Proceedings of the National Academy of Sciences USA* **100**, 364–369.

Hu, C.Y. and Wang, L.Z. (1999) *In planta* soybean transformation technologies developed in China: procedure, confirmation and field performance. *In Vitro Cellular and Developmental Biology, Plant* **35**, 417–420.

Hu, T., Metz, S., Chay, C., Zhou, H.P., Biest, N., Chen, G., Cheng, M., Feng, X., Radionenko, M., Lu, F. and Fry, J. (2003) *Agrobacterium*-mediated large-scale transformation of wheat (*Triticum aestivum* L.) using glyphosate selection. *Plant Cell Reports* **21**, 1010–1019.

Ishida, Y., Saito, H., Ohta, S., Hiei, Y., Komari, T. and Kumashiro, T. (1996) High efficiency transformation of maize (*Zea mays* L.) mediated by *Agrobacterium tumefaciens*. *Nature Biotechnology* **14**, 745–750.

Joersbo, M. (2001) Advances in the selection of transgenic plants using non-antibiotic marker genes. *Physiologia Plantarum* **111**, 269–272.

Joersbo, M., Petersen, S.G. and Okkels, F.T. (1999) Parameters interacting with mannose selection employed for the production of transgenic sugar beet. *Physiologia Plantarum* **105**, 109–115.

Jun, S.I., Kwon, S.Y., Paek, K.Y. and Paek, K.H. (1995) *Agrobacterium*-mediated transformation and regeneration of fertile transgenic plants of Chinese cabbage (*Brassica campestris* ssp. *pekinensis* cv. Spring Flavor). *Plant Cell Reports* **14**, 620–625.

Karimi, M., Inze, D. and Depicker, A. (2002) GATEWAY(TM) vectors for *Agrobacterium*-mediated plant transformation. *Trends in Plant Science* **7**, 193–195.

Khanna, H.K. and Daggard, G.E. (2003) *Agrobacterium tumefaciens*-mediated transformation of wheat using a superbinary vector and a polyamine-supplemented regeneration medium. *Plant Cell Reports* **21**, 429–436.

Kikkert, J.R., Humiston, G.A., Roy, M.K. and Sanford, J.C. (1999) Biological projectiles (phage, yeast, bacteria) for genetic transformation of plants. *In Vitro Cellular and Developmental Biology, Plant* **35**, 43–50.

Klee, H., Horsch, R. and Rogers, S. (1987) *Agrobacterium*-mediated plant transformation and its further applications to plant biology. *Annual Review of Plant Physiology and Plant Molecular Biology* **38**, 467–486.

Kohli, A., Leech, M., Vain, P., Laurie, D.A. and Christou, P. (1998) Transgene organization in rice engineered through direct DNA transfer supports a two-phase integration mechanism mediated by the establishment of integration hot spots. *Proceedings of the National Academy of Sciences USA* **95**, 7203–7208.

Kohli, A., Twyman, R.M., Abranches, R., Wegel, E., Stoger, E. and Christou, P. (2003) Transgene integration, organization and interaction in plants. *Plant Molecular Biology* **52**, 247–258.

Komari, T. (1989) Transformation of callus cultures of nine plant species mediated by *Agrobacterium*. *Plant Science* **60**, 223–229.

Komari, T. (1990) Transformation of cultured cells of *Chenopodium quinoa* by binary vectors that carry a fragment of DNA from the virulence region of Ptibo542. *Plant Cell Reports* **9**, 303–306.

Koncz, C., Martini, N., Mayerhofer, R., Konczkalman, Z., Korber, H., Redei, G.P. and Schell, J. (1989) High frequency T-DNA-mediated gene tagging in plants. *Proceedings of the National Academy of Sciences USA* **86**, 8467–8471.

Krens, F.A., Molendijk, L., Wullems, G.J. and Schilperoort, R.A. (1982) *In vitro* transformation of plant protoplasts with Ti-plasmid DNA. *Nature* **296**, 72–74.

Kunik, T., Tzfira, T., Kapulnik, Y., Gafni, Y., Dingwall, C. and Citovsky, V. (2001) Genetic transformation of HeLa cells by *Agrobacterium*. *Proceedings of the National Academy of Sciences USA* **98**, 1871–1876.

Li, X.G., Volrath, S.L., Nicholl, D.B.G., Chilcott, C.E., Johnson, M.A., Ward, E.R. and Law, M.D. (2003) Development of protoporphyrinogen oxidase as an efficient selection marker for *Agrobacterium tumefaciens*-mediated transformation of maize. *Plant Physiology* **133**, 736–747.

Lindsey, K., Wei, W.B., Clarke, M.C., McArdle, H.F., Rooke, L.M. and Topping, J.F. (1993) Tagging genomic sequences that direct transgene expression by activation of a promoter trap in plants. *Transgenic Research* **2**, 33–47.

Liu, F., Cao, M.Q., Yao, L., Li, Y., Robaglia, C. and Tourneur, C. (1998) *In planta* transformation of pakchoi (*Brassica campestris* L. ssp. *chinensis*) by infiltration of adult plants with *Agrobacterium*. *Acta Horticulturae* **467**, 187–192.

Liu, H.K., Yang, C. and Wei, Z.M. (2004) Efficient *Agrobacterium tumefaciens*-mediated transformation of soybeans using an embryonic tip regeneration system. *Planta* **219**, 1042–1049.

Liu, Y.G., Shirano, Y., Fukaki, H., Yanai, Y., Tasaka, M., Tabata, S. and Shibata, D. (1999) Complementation of plant mutants with large genomic DNA fragments by a transformation-competent artificial chromosome vector accelerates positional cloning. *Proceedings of the National Academy of Sciences USA* **96**, 6535–6540.

Lucca, P., Ye, X.D. and Potrykus, I. (2001) Effective selection and regeneration of transgenic rice plants with mannose as selective agent. *Molecular Breeding* **7**, 43–49.

Machado, M.L.D., Machado, A.D., Hanzer, V., Weiss, H., Regner, F., Steinkellner, H., Mattanovich, D., Plail, R., Knapp, E., Kalthoff, B. and Katinger, H. (1992) Regeneration of transgenic plants of *Prunus armeniaca* containing the coat protein gene of Plum Pox virus. *Plant Cell Reports* **11**, 25–29.

Maheswaran, G., Welander, M., Hutchinson, J.F., Graham, M.W. and Richards, D. (1992) Transformation of apple rootstock M26 with *Agrobacterium tumefaciens*. *Journal of Plant Physiology* **139**, 560–568.

Maliga, P. (2004) Plastid transformation in higher plants. *Annual Review of Plant Biology* **55**, 289–313.

Manoharan, M., Vidya, C.S.S. and Sita, G.L. (1998) *Agrobacterium*-mediated genetic transformation in hot chilli (*Capsicum annuum* L. var. Pusa jwala). *Plant Science* **131**, 77–83.

Matthews, P.R., Wang, M.B., Waterhouse, P.M., Thornton, S., Fieg, S.J., Gubler, F. and Jacobsen, J.V. (2001) Marker gene elimination from transgenic barley, using co-transformation with adjacent 'twin T-DNAs' on a standard *Agrobacterium* transformation vector. *Molecular Breeding* **7**, 195–202.

McCabe, D.E. and Martinell, B.J. (1993) Transformation of elite cotton cultivars via particle bombardment of meristems. *Bio-Technology* **11**, 596–598.

McCormick, S., Niedermeyer, J., Fry, J., Barnason, A., Horsch, R. and Fraley, R. (1986) Leaf disk transformation of cultivated tomato (*Lycopersicon esculentum*) using *Agrobacterium tumefaciens*. *Plant Cell Reports* **5**, 81–84.

Miguel, C.M. and Oliveira, M.M. (1999) Transgenic almond (*Prunus dulcis* Mill.) plants obtained by *Agrobacterium*-mediated transformation of leaf explants. *Plant Cell Reports* **18**, 387–393.

Miljusdjukic, J., Neskovic, M., Ninkovic, S. and Crkvenjakov, R. (1992) *Agrobacterium*-mediated transformation and plant regeneration of buckwheat (*Fagopyrum esculentum* Moench). *Plant Cell Tissue and Organ Culture* **29**, 101–108.

Mohanty, A., Sarma, N.P. and Tyagi, A.K. (1999) *Agrobacterium*-mediated high frequency transformation of an elite indica rice variety Pusa Basmati 1 and transmission of the transgenes to R2 progeny. *Plant Science* **147**, 127–137.

Mourgues, F., Chevreau, E., Lambert, C. and deBondt, A. (1996) Efficient *Agrobacterium*-mediated transformation and recovery of transgenic plants from pear (*Pyrus communis* L.). *Plant Cell Reports* **16**, 245–249.

Murray, F., Brettell, R., Matthews, P., Bishop, D. and Jacobsen, J. (2004) Comparison of *Agrobacterium*-mediated transformation of four barley cultivars using the GFP and GUS reporter genes. *Plant Cell Reports* **22**, 397–402.

Negrotto, D., Jolley, M., Beer, S., Wenck, A.R. and Hansen, G. (2000) The use of phosphomannose isomerase as a selectable marker to recover transgenic maize plants (*Zea mays* L.) via *Agrobacterium* transformation. *Plant Cell Reports* **19**, 798–803.

O'Kennedy, M.M., Burger, J.T. and Botha, F.C. (2004) Pearl millet transformation system using the positive selectable marker gene phosphomannose isomerase. *Plant Cell Reports* **22**, 684–690.

Ooms, G., Burrell, M.M., Karp, A., Bevan, M. and Hille, J. (1987) Genetic transformation in two potato cultivars with T-DNA from disarmed *Agrobacterium*. *Theoretical and Applied Genetics* **73**, 744–750.

Park, S.M. and Kim, D.H. (2004) Transformation of a filamentous fungus *Cryphonectria parasitica* using *Agrobacterium tumefaciens*. *Biotechnology and Bioprocess Engineering* **9**, 217–222.

Parrott, W.A., Hoffman, L.M., Hildebrand, D.F., Williams, E.G. and Collins, G.B. (1989) Recovery of primary transformants of soybean. *Plant Cell Reports* **7**, 615–617.

Paszkowski, J., Shillito, R.D., Saul, M., Mandak, V., Hohn, T., Hohn, B. and Potrykus, I. (1984) Direct gene transfer to plants. *EMBO Journal* **3**, 2717–2722.

Patel, M., Johnson, J.S., Brettell, R.I.S., Jacobsen, J. and Xue, G.P. (2000) Transgenic barley expressing a fungal xylanase gene in the endosperm of the developing grains. *Molecular Breeding* **6**, 113–123.

Pavingerova, D., Briza, J., Kodytek, K. and Niedermeierova, H. (1997) Transformation of *Rhododendron* spp. using *Agrobacterium tumefaciens* with a GUS-intron chimeric gene. *Plant Science* **122**, 165–171.

Pawlowski, W.P. and Somers, D.A. (1998) Transgenic DNA integrated into the oat genome is frequently interspersed by host DNA. *Proceedings of the National Academy of Sciences USA* **95**, 12106–12110.

Pena, L., Perez, R.M., Cervera, M., Juarez, J.A. and Navarro, L. (2004) Early events in *Agrobacterium*-mediated genetic transformation of citrus explants. *Annals of Botany* **94**, 67–74.

Peters, N.R., Ackerman, S. and Davis, E.A. (1999) A modular vector for *Agrobacterium* mediated transformation of wheat. *Plant Molecular Biology Reporter* **17**, 323–331.

Pius, P.K. and Achar, P.N. (2000) *Agrobacterium*-mediated transformation and plant regeneration of *Brassica oleracea* var. capitata. *Plant Cell Reports* **19**, 888–892.

Pua, E.C., Mehrapalta, A., Nagy, F. and Chua, N.H. (1987) Transgenic plants of *Brassica napus* L. *Bio-Technology* **5**, 815–817.

Rachmawati, D., Hosaka, T., Inoue, E. and Anza, H. (2004) *Agrobacterium*-mediated transformation of javanica rice cv. Rojolele. *Bioscience Biotechnology and Biochemistry* **68**, 1193–1200.

Rao, K.S. and Rohini, V.K. (1999) *Agrobacterium*-mediated transformation of sunflower (*Helianthus annus* L.): a simple protocol. *Annals of Botany* **83**, 347–354.

Rashid, H., Yokoi, S., Toriyama, K. and Hinata, K. (1996) Transgenic plant production mediated by *Agrobacterium* in indica rice. *Plant Cell Reports* **15**, 727–730.

Rasmussen, J.L., Kikkert, J.R., Roy, M.K. and Sanford, J.C. (1994) Biolistic transformation of tobacco and maize suspension cells using bacterial cells as microprojectiles. *Plant Cell Reports* **13**, 212–217.

Reed, J., Privalle, L., Powell, M.L., Meghji, M., Dawson, J., Dunder, E., Suttie, J., Wenck, A., Launis, K., Kramer, C., Chang Y-F., Hansen, G. and Wright, M. (2001) Phosphomannose isomerase: an efficient selectable marker for plant transformation. *In Vitro Cellular and Developmental Biology-Plant* **37**, 127–132.

Rhodes, C.A., Pierce, D.A., Mettler, I.J., Mascarenhas, D. and Detmer, J.J. (1988) Genetically transformed maize plants from protoplasts. *Science* **240**, 204–207.

Rommens, C.M., Humara, J.M., Ye, J.S., Yan, H., Richael, C., Zhang, L., Perry, R. and Swords, K. (2004) Crop improvement through modification of the plant's own genome. *Plant Physiology* **135**, 421–431.

Ruvkun, G.B. and Ausubel, F.M. (1981) A general method for site-directed mutagenesis in prokaryotes. *Nature* **289**, 85–88.

Sanford, J.C. (1988) The biolistic process. *Trends in Biotechnology* **6**, 299–302.

Sanford, J.C. (1990) Biolistic plant transformation. *Physiologia Plantarum* **79**, 206–209.

Sanford, J.C., Smith, F.D. and Russell, J.A. (1993) Optimizing the biolistic process for different biological applications. *Methods in Enzymology* **217**, 483–509.

Schaefer, D.G. and Zryd, J.P. (1997) Efficient gene targeting in the moss *Physcomitrella patens*. *The Plant Journal* **11**, 1195–1206.

Shimamoto, K., Terada, R., Izawa, T. and Fujimoto, H. (1989) Fertile transgenic rice plants regenerated from transformed protoplasts. *Nature* **338**, 274–276.

Smithies, O. (2001) Forty years with homologous recombination. *Nature Medicine* **7**, 1083–1086.

Svitashev, S., Ananiev, E., Pawlowski, W.P. and Somers, D.A. (2000) Association of transgene integration sites with chromosome rearrangements in hexaploid oat. *Theoretical and Applied Genetics* **100**, 872–880.

Svitashev, S.K., Pawlowski, W.P., Makarevitch, I., Plank, D.W. and Somers, D.A. (2002) Complex transgene locus structures implicate multiple mechanisms for plant transgene rearrangement. *The Plant Journal* **32**, 433–445.

Terada, R., Urawa, H., Inagaki, Y., Tsugane, K. and Iida, S. (2002) Efficient gene targeting by homologous recombination in rice. *Nature Biotechnology* **20**, 1030–1034.

Tingay, S., McElroy, D., Kalla, R., Fieg, S., Wang, M.B., Thornton, S. and Brettell, R. (1997) *Agrobacterium tumefaciens*-mediated barley transformation. *The Plant Journal* **11**, 1369–1376.

Topping, J.F., Wei, W.B. and Lindsey, K. (1991) Functional tagging of regulatory elements in the plant genome. *Development* **112**, 1009–1019.

Toriyama, K., Arimoto, Y., Uchimiya, H. and Hinata, K. (1988) Transgenic rice plants after direct gene transfer into protoplasts. *Bio-Technology* **6**, 1072–1074.

Trieu, A.T., Burleigh, S.H., Kardailsky, I.V., Maldonado-Mendoza, I.E., Versaw, W.K., Blaylock, L.A., Shin, H.S., Chiou, T.J., Katagi, H., Dewbre, G.R., Weigel, D. and Harrison, M.J. (2000) Transformation of *Medicago truncatula* via infiltration of seedlings or flowering plants with *Agrobacterium*. *The Plant Journal* **22**, 531–541.

Trifonova, A., Madsen, S. and Olesen, A. (2001) *Agrobacterium*-mediated transgene delivery and integration into barley under a range of *in vitro* culture conditions. *Plant Science* **161**, 871–880.

Tyagi, A.K. and Mohanty, A. (2000) Rice transformation for crop improvement and functional genomics. *Plant Science* **158**, 1–18.

Tzfira, T., Li, J.X., Lacroix, B. and Citovsky, V. (2004) *Agrobacterium* T-DNA integration: molecules and models. *Trends in Genetics* **20**, 375–383.

Vanhaute, E., Joos, H., Maes, M., Warren, G., Schell, J. and Vanmontagu, M. (1983) Intergeneric transfer and exchange recombination of restriction fragments cloned in Pbr322—a novel strategy for the reversed genetics of the Ti plasmids of *Agrobacterium tumefaciens*. *EMBO Journal* **2**, 411–417.

Vasil, V., Castillo, A.M., Fromm, M.E. and Vasil, I.K. (1992) Herbicide resistant fertile transgenic wheat plants obtained by microprojectile bombardment of regenerable embryogenic callus. *Bio-Technology* **10**, 667–674.

Wang, A.S., Evans, R.A., Altendorf, P.R., Hanten, J.A., Doyle, M.C. and Rosichan, J.L. (2000) A mannose selection system for production of fertile transgenic maize plants from protoplasts. *Plant Cell Reports* **19**, 654–660.

Wang, M.B., Abbott, D.C., Upadhyaya, N.M., Jacobsen, J.V. and Waterhouse, P.M. (2001) *Agrobacterium tumefaciens*-mediated transformation of an elite Australian barley cultivar with virus resistance and reporter genes. *Australian Journal of Plant Physiology* **28**, 149–156.

Wang, W.C., Menon, G. and Hansen, G. (2003) Development of a novel *Agrobacterium*-mediated transformation method to recover transgenic *Brassica napus* plants. *Plant Cell Reports* **22**, 274–281.

Ward, D.V., Zupan, J.R. and Zambryski, P.C. (2002) *Agrobacterium* VirE2 gets the VIP1 treatment in plant nuclear import. *Trends in Plant Science* **7**, 1–3.

Weber, S., Friedt, W., Landes, N., Molinier, J., Himber, C., Rousselin, P., Hahne, G. and Horn, R. (2003) Improved *Agrobacterium*-mediated transformation of sunflower (*Helianthus annuus* L.): assessment of macerating enzymes and sonication. *Plant Cell Reports* **21**, 475–482.

Weir, B., Gu, X., Wang, M.B., Upadhyaya, N., Elliott, A.R. and Brettell, R.I.S. (2001) *Agrobacterium tumefaciens*-mediated transformation of wheat using suspension cells as a model system and green fluorescent protein as a visual marker. *Australian Journal of Plant Physiology* **28**, 807–818.

Wilson, F.M. and James, D.J. (2003) Regeneration and transformation of the premier UK apple (*Malus* × *pumila* Mill.) cultivar Queen Cox. *Journal of Horticultural Science and Biotechnology* **78**, 656–662.

Wordragen, M.F. and Dons, J.J.M. (1992) *Agrobacterium tumefaciens*-mediated transformation of recalcitrant crops. *Plant Molecular Biology Reporter* **10**, 12–36.

Wright, M., Dawson, J., Dunder, E., Suttie, J., Reed, J., Kramer, C., Chang, Y., Novitzky, R., Wang, H. and Artim-Moore, L. (2001) Efficient biolistic transformation of maize (*Zea mays* L.) and wheat (*Triticum aestivum* L.) using the phosphomannose isomerase gene, *pmi*, as the selectable marker. *Plant Cell Reports* **20**, 429–436.

Wu, H., Sparks, C., Amoah, B. and Jones, H.D. (2003) Factors influencing successful *Agrobacterium*-mediated genetic transformation of wheat. *Plant Cell Reports* **21**, 659–668.

Wu, H.X., McCormac, A.C., Elliott, M.C. and Chen, D.F. (1998) *Agrobacterium*-mediated stable transformation of cell suspension cultures of barley (*Hordeum vulgare*). *Plant Cell Tissue and Organ Culture* **54**, 161–171.

Xiang, C.B., Han, P., Lutziger, I., Wang, K. and Oliver, D.J. (1999) A mini binary vector series for plant transformation. *Plant Molecular Biology* **40**, 711–717.

Yan, B., Reddy, M.S.S., Collins, G.B. and Dinkins, R.D. (2000) *Agrobacterium tumefaciens*-mediated transformation of soybean *Glycine max* (L.) Merrill. using immature zygotic cotyledon explants. *Plant Cell Reports* **19**, 1090–1097.

Ye, G.N., Stone, D., Pang, S.Z., Creely, W., Gonzalez, K. and Hinchee, M. (1999) *Arabidopsis* ovule is the target for *Agrobacterium in planta* vacuum infiltration transformation. *The Plant Journal* **19**, 249–257.

Zambryski, P., Joos, H., Genetello, C., Leemans, J., van Montagu, M. and Schell, J. (1983) Ti plasmid vector for the introduction of DNA into plant cells without alteration of their normal regeneration capacity. *EMBO Journal* **2**, 2143–2150.

Zapata, C., Park, S.H., El-Zik, K.M. and Smith, R.H. (1999) Transformation of a Texas cotton cultivar by using *Agrobacterium* and the shoot apex. *Theoretical and Applied Genetics* **98**, 252–256.

Zhang, H.M., Yang, H., Rech, E.L., Golds, T.J., Davis, A.S., Mulligan, B.J., Cocking, E.C. and Davey, M.R. (1988) Transgenic rice plants produced by electroporation-mediated plasmid uptake into protoplasts. *Plant Cell Reports* **7**, 379–384.

Zhang, W., Subbarao, S., Addae, P., Shen, A., Armstrong, C., Peschke, V. and Gilbertson, L. (2003) Cre/lox-mediated marker gene excision in transgenic maize (*Zea mays* L.) plants. *Theoretical and Applied Genetics* **107**, 1157–1168.

Zhang, W. and Wu, R. (1988) Efficient regeneration of transgenic plants from rice protoplasts and correctly regulated expression of the foreign gene in the plants. *Theoretical and Applied Genetics* **76**, 835–840.

Zhao, Z.Y., Cai, T.S., Tagliani, L., Miller, M., Wang, N., Pang, H., Rudert, M., Schroeder, S., Hondred, D., Seltzer, J. and Pierce, D. (2000) *Agrobacterium*-mediated sorghum transformation. *Plant Molecular Biology* **44**, 789–798.

Zhao, Z.Y., Gu, W.N., Cai, T.S., Tagliani, L., Hondred, D., Bond, D., Schroeder, S., Rudert, M. and Pierce, D. (2002) High throughput genetic transformation mediated by *Agrobacterium tumefaciens* in maize. *Molecular Breeding* **8**, 323–333.

Zupan, J., Muth, T.R., Draper, O. and Zambryski, P. (2000) The transfer of DNA from *Agrobacterium tumefaciens* into plants: a feast of fundamental insights. *The Plant Journal* **23**, 11–28.

Zupan, J.R., Ward, D. and Zambryski, P. (1998) Assembly of the VirB transport complex for DNA transfer from *Agrobacterium tumefaciens* to plant cells. *Current Opinion in Microbiology* **1**, 649–655.

2.2

Enhanced Nutritional Value of Food Crops

Dietrich Rein

BASF Plant Science Holding GmbH, Agricultural Center, D-67117 Limburgerhof, Germany

Karin Herbers

Sungene GmbH, D-06466 Gatersleben, Germany. Current address: BASF Plant Science GmbH, Agricultural Center, D-67117 Limburgerhof, Germany

Introduction

Agricultural technology has achieved major successes in increasing and securing output as well as improving product quality. Until recently, emphasis of breeders was on increasing macronutrients and yield. Developed countries have achieved an over-supply in food energy. Now agriculture strives toward improving food quality, micronutrient composition and functional ingredients. The advent of plant biotechnology allows targeted improvement of food crops toward better agronomic performance as well as nutrient composition. Nutritionally valuable components can be increased or newly introduced into crops. These components find nutritional applications directly in a food or, as extracts from the plant, added to foods. Antinutritive or allergenic components can be reduced or removed through plant biotechnology. With the increase in life expectancy, major chronic disease burdens include cardiovascular

Abbreviations: AMD, age related macula degeneration; ARA, arachidonic acid; DHA, docosahexaenoic acid; EPA, eicosapentaenoic acid; FDA, U.S. Food and Drug Administration; GM, genetically modified; GRAS, generally recognized as safe; LDL, low density lipoprotein; PUFAs, polyunsaturated fatty acids

disease, cancer, type 2 diabetes, obesity, osteoporosis, immune dysfunction and mental disability. These diseases can be affected by lifestyle and diet. Thus, nutrition research has focused on the role of functional ingredients letting people age healthy and perform at their best. Here, we discuss opportunities to adapt crops to human needs focusing on nutraceuticals with proven health benefits. Different options and their respective technical status will be presented, that is (i) to enrich functional ingredients in food (high xanthophylls in fruit, plant sterols in cereals, and pro-vitamin A enhanced crops), (ii) to have the plant produce valuable ingredients (vitamin E and polyunsaturated fatty acids (PUFAs) in oil crops) and (iii) to modify its composition (high-resistance starch).

Each of us can optimize diet or enhance nutritional health benefits by considerate food choices with respect to quality and quantity. Still, healthy foods may be delivered more effectively, in a more tasty and a more sustainable manner through plant biotechnology. In addition, plant biotechnology can provide foods not normally tolerated by people with specific sensitivities. Today's dietary recommendations take into consideration many individual physiological and lifestyle parameters as well as disease risk factors to determine an optimal food composition and diet pattern. Recommendations for food choices are based on the accumulated knowledge in nutrition and medical research. Nutrition recommendations are constantly updated, taking into account verified scientific findings on diet–disease relationships and healthy ingredients. The difference between 'healthy' versus 'unhealthy' foods by today's standards lies in the amounts of different foods and ingredients consumed, the way of preparation, the freshness and the concentration of contaminants. In the average Western diet we recognize those ingredients that have health benefits but faded from modern foods such as flavonoids, fiber and n-3 (or omega-3) fatty acids (O'Keefe, Jr. and Cordain, 2004; Simopoulos, 1999). Other ingredients such as plant sterols have been recognized (Katan et al., 2003) or suggested, such as resistant starch (Behall and Howe, 1995), zeaxanthin and lutein (Krinsky, Landrum and Bone, 2003), by scientific evidence for their health benefits and it would be desirable to increase their intake from foods. Finally there are components we want to reduce in our diet. High amounts of saturated and trans fatty acids (Grundy, Abate and Chandalia, 2002; Sacks and Katan, 2002) or high glycemic foods (Jenkins et al., 2002) are not considered healthy. Antinutritive, toxic or allergenic components such as phytic acid (Davidsson, 2003), mycotoxins (Cleveland et al., 2003; Food and Agriculture Organization of the United Nations, 1979), wheat gluten (Green and Jabri, 2003) and food allergens (Breiteneder and Radauer, 2004) may be harmful to individuals.

GM crops provide the opportunity to improve human health by supporting established health concepts. However, limitations in the development of GM nutritionally enhanced crops exist: (i) in some cases new health-promoting components are still classified as pharmaceuticals despite their recognized health benefit to a general population (e.g. aspirin (Schafer, 1996) and lovastatin (Smith, Jr., 2000)), (ii) the health-promoting effects of certain compounds have not sufficiently been demonstrated in order to allow for the expensive commercial development of crops with the active substance enriched (e.g. resveratrol (Pervaiz, 2003; Stewart, Artime and O'Brian, 2003)), (iii) transgenic plant material or transgenically produced ingredients do not meet broad public acceptance, especially in European foods, (iv) commercially relevant levels of target compounds may be difficult to achieve by molecular engineering, or (v) the limited market may not justify the high upfront cost of development.

This chapter will provide an overview of the current status, and future developments in plant biotechnology with respect to functional foods and functional ingredients for human health and performance. Our focus is on traits with sufficient health evidence to possibly capture economic opportunities. These traits include pro-vitamin A, lutein and zeaxanthin, plant sterols, resistant starch and PUFAs.

Definition of Enhanced Nutrition

Enhanced nutrition means (i) supplying essential nutrients through foods not conventionally serving as source for the respective nutrient, (ii) removing food components from or (iii) adding them to the diet where they serve nutrient functions beyond energy needs. Enhanced nutrition promotes health and/or performance, or prevents from diseases. The specific functional ingredients are termed 'nutraceuticals', marketed either as dietary supplements or as 'functional foods'. Functional foods are foods fortified with added, concentrated or (bio)technologically introduced nutraceuticals (ingredients) to a functional level.

Economic Development of Nutraceuticals

A rising demand for functional foods can be seen in developed countries. In the US it claimed around $22 bn in 2003, with an annual growth estimated to continue at 7 % (Nutrition Business International LLC, 2003). This constitutes approximately 4 % of the $550 bn US food market. The US dietary supplement market seems more mature with around $19 bn sales in 2003 and a smaller predicted growth of 2–4 % until 2013 (Nutrition Business International LLC, 2003). Of the estimated $21 bn Japanese nutraceutical products market in 2002, 60 % were functional foods. In Japan, per capita spending on functional foods is the world's highest at about $100 a year, compared with less than $70 in the US (Yamaguchi, 2003; FOSHU Approval—Is It Worth The Price? www.npicenter.com). In Japan, the 454 approved FOSHU (Foods for Specified Health Uses) products (Sep 2004) accounted for more than $4 bn sales in 2003 (www.naturalproductsasia.com, www.npicenter.com). In contrast to the US, Europe prefers its nutraceuticals in form of functional foods rather than supplements. Retail spending for functional foods in the old EU 15 countries was estimated to be around $17 bn plus less than $15 bn for dietary supplements.

The aging demographics of the 21st century will put more self-care responsibility on everybody. With the number of people over the age of 65 rising, for example in the US, to about 70 million by 2035, the medical care costs of people with chronic diseases will greatly increase (www.cdc.gov). The influence of diet on chronic disease is largest for obesity, followed by cardiovascular diseases, diabetes, dental caries and cancer. In 2004, the direct and indirect cost of cardiovascular disease in the US population is estimated to be $368 bn (American Heart Association, 2003). The US Center for Disease Control and Prevention (www.cdc.gov) estimates that each year over $33 bn in medical costs and $9 bn in lost productivity due to heart disease, cancer, stroke, and diabetes are attributed to diet.

Population rises in developing countries and with it nutrition needs. Enhanced nutrition implies reducing malnutrition with basic nutrients including vitamins, proteins and minerals. For the year 2002, the World Health Organization claims 34.4 million years of lost healthy life due to nutritional deficiencies in its 192 member states (World Health Organization, 2004). Protein-energy malnutrition accounts for 16.9 million and vitamin A deficiency for 800 000 years. The more people to be fed from a limited agricultural surface, the more important plant-based nutrition will become. Animal-sourced protein, although higher in nutritional value, will become too inefficient for some areas of the world. High-quality protein and optimized nutrient density in staple crops may become an economic and nutritious alternative. Development in these areas is an important investment in human health and nutrition of nonprofit and government programs as well as private corporations. Breeding programs and molecular engineering will increasingly address the production of valuable supplements for extractions from plants as well as production of health-promoting substances in staple crops and specialty plants for direct human consumption. Table 2.2.1 gives an overview on important enhanced nutrition products and crops that are being addressed in academic research. A few of these traits have been adopted by plant biotechnology industries to develop the respective traits for the market. The table is not understood to be complete. Areas such as allergen-free food, flavonoids as putative nutraceuticals and others are not covered.

β-Carotene, Precursor for Vitamin A

Vitamin A deficiency causes one of the largest nutritional disease burdens together with protein-, iron- and iodine-deficiency, mainly occurring in developing countries (World Health Organization, 2004). The FAO/WHO estimates mean requirements and safe levels of intake to be 500–850 µg retinol equivalents (RE)/d (or 3–5 mg β-carotene/d) for adults at a conversion of 1 µg β-carotene $= 0.167$ µg RE (Food and Agriculture Organization of the United Nations, 2002). Many developing countries in Africa and Asia have a high occurrence of vitamin A deficiency (World Health Organization, 1998), apparently lacking access to an affordable and sustained source of the vitamin or its precursor β-carotene. Good sources of vitamin A are animal products, however, they are expensive. Thus the development and supply of plants to support vitamin A intake in these countries is highly needed.

Nutritional Value, Prevention of Disease by β-Carotene

Vitamin A is found in the body primarily in the form of retinol and retinyl ester, the majority being stored in the liver. It functions in visual perception, cellular differentiation and the immune response. In vision, vitamin A participates as retinal in the visual cycle, whereas vitamin A plays an important role in gene expression in the form of retinoic acid, maintaining differentiation of epithelial cells in intestine, skin and lung (Solomons, 2001).

Early signs of vitamin A deficiency include night blindness due to the role of retinal in vision. The roles of retinol and retinoic acid in immunity and gene expression become apparent with more profound deficiency (Food and Agriculture Organization of the

Table 2.2.1 *Overview of important nutraceuticals addressed by Plant Biotechnology. ARA, arachidonic acid; DHA, docosahexaenoic acid; EPA, eicosapentaenoic acid; LDL, low density lipoprotein.*

Trait	Proposed crop	Health impact	Comments
Vitamin E	Oil crops	Supplementary tocopherols or tocotrienols may reduce cardiovascular disease (Brigelius-Flohe et al., 2002; Harris, Devaraj and Jialal, 2002) and may protect against cancer (Dutta and Dutta, 2003; Kuchide et al., 2003) and diabetes (Vega-Lopez, Devaraj and Jialal, 2004) through antioxidant effects and gene regulation (Zingg and Azzi, 2004)	Benefits of tocopherols and tocotrienols beyond vitamin E function are clinically not well established; mechanisms are insufficiently understood; relative efficacies of tocopherols and tocotrienols are still debated
β-Carotene	Rice endosperm, banana, other grains, tomato and oil crops	Converted to vitamin A, β-carotene prevents blindness and maintains the immune system and general health; shortage is still prominent in developing countries (Potrykus, 2003; World Health Organization, 2004; Zimmermann and Hurrell, 2002)	Recent scientific success has demonstrated that nutritionally relevant concentrations of β-carotene can be expressed in rice (Paine et al., 2005). Commercial products await development.
Iron and zinc	Staple crops including rice	Nutritional deficiencies (e.g. protein, iron, zinc, iodine and vitamin A) affect more than 3 bn people worldwide and account for almost two thirds of childhood deaths (World Health Organization, 2004); once a micronutrient enhanced crop system is in place it may be more sustainable than current supplementation approaches (Bouis, 2003; Welch and Graham, 2004)	Conventional breeding techniques may be more effective to increase mineral content of staple crops; 'biofortified' staple crops may mask a lack of dietary diversity and other nutritional deficiencies

(*continued*)

Table 2.2.1   *(Continued)*

Trait	Proposed crop	Health impact	Comments
Resistant starch	Starch storing crops such as corn, wheat and potato	High-amylose starch as source for resistant starch in bakery products may lower blood glucose and insulin response (Behall et al., 1989; Behall and Hallfrisch, 2002) and energy density; resistant starch supports individuals affected by diabetes or metabolic syndrome (Brand-Miller, 2003)	Physiologically effective only in specific food preparations; gastrointestinal problems in some individuals
Carotenoids, zeaxanthin and lutein	Tomato, potato, carrot and other crops	Dietary zeaxanthin and lutein can increase macular pigment density in humans (Richer et al., 2004), which might be protective against the development of age-related macular degeneration (AMD) (Krinsky et al., 2003); carotenoids may also protect light-exposed skin and tissue (Sies and Stahl, 2003)	A causative relationship between carotenoid intake and reduced AMD has yet to be demonstrated; moreover, it is still uncertain which carotenoid is most effective
Carotenoid, astaxanthin	Tomato, carrot and other crops	Potent antioxidant carotenoid that may be effective in prevention of diseases (Guerin, Huntley and Olaizola, 2003) e.g. immune disorders, tumor induction or growth, inflammation and infection (Bennedsen et al., 1999; Chew and Park, 2004; Jyonouchi et al., 2000)	Evidence for human health promotion is only suggestive, since most studies were performed on cell culture or animal models; astaxanthin is associated with several proposed health functions, the mechanisms of which are little understood
Carotenoid, lycopene	Tomato, carrot and others	Powerful antioxidant, may reduce risk of cancers (e.g. prostate) (Hadley et al., 2002; Kristal, 2004; Muller, Altehed and Stehle, 2003)	Clinical intervention studies to show protection from prostate cancer still lacking; difference between synthetic lycopene and lycopene-rich carotenoid mixtures not clear

Table 2.2.1 (Continued)

Trait	Proposed crop	Health impact	Comments
Fish oil type fatty acids, EPA and DHA	Oil-rich crops such as soybean or linseed	Sustainable vegetable source of EPA and DHA with healthier background lipids, less contaminants and better taste for reduction of cardiovascular risk (Bang, Dyerberg and Hjoorne, 1976; Geelen *et al.*, 2004; Kromann and Green, 1980; MacLean *et al.*, 2004; von Schacky, 2004), mental disability and cancer (Larsson *et al.*, 2004)	Cardiovascular health benefits are well recognized, but benefits toward diabetes, inflammatory diseases or cancer are less established; mechanism of action insufficiently understood; stability and taste of EPA and DHA in foods still challenging
Phytosterols	Oil crops, vegetables and grain	Proven to lower plasma LDL-cholesterol as drug and as food ingredient (Katan *et al.*, 2003; Ostlund, Jr., 2004); approved as nutraceutical in the US and EU with the option to carry a US health claim (www.cfsan.fda.gov/~dms/flg-6c.html)	Currently still abundant conventional production capacity preventing serious investments into GM crops; consumer interest slower growing than expected despite proven health benefits
Resveratrol	Tomato and other fruits, rapeseed	May prevent specific cancers such as prostate (Cal *et al.*, 2003; Stewart *et al.*, 2003) and may provide health benefits as antioxidant (Pervaiz, 2003)	Benefits not well established in clinical studies but rather extrapolated from cell culture and animal research; pharmaceutical-like anticancer effects proposed, which may conflict with nutraceutical safety requirements
Lovastatin	Specialty crops	Well-proven HMG-CoA reductase inhibitor with broad cardiovascular benefits and excellent safety record (Davidson, 2001), originally isolated from fungus (Tobert, 2003), has reached 'over the counter' drug status in the UK in 2004 (http://medicines.mhra.gov.uk)	Not yet approved as supplement or food ingredient; side effects in some individuals

(continued)

Table 2.2.1   *(Continued)*

Trait	Proposed crop	Health impact	Comments
		and is being discussed as over the counter drug in the US (Smith, Jr., 2000); suitable for incorporation into functional foods after approval	
ARA	Oil crops	Required for fetal development in its function as eicosanoid precursor, developmental regulator and structural membrane component, dietary essentiality for infants debated (Carlson *et al.,* 1993; Innis, 2003; Larque, Demmelmair and Koletzko, 2002).	Only a limited market for infant formula; few applications for adults
Tailored fats	Oil crops	1,3-Diacylglycerol provides significantly less energy for fat synthesis than triacylglycerols despite similar taste and technological properties (Maki *et al.,* 2002; Nagao *et al.,* 2000; Tada and Yoshida, 2003)	Clinical long-term effects of 1,3-diacylglycerol consumption have not been studied adequately, thus a chance for undesirable side effects cannot yet be ruled out
Lactoferrin	Cereals	Wheat or rice expressing the high-quality protein lactoferrin with biological activities. Activities include antimicrobial (Farnaud and Evans, 2003), immunomodulatory, antioxidant, anti-inflammatory (Weinberg, 2001) and probiotic effects; it may have a role in mediating iron homeostasis (Ward and Conneely, 2004); 'activated' lactoferrin may be useful for food sanitary applications (Naidu and Nimmagudda, 2003)	Human genes needed to produce plant derived 'Human lactoferrin'; nevertheless, plant-derived lactoferrin will remain distinct from human lactoferrin in its carbohydrate moieties, implications unclear

Table 2.2.1 (Continued)

Trait	Proposed crop	Health impact	Comments
Celiac disease sensitive protein	Wheat	Bakery and other wheat products tolerable for celiac disease patients could alleviate nutritional deficiency symptoms for the large number of affected individuals (Fasano et al., 2003; Green et al., 2003)	Freedom of symptoms can never be guaranteed to all celiac individuals; food technological problems may arise from alteration or removal of celiac disease-sensitive proteins
Better tasting beans	Soybean	Reduction or elimination of the 'bean like' aroma defects of soy (MacLeod and Ames, 1988) could move the minor food ingredient soy from a feed to a healthy food crop	Soy may be less obviously recognized in cases where allergic individuals are involved or where 'adulterations' of traditional meat products is an issue (although soy is usually the healthier alternative)
Tailored fatty acids	Oil crops (corn, soy, safflower, sunflower and canola)	High-oleic acid oils for improved stability and viscosity as healthy alternative to hydrogenation or use of saturated fat (http://web.aces.uiuc.edu/value/factsheets/); nutrition labeling advantage in the US starting in 2006 (http://www.cfsan.fda.gov/~dms/transfat.html); similar labeling anticipated in EU	Health-promoting omega-3 fatty acids are replaced by nutritionally neutral oleic acid instead of finding technological solutions to stabilize sensitive omega-3s; high oleic traits have already been generated transgenically and nontransgenically

United Nations, 2002; Sommer and Davidson, 2002). Bacterial invasion and permanent scarring of the cornea of the eye (xerophthalmia) are later symptoms. Vitamin A deficiency has also been associated with increased child mortality and vulnerability to infection, particularly measles and diarrhea. Severe deficiency results in blindness, and in altered appearance and function of intestine, skin and lung. Risk for vitamin A deficiency is greatest in children because adequate liver vitamin A stores have yet to be built.

Sommer and Davidson (2002) estimate that 140 million preschool-aged children and at least 7.2 million pregnant women are vitamin A deficient. Most of these women suffer not only clinical complications, primarily xerophthalmia, but also increased mortality. Of the estimated 0.5 million children worldwide, which become blind each year, 70 % is due to vitamin A deficiency. Of children who are blind from keratomalacia or who have corneal disease, more than 50 % are reported to die (Food and Agriculture Organization of the United Nations, 2002). Thus, the problem of micronutrient deficiency accounts for a tremendous loss in 'years of healthy' life, a humanitarian and economical problem that is worth addressing through GM technology.

Sources of β-Carotene

β-Carotene is found in green leafy vegetables (e.g. spinach and young leaves from various sources), yellow vegetables (e.g. pumpkins and carrots), noncitrus fruits (e.g. mangoes, apricots and tomato) and palm oil or fruit (Food and Agriculture Organization of the United Nations, 2002; www.nal.usda.gov/fnic/foodcomp/Data/car98/car_tble.pdf). Pre-formed vitamin A is only found in animal products (e.g. liver and eggs) or in fortified processed foods (Solomons, 2001). Both β-carotene and vitamin A are rare in diets of economically deprived populations that often have to survive on starchy staples with little fruit or vegetables. Thus, the availability of high β-carotene staples (grains and tuberous roots) should help to alleviate the nutritional deficiency in some areas of the world.

It is being tried to engineer crops containing high levels of β-carotene. Successful modifications have been reported from transgenic rice, tomatoes and canola (for an overview see Table 2.2.1, Herbers, 2003; the biosynthetic pathway for carotenoids in plants is depicted in Figure 2.2.1). 'Golden Rice' is the most prominent example for genetic engineering of pro-vitamin A in plants. A *de novo* carotenoid biosynthetic pathway had to be introduced into rice endosperm to yield β-carotene (summarized by Beyer *et al.*, 2002). Although promising results were obtained, the levels of β-carotene achieved in these rice plants (1.6 mg/kg dry weight rice endosperm) had not been

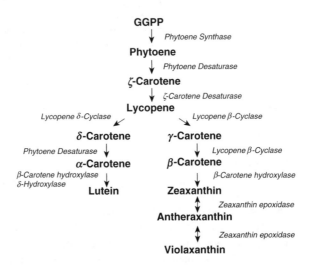

Figure 2.2.1   *Carotenoid biosynthesis in plants.*

sufficient to provide full vitamin A intake. Recent efforts succeeded in increasing the amount of β-carotene in transgenic rice substantially to concentrations of up to 37 mg β-carotene/kg dry weight rice endosperm (Paine *et al.*, 2005). The increase was due to using a maize phytoene synthase instead of the gene from daffodil. With this β-carotene concentration the staple rice could provide a substantial part of daily pro-vitamin A requirements.

In canola seeds β-carotene content was increased by 50-fold by the expression of the phytoene synthase (*crt*B) gene from *E. uredovora* behind the *Brassica* napin promoter. This resulted in levels of up to 0.7 g/kg seed β-carotene and 0.4 g/kg α-carotene (Shewmaker *et al.*, 1999). The β/α carotene ratio of 2:1 could be shifted to about 3:1 by the simultaneous expression of *crt*B, phytoene desaturase (*crt*I) and β-cyclase (*crt*Y) (Ravanello *et al.*, 2003). In Arabidopsis a 43-fold average increase of β-carotene was reached by using the Arabidopsis phytoene synthase under control of the napin A promoter (Lindgren *et al.*, 2003).

Recently, impressive levels of β-carotene have been obtained by expressing tomato lycopene β-cyclase (t*lcy*-b) behind the CaMV 35S promoter in transgenic tomato plants (D'Ambrosio *et al.*, 2004). Fruits reached up to 0.2 g/kg FW due to a total conversion of lycopene into β-carotene as well as to a roughly two-fold increase of total carotenoids. Given the recommendation for daily intakes to be in the range of between 3 and 5 mg β-carotene, about 20 g of fresh tomatoes might be sufficient to support the respective intake.

Colorful Bioactive Carotenoids, Zeaxanthin and Lutein

The nonvitamin A carotenoids, zeaxanthin and lutein, may help preventing degeneration of the eye. Age-related macula degeneration (AMD) is accompanied by a loss of yellow carotenoid pigments that humans cannot synthesize in the macula lutea. Increased consumption of zeaxanthin and lutein is associated with a higher macular pigment density (Landrum and Bone, 2001), but a causative relationship has not yet been experimentally demonstrated (Krinsky *et al.*, 2003). Despite the uncertainty with respect to protection from AMD by dietary zeaxanthin and lutein (Mozaffarieh, Sacu and Wedrich, 2003), reports summarizing research on the dietary effected increase of serum xanthophylls and macular pigment density, and the possible prevention of light-exposed tissue damage, promote their popularity (Alves-Rodrigues and Shao, 2004; Sies and Stahl, 2003). Physiologically there is still debate about which of the two dietary xanthophylls may be the more significant for human health (Krinsky *et al.*, 2003).

Vegetables with relatively high concentrations of zeaxanthin and lutein constitute a minor portion of the Western diet and their bioavailability needs to be considered (Castenmiller *et al.*, 1999). Introduction of, or increase in, zeaxanthin and lutein in crops using GM would make the food supply more versatile to achieve higher plasma and tissue levels.

Nutritional Value, Prevention of Disease and Recommendations

AMD is the leading cause of severe visual impairment and blindness in the elderly. The prevalence of neovascular AMD and/or geographic atrophy in the US population of 40 years and older is estimated to be 1.5 % with 0.6–0.7 % of the citizens having late AMD (Friedman *et al.*, 2004). Due to the rapidly aging Western populations, more than 1 % will be affected by AMD in 2020. AMD is a condition that primarily affects the part of the retina responsible for sharp central vision. Two major types of disease are differentiated (National Eye Institute, 2002): (i) Dry AMD (nonexudative, ∼80 % of cases) involves the presence of drusen, fatty deposits under the light-sensing cells in the retina, with atrophy in later stages. Vision loss in early dry AMD is moderate and progresses slowly. (ii) Wet AMD (exudative, ∼20 %) is more threatening to vision because of neovascularization under the retina with vessels breaking open and leaking fluid. This distorts vision and causes scar tissue to form.

Dietary supplementation with zeaxanthin and lutein was suggested as treatment for AMD long after macular pigment was identified to contain xanthophylls (Krinsky, 2002). The center of the human retina, the macula lutea, contains an enrichment of the carotenoids 3R,3′R-zeaxanthin, 3R,3′S(meso)-zeaxanthin and lutein up to a concentration of 1 mM, almost three magnitudes above normal plasma concentration (Landrum and Bone, 2001). The specific enrichment together with the ability of the xanthophylls to protect from high-energy blue light damage in the eye, and their antioxidant functions as radiation quencher and radical chain-breaking antioxidant (Krinsky, 1989) led to the proposition that dietary zeaxanthin and lutein may protect from maculapathy (including AMD) (Landrum and Bone, 2001).

Dietary zeaxanthin and lutein are absorbed well from foods or supplements and distributed through plasma lipoproteins to peripheral tissues. Lutein can be converted to zeaxanthin (Khachik, Bernstein and Garland, 1997) and zeaxanthin to all-E-3-dehydro-lutein (Hartmann *et al.*, 2004) *in vivo*. However, it is not yet known which of the carotenoids forms meso-zeaxanthin in the retina (Krinsky *et al.*, 2003). Xanthophyll supplementation can increase macular pigment density (Richer *et al.*, 2004). Still, the proposed function of macular lutein, zeaxanthin and meso-zeaxanthin to support photoprotectors within the retina has yet to be proven (Landrum and Bone, 2001; Mozaffarieh *et al.*, 2003). Considering the increasing prevalence of AMD and the large economic impact of severe impairment of vision, benefits from zeaxanthin and lutein in prevention of the disease could have an enormous social and economic impact.

Sources of Zeaxanthin and Lutein

Zeaxanthin and lutein can primarily be found in vegetables. These include, for example (lutein + zeaxanthin mg/100 g edible portion, (www.nal.usda.gov/fnic/foodcomp/Data/car98/car_tble.pdf; Elmadfa *et al.*, 2004)): corn (2), broccoli (2), zucchini (2), collards (8), turnip greens (8), spinach (7–12), Savoy cabbage (22), kale (15–40). The concentrations in these vegetables are high considering that the average Western diet provides 1–3 mg/d zeaxanthin and lutein (Nebeling *et al.*, 1997). The low average intake, despite the availability of xanthophyll-rich vegetables, indicates the relatively low popularity of these vegetables.

For supplements and functional foods, lutein can be enriched from vegetables and from marigold (*Tagetes erecta*, mainly producing lutein). In the US, FDA has no questions regarding 'crystalline lutein', a blend of zeaxanthin and lutein to be added to a variety of food items in the range of 0.3–3.0 mg per eating occasion, to be considered

GRAS (GRN 140). 'Crystalline lutein' is produced and purified from marigold oleoresin, which is a hydrophobic solvent extract from dried marigold petals. Zeaxanthin can also be produced synthetically, and it can be produced from mutated *Flavobacterium* cultures for which the carotenoid biosynthesis genes are known (Pasamontes *et al.*, 1997). There are several bacterial production systems supplying the supplement and functional food market.

GM approaches to elevate endogenous levels of zeaxanthin (see Figure 2.2.1 for biosynthetic scheme) have been performed in tomatoes and potato tubers. An increase of zeaxanthin was achieved for the first time by the combined over-expression of Arabidopsis β-cyclase and pepper β-carotene hydroxylase under the control of the phytoene desaturase promoter in tomato fruits (Dharmapuri *et al.*, 2002). The double transformants accumulated up to 11 mg/kg FW cryptoxanthin and up to 13 mg/kg FW zeaxanthin (Dharmapuri *et al.*, 2002). The latter pigments are below detection limits in wild-type fruits, which usually accumulate mainly lycopene. Another transgenic strategy was successfully employed by Römer *et al.* (2002) in transgenic potato tubers. In these organs the most abundant carotenoids are violaxanthin (see Figure 2.2.1) followed by lutein; total carotenoids yielding altogether between 10 and 25 mg/kg DW tuber. Violaxanthin is formed by the action of zeaxanthin epoxidase via antheraxanthin, the monoepoxy intermediate. Römer *et al.* (2002) reduced the expression of zeaxanthin epoxidase. This approach yielded levels of zeaxanthin elevated up to 130-fold, reaching 40 mg zeaxanthin/kg DW. In addition, most of the tubers containing higher zeaxanthin levels showed increased levels of total carotenoids (up to 5.7-fold). Company statements give recommendations of >6 mg for daily intakes of lutein and zeaxanthin, levels acceptable to FDA in GRAS notifications GRN 110 and 140. In relation to these amounts it becomes obvious that both the transgenic tomatoes and potatoes have been significantly biofortified for zeaxanthin by genetic modifications.

Plant Sterol Rich Foods for Healthy Blood Lipids

With optimized dietary intake, plant sterols (phytosterols) and/or stanols (1.5–2 g/d) can lower human serum total- and low-density lipoprotein (LDL) or 'bad' cholesterol by about 10 % (Katan *et al.*, 2003). The associated reduction in clinical manifestation of coronary heart disease is expected to be around 20 % (Miettinen and Gylling, 2003). Properly solubilized free or esterified plant sterols/stanols or equivalent esters at 0.8–1.0 g per day, in fortified and unfortified food vehicles, lower LDL cholesterol and maintain good heart health (Berger *et al.*, 2004). Average plant sterol intake with Westernized diets is only 0.2–0.4 g/d (Normen *et al.*, 2001). Historically, human plant sterol intake must have been higher, because the introduction of oil and fat refining technology not only improved oil taste and stability but also lowered dietary plant sterols. The deodorization step in today's oil refinement removes, depending on the process applied, approximately half of the sterols from the oil and thus from the diet (Belitz and Grosch 1999; Przybylski, 2001).

Plant sterols accumulate in the lipid fraction of seeds and fruit. Substantial amounts of the major sterols, β-sitosterol, campesterol and stigmasterol, can also be found in cereals including wheat (Normen *et al.*, 2001). Plant stanols naturally occur as the hydrogenated counterparts of the respective sterols. The major stanols, β-sitostanol and campestanol, contribute <10 % to dietary plant sterol intake (Normen et al., 2001). The steroid alcohols share a ring structure with cholesterol but differ in respective side chains. Plant sterols may be present either free, as long-chain fatty acyl esters (about 50 %) or as glycosides (minor).

Higher dietary plant sterol intake can be achieved through adding sterols to specific items in the food chain or through genetically increasing the production in the crop plant. Even low fat foods can be efficient in reducing serum cholesterol, if plant sterols were appropriately emulsified (Ostlund, Jr., 2004). The relatively high sterol content of wheat (60–80 mg/100 g) and other cereals (Piironen et al., 2002) compared to about 250 mg/100 g refined oil in European food products shows that low-fat food items contribute significantly to plant sterol intake (Normen et al., 2001). Thus, in addition to oil crops, crops grown for protein or starch such as soy, peas, wheat, corn, bananas or tomatoes serve an important role in dietary sterol intake and may be altered to overexpress sterols in the future.

A high plant sterol trait in edible plants would offer an effective dose of plant sterol/stanol in a serving size. The US FDA approved a health claim for plant sterols or stanols to reduce risk of coronary heart disease. A food, of which two servings can be consumed per day with 0.4 g free sterols per serving, can carry the claim that it 'may reduce the risk of heart disease' if included into a diet low in saturated fat and cholesterol (21 CFR 101.83).

Nutritional Value, Prevention of Disease and Recommendations

Plant sterol/stanol esters may reduce the risk of atherosclerosis and cardiovascular disease by lowering blood cholesterol (US Health Claim, (21 CFR 101.83). www.cfsan.fda.gov/~dms/hclaims.html). To understand cholesterol-lowering mechanisms it is important to recognize that the human intestine discriminates between cholesterol and plant sterol for absorption and metabolism. Cholesterol is absorbed efficiently and recovered in lipoproteins (35–70 %), whereas little plant sterols/stanols are recovered in plasma (0.02–3.5 %) (de Jong et al., 2003) although plant sterols and cholesterol have similar physicochemical properties.

There are two hypotheses trying to explain the cholesterol-lowering effect of plant sterols. First, due to their similar physicochemical properties plant sterols displace cholesterol from the intestinal micelles without being absorbed into plasma in significant quantities. The second hypothesis is that plant sterols are rapidly taken up by enterocytes together with cholesterol. Once in the enterocyte plant sterols increase expression of the adenosine triphosphate-binding cassette A1 sterol transporter (Ostlund, Jr., 2004; Plat and Mensink, 2002). Intestinal cell sensors (SREBP-2 and LXR) may not differentiate between cholesterol and plant sterols. Consequently, the transporter mediates excretion of the sterol components out of intestinal cells back to the lumen (Plat and Mensink, 2002).

Already the low current dietary intake of 0.2–0.4 g plant sterols/d is bioactive by reducing cholesterol absorption (Normen et al., 2001; Ostlund, Jr., 2002). Due to their proven cholesterol-lowering action, FDA approved a health claim (2000) for plant sterols and risk of coronary heart disease for foods containing at least 0.65 g plant sterol ester or 1.7 g plant stanol ester per reference amount (21 CFR 101.83) (current research suggests that there is little difference in cholesterol-lowering efficacy between sterols and stanols

(Ostlund, Jr., 2002)). Also the US National Cholesterol Education Program recommends plant stanols/sterols of 2 g per day as a therapeutic option for lowering LDL (National Cholesterol Education Program (NCEP) Expert Panel, 2004). This amount is in the maximum LDL cholesterol-lowering range (10 %) achievable by diet (1.5–2 g free or 2.5–3.3 g esterified plant sterols/stanols per day) (Katan *et al.*, 2003; Lichtenstein, 2002). Although nearly all foods contribute appreciably to plant sterol/stanol intake, except for refined carbohydrates and animal products (Normen *et al.*, 2001), recommendations in the gram per day range can only be achieved by consuming fortified foods.

Sources for Plant Sterols and Stanols

Plant sterols are obtained as a side product during vegetable oil processing, whereas stanol-rich products are derived from tall oil, a waste product from the paper industry. Crude vegetable oil is refined through a series of unit operations known as physical or alkaline refining. The processing can include the steps of degumming, caustic refining, bleaching and deodorization (Belitz and Grosch, 1999). The primary purpose of deodorization is to improve taste, odor and stability via the removal of undesirable volatiles and pigments. Sterols are recovered in the unsaponifiable fraction from deodorizer distillate of soy, corn and rapeseed oil. Sterols are further concentrated, for example by distillation techniques.

Tall oil is produced through organic solvent extraction from tall oil soap, a by-product of the pulping process used for coniferous trees during paper manufacturing. The tall oil is extracted with alcohol and heated to give primary plant sterol crystals. After cooling, these crystals are washed with water and filtered to separate residues. The plant sterols are recrystallized, filtered, dried under vacuum, milled and sieved. The resulting crystalline product is predominantly a mixture of β-sitosterol, β-sitostanol, campesterol and campestanol (adapted from US Generally Recognized as Safe Notifications, GRN 39 and 112, www.cfsan.fda.gov/~rdb/opa-gras.html). Plant sterol- and stanol-enriched products include vegetable oil spreads, dressings for salad, health drinks, health bars and yogurt-type products.

Metabolic engineering approaches have shown that it is possible to increase sterol levels in plants by expressing single genes coding for key regulatory enzymes. So far, maximum increases of plant sterols were reported to be in the range of 3–6-fold by the expression of HMG-CoA reductase in tobacco plants (Chappel *et al.*, 1995; Schaller *et al.*, 1995). The expression of C-24 methyltransferase type 1 in tobacco yielded levels elevated by about 1.4-fold (Holmberg *et al.*, 2002). Venkatramesh *et al.* (2003) were able to show that plant sterols could be converted into their respective hydrogenated forms, the stanols, in transgenic *B. napus* and soybean by the expression of a gene encoding 3-hydroxysteroid oxidase from *Streptomyces*. Thus, genetic modification gives the potential to enrich plant sterols and stanols in crop plants for enhanced nutrition purposes.

Resistant Starch, a Valuable Food Fiber

Carbohydrate quantity and quality have become a rising issue in the development of type 2 diabetes and obesity, both dominant problems in the US (Mokdad *et al.*, 2001) and

Europe. A key factor in carbohydrate quality is the time of digestion and glucose absorption into blood circulation (glycemic response). Consumption of carbohydrates that are slowly digested or not digested in the small intestine, for example resistant starch or dietary fiber, provides health benefits (Augustin *et al.*, 2003; Brand-Miller, 2003). Benefits include an improved glycemic control in diabetes (Willett, Manson and Liu, 2002), and reductions in blood insulin and in postprandial lipids (Behall and Hallfrisch, 2002; Behall *et al.*, 1989). In addition, resistant starch favorably affects the large intestine microflora and generates butyric acid serving the colon as energy source. These benefits are associated with a reduced risk for development of type 2 diabetes (Salmeron *et al.*, 1997a,b).

Increasing the resistant starch portion in a starchy food crop may be valuable in commonly consumed high glycemic foods such as potatoes, corn or rice (Foster-Powell *et al.*, 2002). Starchy foods usually contain 75 % of the starch as amylopectin with the remainder as amylose. Amylose is more resistant to intestinal digestion than amylopectin and sugars because of slower intestinal degradation by α-amylase. White potatoes account for a large proportion of starchy vegetable consumption in the US (US Department of Agriculture—Agriculture Research Service, 2000). Similarly, corn and rice are popular staple foods with high glycemic indices in most applications (Foster-Powell *et al.*, 2002). Reversing the amylopectin to amylose ratio in favor of amylose, as was achieved in high-amylose corn, lowers glycemic response to foods containing the ingredient (Vonk *et al.*, 2000). High amylose in the starch fraction of other starch-containing crops, such as potato or rice could also improve the resistance to rapid digestion and absorption of these high glycemic foods.

Nutritional Value and Prevention of Disease

Most of the potential health benefits of amylose from different food sources are associated with the relatively slow digestion of this starch type and the delayed blood glucose and insulin response after absorption (Behall *et al.*, 1989; Goddard *et al.*, 1984). High amounts of amylose double helices and amylose–lipid complexes are believed to be responsible for reduced starch digestion by pancreatic α-amylase in the human small intestine. Amylose is thus considered 'resistant starch' and produces a slow and comparatively smaller but more sustained rise in blood glucose and insulin than amylopectin or sugars. The insulin requirement for amylose digestion is relatively low (Behall *et al.*, 1988), whereas amylopectin raises blood glucose, insulin and glucagon to a larger extent, thus stressing glucose regulation.

One approach to rate carbohydrate quality is by its glycemic index. It measures the blood glucose response of specific food carbohydrates relative to dietary glucose or white bread (Ludwig and Eckel, 2002). Although the glycemic index is still controversial (Pi-Sunyer, 2002), in large prospective epidemiologic studies a high-glycemic index of the overall diet has been associated with a greater risk of type 2 diabetes (Willett *et al.*, 2002). Diabetes diets aim at keeping constant blood glucose level by avoiding foods that cause a rapid rise after consumption. Several studies demonstrated a lower glycemic response after hyperinsulinemic and normal subjects consumed diets with amylose replacing amylopectin (Behall *et al.*, 2002; Behall and Howe, 1995). Diabetes prevention programs utilizing resistant starch to reduce glycemic response may be more cost-effective than medical treatment (Brand-Miller, 2004).

High amylose may also benefit cardiovascular health. Postprandial lipemia, hyper-insulinemia and diabetes contribute to cardiovascular risk. A low-glycemic load (the glycemic index multiplied by the amount of carbohydrate) diet has been associated with lower risk of type 2 diabetes and cardiovascular disease in a prospective cohort study (Liu and Willett, 2002). Low-glycemic load was associated with elevated 'good' high-density lipoprotein-cholesterol in participants of the Nurses' Health Study (Liu *et al.*, 2001). Thus, diets high in resistant starches support heart health through a more constant blood glucose level, although effects on plasma lipoproteins are limited. Highly resistant starch diets may also affect food intake and weight control. Lack of the sensation of satiety and a fluctuating blood glucose and insulin level may be critical in body weight regulation (Ludwig, 2000). The rapid absorption of glucose after high glycemic meals induces a sequence of hormonal and metabolic changes that could promote more food intake.

Taken together, increasing the resistant starch fraction in foods can help lower postprandial glucose and insulin response as well as food intake. The reduced glycemic response may lower the risk of diabetes, secondary cardiovascular disease and obesity. The replacement of significant amounts of rapidly digestible carbohydrates with resistant starch in popular starchy crops (potato, corn and rice) can be considered a nutritional enhancement.

Sources for Resistant Starches

Food labels claiming a product to have a low glycemic index have been permitted in Australia provided the food product meets certain nutrient criteria. Most other countries have not yet allowed favorable labeling indicating the glycemic index. High-amylose starch is currently derived mainly from non-GM high-amylose corn (maize) mutants. Some elevated-amylose rice varieties have also been shown to provide low glycemic indices (Foster-Powell *et al.*, 2002) as does specifically processed tapioca starch.

High-amylose resistant corn starches available to food manufacturers include Hylon, Hi-Maize™ and Novelose® (National Starch and Chemical Company, Bridgewater, NJ), AmyloGel (Cargill, Cedar Rapids, IA), Gelose® (Penford Australia, Lane Cove, Australia) and Eurylon® (Roquette America, Inc.). These are processed resistant starches based on non-GMO high amylose (50–90 %) corn (Vonk *et al.*, 2000). Cerestar (Mechelen, Belgium) offers C*ActiStar a partially hydrolyzed and retrogradated non-GMO tapioca starch containing >50 % resistant starch. Above starches can be labeled to contain 30–60 % total dietary fiber for the starch fraction in a food. Products formulated with resistant starch may carry a 'Good source of fiber' in the US, when formulated to deliver 2.5 g of fiber per serving or 'High source of fiber', at 5.0 g. An additional benefit of resistant starches is the lower amount of energy available to human nutrition, 2–3 instead of 4 kcal/g depending on the product (Behall and Howe, 1996). Food categories include diabetic bread products, breakfast cereals, pasta and extruded products.

A recent review (Jobling, 2004) describes the technical achievements for yielding high-amylose starch by genetic engineering. So far, the most successful attempts to increase amylose levels from 20 % to 30 % in normal starch to >60 % have been by the simultaneous downregulation of starch branching enzymes I and II (SBE I and II, Schwall *et al.*, 2000) or by expressing antibodies against SBEII (Jobling *et al.*, 2003). Both approaches were performed in transgenic potato plants. These approaches should be transferable also to other starchy crops in order to increase the amount of amylose.

PUFAs

Long-chain polyunsaturated *n*-3 (omega-3) fatty acids are vital constituents of human metabolism. For adults, eicosapentaenoic acid (EPA) and docosahexaenoic acid (DHA) have physiological benefits mainly toward cardiovascular protection, reduction of inflammation and mental performance. Infants benefit from adding DHA and arachidonic acid (ARA) to infant formula. Hereby they receive nutrients close to mother's milk. This ensures access of the developing organism to DHA, the primary structural fat in the brain and eyes, and to ARA, a precursor of a hormone-like growth mediator and the most prevalent omega-6 fatty acid in the brain (Jensen, 1999). In this section we will consider EPA and DHA for improved adult nutrition. The accumulating scientific evidence indicating adult health benefits causes an increase in demand for dietary EPA and DHA, and requires more sustainable and safe production systems for the nutrients.

EPA and DHA are currently sourced from fish, marine animals and marine microalgae. Oil crops lack the genes to desaturate and elongate their fatty acids to yield EPA and DHA. The production of EPA and DHA in oil crops should have substantial advantages over fish oils (Wu *et al.*, 2005). Fish oils accumulate contaminants such as heavy metals, dioxins and polychlorinated biphenyls. Background fatty acids in fish oils are typically those of saturated animal fat and thus less healthy than those from plant oils. Importantly, the lack of odorous amines and metal catalysts degrading the fragile fatty acids in future crop oils may substantially improve taste and stability. Finally, the sustainability and the economics of modern crops will allow the incorporation of EPA and DHA into a wide range of foods.

Nutritional Value, Prevention of Disease and Recommendations

EPA and DHA are important components of human brain and retina, and precursors for hormone-like mediators (eicosanoids) affecting chronic diseases. Although only linoleic- (*n*-6) and α-linolenic acids (*n*-3) are considered essential for humans (Food and Nutrition Board (FNB), 2002), evidence accumulates that dietary supplementation with EPA and DHA has cardiovascular and other benefits for a broad population (von Schacky, 2004). Dietary EPA and DHA reduce blood coagulation, blood lipids and blood pressure, may improve the lipoprotein profile and have anti-arrhythmic effects (Calder, 2004; Geelen *et al.*, 2004). EPA and DHA were shown to suppress inflammatory and allergic processes (Calder 2002), benefit mental disorders (Horrocks and Farooqui, 2004; Morris *et al.*, 2003) and depression (Hibbeln, 2002). The significance of cardiovascular health benefits has been discussed for years and the recent scientific substantiation may convince the US FDA to eventually upgrade the current 'Qualified Health Claim' on 'Omega-3 Fatty Acids and Coronary Heart Disease' (Docket No. 2004Q-0401) into a full claim.

During the last 30 years, evidence has accumulated that dietary supplementation with EPA and DHA helps prevent chronic diseases (Bang *et al.*, 1976; Kromann *et al.*, 1980). Studies by Burr *et al.* (1989) suggested that modest intake of fatty fish reduces mortality in men who have recovered from myocardial infarction. The GISSI Prevenzione trial confirmed protective findings in a large intervention study in patients surviving myocardial infarction (GISSI-Prevenzione Investigators, 1999). Treatment with about 0.9 g EPA and DHA ethyl esters per day significantly lowered the risk of death and cardiovascular death. The effect of EPA and DHA on insulin sensitivity in type 2 diabetes is currently an area of considerable interest. Epidemiological studies suggest that improved plasma insulin or insulin sensitivity is not observed in diabetics (MacLean *et al.*, 2004), but that fish consumption protects against cardiovascular disease and total mortality in diabetics (Hu *et al.*, 2003). Clinical data with respect to diabetes are still heterogeneous and require more research (Julius, 2003).

International dietary fatty acid recommendations show a trend toward recognizing the significance of fish oil type fats or EPA and DHA (PUFA Newsletter, September 2003, www.fatsoflife.com). In the US and most European countries, omega-3 fatty acid intake for adults ranges between 1 and 2 g/d (0.5–1.0 % of energy), the majority is derived from α-linolenic acid and only 0.1–0.3 g/d from EPA plus DHA (Kris-Etherton *et al.*, 2000; Sanders, 2000). A few industrialized countries have higher fish oil intakes such as Portugal, Spain, France and Japan. Vegetarians, especially vegans, often show depressed tissue EPA and DHA levels (Davis and Kris-Etherton, 2003). Before the introduction of today's *n*-6-rich oil crops, *n*-3 fatty acids had a more significant role in our diets, with a dietary *n*-6/*n*-3 fatty acid ratio of about 2:1 (Simopoulos, 1999). Today the ratio may exceed 10:1 in most Western diets. Of concern is the rise in pro-inflammatory and regulatory eicosanoids produced from ARA metabolites. Thus, an improved supply from plant oil derived EPA and DHA would be desirable.

Taking together scientific evidence on the benefits of EPA and DHA for adult health and the levels of supplementation at which benefits were achieved, a prudent goal would be an intake of 0.5–1.0 g EPA and DHA per day. Thus, the current intake should be increased 3–5-fold, for which GM oils may offer a healthy and sustainable alternative.

Sources for PUFAs

Current production systems are either not sustainable or not economic to supply large quantities of the active ingredients that will be needed in future broad food supplementation. The predominant sources of EPA and DHA are fish and fish oils. Some fatty fish, particularly halibut, mackerel, herring, and salmon, are rich sources of EPA and DHA. The content of *n*-3 fatty acids may vary significantly depending on the type of fish, the environment and if it was farm-raised or wild. Typically, fatty acid patterns of farm-raised salmon are adjusted to its wild counterpart by adding marine oils to the diet during the finishing of the animals. This method has limits since other factors such as physical activity and environment also affect fish fatty acid pattern. Wild fish exploitation has raised serious concerns about the ecological effects of industrialized ocean fishing (Myers and Worm, 2003) with a plea for more sustainable management (Pauly *et al.*, 2002).

Alternatively DHA can be produced by fermentation of microalgae and subsequent extraction of the oil (Australia New Zealand Food Authority C, 2002). Fermentative production systems have been optimized for oil yield as well as for feed and energy efficiency. Due to its high costs the technology will remain economical only for niche products. A cost-efficient production in transgenic oil crops might constitute an alternative providing diverse advantages over aquaculture, wild fish exploitation and microbial fermentation. For technical achievements in the area of genetic engineering the reader is referred to Wu *et al.* (2005) and to Chapter 2.3.

Conclusions

The concept of enhanced nutrition to maintain health and performance of an aging society will become increasingly important in the coming years, especially in the industrialized nations. Nutraceuticals, which underpin disease prevention strategies, are either contained in functional food or are consumed as supplements. Often, these components are not 'essential' by a nutrient definition. We have discussed several compounds with health benefits taken from diverse biochemical fields and with varying beneficial effects. In short, the scientific literature proposes that plant sterols support a healthy blood lipoprotein profile, PUFAs reduce cardiovascular disease risk and mental degradation, pro-vitamin A has essential visual functions, zeaxanthin and lutein may prevent visual degradation and the intake of resistant starch reduces glycemic response and may lower the risk of type 2 diabetes and obesity.

The discussion of each of these nutraceuticals has shown that several options exist to allow for their production, including chemical synthesis, fermentation, extraction from available natural sources, breeding and plant biotechnology. In all cases plant biotechnology has progressed significantly, yielding elevated levels of the respective compound in diverse transgenic plant species.

Given these technical achievements it can be concluded that GM crops in general offer the opportunity to enrich components with proven health benefits and thereby improve specific food compositions. Moreover, metabolic engineering can also transfer healthy components from a rarely consumed food or a food with nutritional disadvantages to popular staple food to make it more beneficial. Finally, an important topic which has not been dealt with in this chapter, GM techniques can help reduce the amount of problematic ingredients in foods such as allergens (Chapter 3.1) and compounds adversary to health. Thus, GM can substantially improve diet health efficacy and quality of food.

Economic considerations determine which option will be taken by private companies to create enhanced nutritive compounds. The development of a trait by GM technology is costly and has long timelines until market introduction. To be profitable (i) the quality and contents of the nutraceutical have to be substantial, (ii) cultivation/production/extraction costs have to be relatively low compared to competitive systems and (iii) high-value markets in developed countries have to be targeted in order to obtain the return on investment. Commercialization has been profitable in a few GM crops including corn, cotton, rapeseed and soybean with so far little effort to improve food quality (Liu and Willett, 2002; Stoutjesdijk *et al.*, 2000). The next generation of GM crops is likely to serve consumer demand for nutritional benefits with crop traits sufficiently large to be profitable, for example PUFAs from oil crops (Drexler *et al.*, 2003; Wu *et al.*, 2005; Qi *et al.*, 2004).

The concept of enhanced nutrition by GM technology may also serve developing countries. Co-operative research developments by publicly funded institutions support the exploitation of GM results in developing countries to improve not only agronomic performance, but also nutrient availability in food. To this end, genetically engineered rice containing β-carotene in the endosperm was developed to provide an alternative intervention to combat vitamin A deficiency (Toenniessen, 2002). Recently, β-carotene concentrations in rice endosperm have reached levels to provide a reasonable source for vitamin A needs (Paine *et al.*, 2005). Micronutrient deficiencies including iron and zinc, which cause severe human health problems in developing countries are also being addressed both by breeding (Welch and Graham, 2002) as well as by genetic engineering (Lucca *et al.*, 2001; Zimmermann and Hurrell, 2002).

Acknowledgment

We would like to thank Bernd Sonnenberg for critically reading the manuscript.

References

Alves-Rodrigues, A. and Shao, A. (2004) The science behind lutein. *Toxicology Letters* 150, 57–83.
American Heart Association (2003) *Heart Disease and Stroke Statistics (2004) Update*, Dallas, TX, pp. 1–52.
Augustin, L.S., Gallus, S., Bosetti, C., Levi, F., Negri, E., Franceschi, S., Dal Maso, L., Jenkins, D.J., Kendall, C.W. and La Vecchia, C. (2003) Glycemic index and glycemic load in endometrial cancer. *International Journal of Cancer* 105, 404–407.
Australia New Zealand Food Authority C. Review of Infant Formula, Supplementary Final Assessment (Inquiry—S.24) Report. ANZFA—Review Of Infant Formula Proposal P93, 1–276. 8-5-2002.
Bang, H.O., Dyerberg, J. and Hjoorne, N. (1976) The composition of food consumed by Greenland Eskimos. *Acta Medica Scandinavica* 200, 69–73.
Behall, K.M. and Hallfrisch, J. (2002) Plasma glucose and insulin reduction after consumption of breads varying in amylose content. *European Journal of Clinical Nutrition* 56, 913–920.
Behall, K.M. and Howe, J.C. (1995) Effect of long-term consumption of amylose vs. amylopectin starch on metabolic variables in human subjects. *American Journal of Clinical Nutrition* 61, 334–340.
Behall, K.M. and Howe, J.C. (1996) Resistant starch as energy. *Journal of the American College of Nutrition* 15, 248–254.
Behall, K.M., Scholfield, D.J., Yuhaniak, I. and Canary, J. (1989) Diets containing high amylose vs. amylopectin starch: effects on metabolic variables in human subjects. *American Journal of Clinical Nutrition* 49, 337–344.
Behall, K.M., Scholfield, D.J. and Canary, J. (1988) Effect of starch structure on glucose and insulin responses in adults. *American Journal of Clinical Nutrition* 47, 428–432.
Belitz, H-D. and Grosch, W. (1999) *Food Chemistry*, Springer-Verlag, Berlin, Germany.
Bennedsen, M., Wang, X., Willen, R., Wadstrom, T. and Andersen, L.P. (1999) Treatment of *H. pylori* infected mice with antioxidant astaxanthin reduces gastric inflammation, bacterial load and modulates cytokine release by splenocytes. *Immunology Letters* 70, 185–189.
Berger, A., Jones, P.J. and Abumweis, S.S. (2004) Plant sterols: factors affecting their efficacy and safety as functional food ingredients. *Lipids in Health and Disease* 3, 5–23.
Beyer, P., Al-Bibili, S., Ye, X., Lucca, P., Schaub, P., Welsch, R. and Potrykus, I. (2002) Golden rice: introducing the β-carotene biosynthesis pathway into rice endosperm by genetic engineering to defeat vitamin A deficiency. *Journal of Nutrition* 132, 506S–510S.
Bouis, H.E. (2003) Micro-nutrient fortification of plants through plant breeding: can it improve nutrition in man at low cost? *Proceedings of the Nutrition Society* 62, 403–411.
Brand-Miller, J.C. (2003) Glycemic load and chronic disease. *Nutrition Reviews* 61, S49–S55.
Brand-Miller, J.C. (2004) Postprandial glycemia, glycemic index, and the prevention of type 2 diabetes. *American Journal of Clinical Nutrition* 80, 243–244.
Brigelius-Flohe, R., Kelly, F.J., Salonen, J.T., Neuzil, J., Zingg, J.M. and Azzi, A. (2002) The European perspective on vitamin E: current knowledge and future research. *American Journal of Clinical Nutrition* 76, 703–716.

Breiteneder, H. and Radauer, C. (2004) A classification of plant food allergens. *Journal of Allergy and Clinical Immunology* **113**, 821–830.

Burr, M.L., Fehily, A.M., Gilbert, J.F., Rogers, S., Holliday, R.M., Sweetnam, P.M., Elwood, P.C. and Deadman, N.M. (1989) Effects of changes in fat, fish, and fibre intakes on death and myocardial reinfarction: diet and reinfarction trial (DART). *Lancet* **2**, 757–761.

Cal, C., Garban, H., Jazirehi, A., Yeh, C., Mizutani, Y. and Bonavida, B. (2003) Resveratrol and cancer: chemoprevention, apoptosis, and chemo-immunosensitizing activities. *Current Medicinal Chemistry Anticancer Agents* **3**, 77–93.

Calder, P.C. (2002) Dietary modification of inflammation with lipids. *Proceedings of the Nutrition Society* **61**, 345–358.

Calder, P.C. (2004) n-3 Fatty acids and cardiovascular disease: evidence explained and mechanisms explored. *Clinical Science* **107**, 1–11.

Carlson, S.E., Werkman, S.H., Peeples, J.M., Cooke, R.J. and Tolley, E.A. (1993) Arachidonic acid status correlates with first year growth in preterm infants. *Proceedings of the National Academy of Sciences USA* **90**, 1073–1077.

Castenmiller, J.J., West, C.E., Linssen, J.P., van het Hof, K.H. and Voragen, A.G. (1999) The food matrix of spinach is a limiting factor in determining the bioavailability of beta-carotene and to a lesser extent of lutein in humans. *Journal of Nutrition* **129**, 349–355.

Chappell, J., Wolf, F., Prouix, J., Cueller, R. and Saunders, C. (1995) Is the reaction catalyzed by 3-hydroxy-3-methyl-glutaryl coenzyme A reductase a rate limiting step for isoprenoid biosynthesis in plants? *Plant Physiology* **109**, 1337–1343.

Chew, B.P. and Park, J.S. (2004) Carotenoid action on the immune response. *Journal of Nutrition* **134**, 257S–261S.

Cleveland, T.E., Dowd, P.F., Desjardins, A.E., Bhatnagar, D. and Cotty, P.J. (2003) United States Department of Agriculture-Agricultural Research Service research on pre-harvest prevention of mycotoxins and mycotoxigenic fungi in US crops. *Pest Management Science* **59**, 629–642.

D'Ambrosio, C., Giorio, G., Marino, I., Merendino, A., Petrozza, A., Salfi, L., Stigliani, A.L. and Cellini, F. (2004) Virtually complete conversion of lycopene into β-carotene in fruits of tomato plants transformed with the tomato lycopene β-cyclase (tlcy-b) cDNA. *Plant Science* **166**, 207–214.

Davidson, M.H. (2001) Safety profiles for the HMG-CoA reductase inhibitors: treatment and trust. *Drugs* **61**, 197–206.

Davidsson, L. (2003) Approaches to improve iron bioavailability from complementary foods. *Journal of Nutrition* **133**, 1560S–1562S.

Davis, B.C. and Kris-Etherton, P.M. (2003) Achieving optimal essential fatty acid status in vegetarians: current knowledge and practical implications. *American Journal of Clinical Nutrition* **78**, 640S–646S.

Dharmapuri, S., Rosati, C., Pallara, P., Aquilani, R., Bouvier, F., Camara, B. and Giuliano, G. (2002) Metabolic engineering of xanthophylls content in tomato fruits. *FEBS Letters* **519**, 30–34.

Drexler, H., Spiekermann, P., Meyer, A., Domergue, F., Zank, T., Sperling, P., Abbadi, A. and Heinz, E. (2003) Metabolic engineering of fatty acids for breeding of new oilseed crops: strategies, problems and first results. *Journal of Plant Physiology* **160**, 779–802.

Dutta, A. and Dutta, S.K. (2003) Vitamin E and its role in the prevention of atherosclerosis and carcinogenesis: a review. *Journal of the American College of Nutrition* **22**, 258–268.

Elmadfa, I., Aign, W., Muskat, E. and Fritzsche, D. (2004) GU Nährwert-Kalorien-Tabelle 2004/05. Munich, Germany, Gräfe und Unzer Verlag GmbH.

Farnaud, S. and Evans, R.W. (2003) Lactoferrin—a multifunctional protein with antimicrobial properties. *Molecular Immunology* **40**, 395–405.

Fasano, A., Berti, I., Gerarduzzi, T., Not, T., Colletti, R.B., Drago, S., Elitsur, Y., Green, P.H., Guandalini, S., Hill, I.D., Pietzak, M., Ventura, A., Thorpe, M., Kryszak, D., Fornaroli, F., Wasserman, S.S., Murray, J.A. and Horvath, K. (2003) Prevalence of celiac disease in at-risk and not-at-risk groups in the United States: a large multicenter study. *Archives of Internal Medicine* **163**, 286–292.

Food and Agriculture Organization of the United Nations (1979) Recommended practices for the prevention of mycotoxins in food, feed and their products. *FAO Food Nutrition Papers* **10**, 1–71.

Food and Agriculture Organization of the United Nations (2002) Vitamin A, in *Human Vitamin and Mineral Requirements*, FAO and WHO, Rome, Italy.

Food and Agriculture Organization of the United Nations (2004) The State of Food and Agriculture 2003–2004, FAO and WHO, Rome, Italy.

Food and Nutrition Board, Institute of Medicine (2002) Dietary reference intakes for energy, carbohydrate, fiber, fat, fatty acids, cholesterol, protein, and amino acids (macronutrients). http://www.nap.edu/books/0309085373/html/.

Foster-Powell, K., Holt, S.H. and Brand-Miller, J.C. (2002) International table of glycemic index and glycemic load values: 2002. *American Journal of Clinical Nutrition* **76**, 5–56.

Friedman, D.S., O'Colmain, B.J., Munoz, B., Tomany, S.C., McCarty, C., de Jong, P.T., Nemesure, B., Mitchell, P. and Kempen, J. (2004) Prevalence of age-related macular degeneration in the United States. *Archives of Ophthalmology* **122**, 564–572.

Geelen, A., Brouwer, I.A., Zock, P.L. and Katan, M.B. (2004) Antiarrhythmic effects of n-3 fatty acids: evidence from human studies. *Current Opinion in Lipidology* **15**, 25–30.

GISSI-Prevenzione Investigators (1999) Dietary supplementation with n-3 polyunsaturated fatty acids and vitamin E after myocardial infarction: results of the GISSI-Prevenzion trial. Gruppo Italiano per lo Studio della Sopravvivenza nell'Infarto miocardico. *Lancet* **354**, 447–455.

Goddard, M.S., Young, G. and Marcus, R. (1984) The effect of amylose content on insulin and glucose responses to ingested rice. *American Journal of Clinical Nutrition* **39**, 388–392.

Green, P.H. and Jabri, B. (2003) Coeliac disease. *Lancet* **362**, 383–391.

Grundy, S.M., Abate, N. and Chandalia, M. (2002) Diet composition and the metabolic syndrome: what is the optimal fat intake? *American Journal of Medicine* **113** (Suppl 9B) 25S–29S.

Guerin, M., Huntley, M.E. and Olaizola, M. (2003) *Haematococcus astaxanthin*: applications for human health and nutrition. *Trends in Biotechnology* **21**, 210–216.

Hadley, C.W., Miller, E.C., Schwartz, S.J. and Clinton, S.K. (2002) Tomatoes, lycopene, and prostate cancer: progress and promise. *Experimental Biology and Medicine* **227**, 869–880.

Harris, A., Devaraj, S. and Jialal, I. (2002) Oxidative stress, alpha-tocopherol therapy, and atherosclerosis. *Current Atherosclerosis Reports* **4**, 373–380.

Herbers, K. (2003) Vitamin production in transgenic plants. *Journal of Plant Physiology* **160**, 821–829.

Hibbeln, J.R. (2002) Seafood consumption, the DHA content of mothers' milk and prevalence rates of postpartum depression: a cross-national, ecological analysis. *Journal of Affective Disorders* **69**, 15–29.

Holmberg, N., Harker, M., Gibbard, C.L., Wallace, A.D., Clayton, J.C., Rawlins, S., Hellyer, A. and Safford, R. (2002) Sterol C-24 methyltransferase type 1 controls the flux of carbon into sterol biosynthesis in tobacco seed. *Plant Physiology* **130**, 303–311.

Horrocks, L.A. and Farooqui, A.A. (2004) Docosahexaenoic acid in the diet: its importance in maintenance and restoration of neural membrane function. *Prostaglandins, Leukotrienes and Essential Fatty Acids* **70**, 361–372.

Hu, F.B., Cho, E., Rexrode, K.M., Albert, C.M. and Manson, J.E. (2003) Fish and long-chain omega-3 fatty acid intake and risk of coronary heart disease and total mortality in diabetic women. *Circulation* **107**, 1852–1857.

Innis, S.M. (2003) Perinatal biochemistry and physiology of long-chain polyunsaturated fatty acids. *Journal of Pediatrics* **143**, S1–S8.

Jenkins, D.J., Kendall, C.W., Augustin, L.S., Franceschi, S., Hamidi, M., Marchie, A., Jenkins, A.L. and Axelsen, M. (2002) Glycemic index: overview of implications in health and disease. *American Journal of Clinical Nutrition* **76**, 266S–273S.

Jensen, R.G. (1999) Lipids in human milk. *Lipids* **34**, 1243–1271.

Jobling, S. (2004) Improving starch for food and industrial applications. *Current Opinion in Plant Biology* **7**, 210–218.

Jobling, S.A., Jarman, C., The, M.M., Holmberg, N., Blake, C. and Verhoeyen, M.E. (2003) Immunomodulation of enzyme function in plants by single-domain antibody fragments. *Nature Biotechnology* **21**, 77–80.

Julius, U. (2003) Fat modification in the diabetes diet. *Experimental Clinical Endocrinology and Diabetes* **111**, 60–65.

Jyonouchi, H., Sun, S., Iijima, K. and Gross, M.D. (2000) Antitumor activity of astaxanthin and its mode of action. *Nutrition and Cancer* **36**, 59–65.

Katan, M.B., Grundy, S.M., Jones, P., Law, M., Miettinen, T. and Paoletti, R. (2003) Efficacy and safety of plant stanols and sterols in the management of blood cholesterol levels. *Mayo Clinic Proceedings* **78**, 965–978.

Khachik, F., Bernstein, P.S. and Garland, D.L. (1997) Identification of lutein and zeaxanthin oxidation products in human and monkey retinas. *Investigative Ophthalmology and Visual Science* **38**, 1802–1811.

Krinsky, N.I. (1989) Antioxidant functions of carotenoids. *Free Radical Biology and Medicine* **7**, 617–635.

Krinsky, N.I. (2002) Possible biologic mechanisms for a protective role of xanthophylls. *Journal of Nutrition* **132**, 540S–542S.

Krinsky, N.I., Landrum, J.T. and Bone, R.A. (2003) Biologic mechanisms of the protective role of lutein and zeaxanthin in the eye. *Annual Reviews of Nutrition* **23**, 171–201.

Kris-Etherton, P.M., Taylor, D.S., Yu-Poth, S., Huth, P., Moriarty, K., Fishell, V., Hargrove, R.L., Zhao, G. and Etherton, T.D. (2000) Polyunsaturated fatty acids in the food chain in the United States. *American Journal of Clinical Nutrition* **71**, 179S–188S.

Kristal, A.R. (2004) Vitamin A, retinoids and carotenoids as chemopreventive agents for prostate cancer. *Journal of Urology* **171**, S54–S58.

Kromann, N. and Green, A. (1980) Epidemiological studies in the Upernavik district, Greenland. Incidence of some chronic diseases 1950–1974. *Acta Medica Scandinavica* **208**, 401–406.

Kuchide, M., Tokuda, H., Takayasu, J., Enjo, F., Ishikawa, T., Ichiishi, E., Naito, Y., Yoshida, N., Yoshikawa, T. and Nishino, H. (2003) Cancer chemopreventive effects of oral feeding alpha-tocopherol on ultraviolet light B induced photocarcinogenesis of hairless mouse. *Cancer Letters* **196**, 169–177.

Landrum, J.T. and Bone, R.A. (2001) Lutein, zeaxanthin, and the macular pigment. *Archives of Biochemistry and Biophysiology* **385**, 28–40.

Larque, E., Demmelmair, H. and Koletzko, B. (2002) Perinatal supply and metabolism of long-chain polyunsaturated fatty acids: importance for the early development of the nervous system. *Annals of the New York Academy of Sciences* **967**, 299–310.

Larsson, S.C., Kumlin, M., Ingelman-Sundberg, M. and Wolk, A. (2004) Dietary long-chain n-3 fatty acids for the prevention of cancer: a review of potential mechanisms. *American Journal of Clinical Nutrition* **79**, 935–945.

Lichtenstein, A.H. (2002) Plant sterols and blood lipid levels. *Current Opinion in Clinical Nutrition and Metabolic Care* **5**, 147–152.

Lindgren, L.O., Stalberg, K.G. and Höglund, A.S. (2003) Seed-specific over-expression of an endogenous *Arabidopsis* phytene synthase gene results in delayed germination and increased levels of carotenoids, chlorophyll, and abscisic acid. *Plant Physiology* **132**, 779–785.

Liu, S. and Willett, W.C. (2002) Dietary glycemic load and atherothrombotic risk. *Current Atherosclerosis Reports* **4**, 454–461.

Liu, S., Manson, J.E., Stampfer, M.J., Holmes, M.D., Hu, F.B., Hankinson, S.E. and Willett, W.C. (2001) Dietary glycemic load assessed by food-frequency questionnaire in relation to plasma high-density-lipoprotein cholesterol and fasting plasma triacylglycerols in postmenopausal women. *American Journal of Clinical Nutrition* **73**, 560–566.

Lucca, P., Hurrell, R. and Potrykus, I. (2001) Genetic engineering approaches to improve bioavailability and the level of iron in rice grains. *Theoretical and Applied Genetics* **102**, 392–397.

Ludwig, D.S. and Eckel, R.H. (2002) The glycemic index at 20 y. *American Journal of Clinical Nutrition* **76**, 264S–265S.

Ludwig, D.S. (2000) Dietary glycemic index and obesity. *Journal of Nutrition* **130**, 280S–283S.

MacLean, C.H., Mojica, W.A. and Morton, S.C. (2004) Effects of omega-3 fatty acids on lipids and glycemic control in type II diabetes and the metabolic syndrome and on inflammatory bowel disease, rheumatoid arthritis, renal disease, systemic lupus erythematosus, and osteoporosis. 89: 04-E012-1. 2004. AHRQ Publication.

MacLeod, G. and Ames, J. (1988) Soy flavor and its improvement. *Critical Reviews in Food Science and Nutrition* **27**, 19–400.

Maki, K.C., Davidson, M.H., Tsushima, R., Matsuo, N., Tokimitsu, I., Umporowicz, D.M., Dicklin, M.R., Foster, G.S., Ingram, K.A., Anderson, B.D., Frost, S.D. and Bell, M. (2002) Consumption of diacylglycerol oil as part of a reduced-energy diet enhances loss of body weight and fat in comparison with consumption of a triacylglycerol control oil. *American Journal of Clinical Nutrition* **76**, 1230–1236.

Miettinen, T.A. and Gylling, H. (2003) Non-nutritive bioactive constituents of plants: phytosterols. *International Journal for Vitamin and Nutrition Research* **73**, 127–134.

Mokdad, A.H., Bowman, B.A., Ford, E.S., Vinicor, F., Marks, J.S. and Koplan, J.P. (2001) The continuing epidemics of obesity and diabetes in the United States. *Journal of the American Medical Association* **286**, 1195–1200.

Morris, M.C., Evans, D.A., Bienias, J.L., Tangney, C.C., Bennett, D.A., Wilson, R.S., Aggarwal, N. and Schneider, J. (2003) Consumption of fish and n-3 fatty acids and risk of incident Alzheimer disease. *Archives of Neurology* **60**, 940–946.

Mozaffarieh, M., Sacu, S. and Wedrich, A. (2003) The role of the carotenoids, lutein and zeaxanthin, in protecting against age-related macular degeneration: a review based on controversial evidence. *Nutrition Journal* **2**, 20–27.

Muller, N., Alteheld, B. and Stehle, P. (2003) Tomato products and lycopene supplements: mandatory components in nutritional treatment of cancer patients? *Current Opinion in Clinical Nutrition and Metabolic Care* **6**, 657–660.

Myers, R.A. and Worm, B. (2003) Rapid worldwide depletion of predatory fish communities. *Nature* **423**, 280–283.

Nagao, T., Watanabe, H., Goto, N., Onizawa, K., Taguchi, H., Matsuo, N., Yasukawa, T., Tsushima, R., Shimasaki, H. and Itakura, H. (2000) Dietary diacylglycerol suppresses accumulation of body fat compared to triacylglycerol in men in a double-blind controlled trial. *Journal of Nutrition* **130**, 792–797.

Naidu, A.S. and Nimmagudda, R. (2003) Activated lactoferrin. Part 1: a novel antimicrobial formulation. AgroFood Industry Hi-Tech 47–50.

National Cholesterol Education Program (NCEP) (2004) *Expert Panel. Detection, Evaluation, and Treatment of High Blood Cholesterol in Adults (Adult Treatment Panel III)*, Volume 3. National Guideline Clearinghouse, pp. 1–36.

National Eye Institute (2002) *Vision Problems in the U.S., Prevalence of Adult Vision Impairment and Age-Related Eye Disease in America*, National Eye Institute, Bethesda, MD, pp. 1–42.

Nebeling, L.C., Forman, M.R., Graubard, B.I. and Snyder, R.A. (1997) Changes in carotenoid intake in the United States: the 1987 and 1992 National Health Interview Surveys. *Journal of the American Dietetic Association* **97**, 991–996.

Normen, A.L., Brants, H.A., Voorrips, L.E., Andersson, H.A., van Den Brandt, P.A. and Goldbohm, R.A. (2001) Plant sterol intakes and colorectal cancer risk in the Netherlands Cohort Study on Diet and Cancer. *American Journal of Clinical Nutrition* **74**, 141–148.

O'Keefe, J.H., Jr. and Cordain, L. (2004) Cardiovascular disease resulting from a diet and lifestyle at odds with our Paleolithic genome: how to become a 21st-century hunter-gatherer. *Mayo Clinic Proceedings* **79**, 101–108.

Ostlund, R.E. Jr. (2002) Phytosterols in human nutrition. *Annual Reviews of Nutrition* **22**, 533–549.

Ostlund, R.E. Jr. (2004) Phytosterols and cholesterol metabolism. *Current Opinion in Lipidology* **15**, 37–41.

Paine, J.A., Shipton, C.A., Chaggar, S., Howells, R.M., Kennedy, M.J., Vernon, G., Wright, S.Y., Hinchliffe, E., Adams, J.L., Silverstone, A.L. and Drake, R. (2005) Improving the nutritional value of Golden Rice through increased pro-vitamin A content. *Nature Biotechnology* **23**, 482–487.

Pasamontes, L., Hug, D., Tessier, M., Hohmann, H.P., Schierle, J. and van Loon, A.P. (1997) Isolation and characterization of the carotenoid biosynthesis genes of Flavobacterium sp. strain R1534. *Gene* **185**, 35–41.

Pauly, D., Christensen, V., Guenette, S., Pitcher, T.J., Sumaila, U.R., Walters, C.J., Watson, R. and Zeller, D. (2002) Towards sustainability in world fisheries. *Nature* **418**, 689–695.

Pervaiz, S. (2003) Resveratrol: from grapevines to mammalian biology. *FASEB Journal* **17**, 1975–1985.

Piironen, V., Toivo, J. and Lampi, A.M. (2002) Plant sterols in cereals and cereal products. *Cereal Chemistry* **79**, 148–154.

Pi-Sunyer, F.X. (2002) Glycemic index and disease. *American Journal of Clinical Nutrition* **76**, 290S–298S.

Plat, J. and Mensink, R.P. (2002) Increased intestinal ABCA1 expression contributes to the decrease in cholesterol absorption after plant stanol consumption. *FASEB Journal* **16**, 1248–1253.

Potrykus, I. (2003) Nutritionally enhanced rice to combat malnutrition disorders of the poor. *Nutrition Reviews* **61**, S101–S104.

Przybylski, R. (2001) *Canola Oil: Physical and Chemical Properties*, Canola Oil Technical Information Kit, Canola Council of Canada, Winnipeg, Canada, pp. 1–12.

Qi, B., Fraser, T., Mugford, S., Dobson, G., Sayanova, O., Butler, J., Napier, J.A., Stobart, A.K. and Lazarus, C.M. (2004) Production of very long chain polyunsaturated omega-3 and omega-6 fatty acids in plants. *Nature Biotechnology* **22**, 739–745.

Ravanello, M.P., Ke, D., Alvarez, J., Huang, B. and Shewmaker, C.K. (2003) Coordinate expression of multiple bacterial carotenoid genes in canola leading to altered carotenoid production. *Metabolic Engineering* **5**, 255–263.

Richer, S., Stiles, W., Statkute, L., Pulido, J., Frankowski, J., Rudy, D., Pei, K., Tsipursky, M. and Nyland, J. (2004) Double-masked, placebo-controlled, randomized trial of lutein and antioxidant supplementation in the intervention of atrophic age-related macular degeneration: the Veterans LAST study (Lutein Antioxidant Supplementation Trial). *Optometry* **75**, 216–230.

Römer, S., Lübeck, J., Kauder, F., Steiger, S., Adomat, C. and Sandmann, G. (2002) Genetic engineering of a zeaxanthin-rich potato by antisense inactivation and co-suppression of carotenoid epoxidation. *Metabolic Engineering* **4**, 263–272.

Sacks, F.M. and Katan, M.B. (2002) Randomized clinical trials on the effects of dietary fat and carbohydrate on plasma lipoproteins and cardiovascular disease. *American Journal of Medicine* **113**, (Suppl 9B) 13S–24S.

Salmeron, J., Ascherio, A., Rimm, E.B., Colditz, G.A., Spiegelman, D., Jenkins, D.J., Stampfer, M.J., Wing, A.L. and Willett, W.C. (1997a) Dietary fiber, glycemic load, and risk of NIDDM in men. *Diabetes Care* **20**, 545–550.

Salmeron, J., Manson, J.E., Stampfer, M.J., Colditz, G.A., Wing, A.L. and Willett, W.C. (1997b) Dietary fiber, glycemic load, and risk of non-insulin-dependent diabetes mellitus in women. *Journal of the American Medical Association* **277**, 472–477.

Sanders, T.A. (2000) Polyunsaturated fatty acids in the food chain in Europe. *American Journal of Clinical Nutrition* **71**, 176S–178S.

Schafer, A.I. (1996) Antiplatelet therapy. *American Journal of Medicine* **101**, 199–209.

Schaller, H., Grausem, B., Benveniste, P., Chye, M.L., Tan, C.T., Song, Y.H. and Chua, N.H. (1995) Expression of the *Hevea brasiliensis* (H.B.K.) Müll. Arg. 3-hydroxy-3-methylglutaryl-coenzyme A reductase 1 in tobacco results in sterol overproduction. *Plant Physiology* **109**, 761–770.

Schwall, G.P., Safford, R., Westcott, R.J., Jeffcoat, R., Tayal, A., Shi, Y.C., Gidley, M.J. and Jobling, S.A. (2000) Production of very-high-amylose potato starch by inhibition of SBE A and B. *Nature Biotechnology* **18**, 551–554.

Shewmaker, C.K., Sheehy, J.A., Daley, M., Colburn, S. and Ke, D.Y. (1999) Seed-specific overexpression of phytoene synthase: increase in carotenoids and other metabolic effects. *Plant Journal* **20**, 401–412.

Sies, H. and Stahl, W. (2003) Non-nutritive bioactive constituents of plants: lycopene, lutein and zeaxanthin. *International Journal for Vitamin and Nutrition Research* **73**, 95–100.

Simopoulos, A.P. (1999) Essential fatty acids in health and chronic disease. *American Journal of Clinical Nutrition* **70**, 560S–569S.

Smith, S.C., Jr. (2000) Bridging the treatment gap. *American Journal of Cardiology* **85**, 3E–7E.

Solomons, N.W. (2001) Vitamin A and carotenoids, in *Present Knowledge in Nutrition* (eds B.A. Bowman and R.M. Russel), ILSI Press, Washington, DC, pp. 127–145.

Sommer, A. and Davidson, F.R. (2002) Assessment and control of vitamin A deficiency: the Annecy Accords. *Journal of Nutrition* **132**, 2845S–2850S.

Stewart, J.R., Artime, M.C. and O'Brian, C.A. (2003) Resveratrol: a candidate nutritional substance for prostate cancer prevention. *Journal of Nutrition* **133**, 2440S–2443S.

Stoutjesdijk, P.A., Hurlestone, C., Singh, S.P. and Green, A.G. (2000) High-oleic acid Australian *Brassica napus* and *B. juncea* varieties produced by co-suppression of endogenous Delta12-desaturases. *Biochemical Society Transactions* **28**, 938–940.

Tada, N. and Yoshida, H. (2003) Diacylglycerol on lipid metabolism. *Current Opinion in Lipidology* **14**, 29–33.

Toenniessen, G.H. (2002) Crop genetic improvement for enhanced human nutrition. *Journal of Nutrition* **132**, 2943S–2946S.

Tobert, J.A. (2003) Lovastatin and beyond: the history of the HMG-CoA reductase inhibitors. *Nature Reviews Drug Discovery* **2**, 517–526.

US Department of Agriculture—Agriculture Research Service (1998) *Pyramid Servings Intakes by U.S. Children and Adults: 1994–96*, 2000, Agricultural Research Service, U.S. Department of Agriculture, Beltsville, MD, pp. 1–47.

Vega-Lopez, S., Devaraj, S. and Jialal, I. (2004) Oxidative stress and antioxidant supplementation in the management of diabetic cardiovascular disease. *Journal of Investigative Medicine* **52**, 24–32.

Venkatramesh, M., Karunanandaa, B., Sun, B., Gunter, C.A., Boddupalli, S. and Kishore, G.M. (2003) Expression of a *Streptomyces* 3-hydroxysteroid oxidase gene in oilseeds for converting phytosterols to phytostanols. *Phytochemistry* **62**, 39–46.

von Schacky, C. (2004) Omega-3 fatty acids and cardiovascular disease. *Current Opinion in Clinical Nutrition and Metabolic Care* **7**, 131–136.

Vonk, R.J., Hagedoorn, R.E., de Graaff, R., Elzinga, H., Tabak, S., Yang, Y.X. and Stellaard, F. (2000) Digestion of so-called resistant starch sources in the human small intestine. *American Journal of Clinical Nutrition* **72**, 432–438.

Ward, P.P. and Connelly, O.M. (2004) Lactoferrin: role in iron homeostasis and host defense against microbial infection. *Biometals* **17**, 203–208.

Welch, R.M. and Graham, R.D. (2004) Breeding for micro-nutrients in staple food crops from a human nutrition perspective. *Journal of Experimental Botany* **55**, 353–364.

Weinberg, E.D. (2001) Human lactoferrin: a novel therapeutic with broad spectrum potential. *Journal of Pharmacy and Pharmacology* **53**, 1303–1310.

Welch, R.M. and Graham, R.D. (2002) Breeding crops for enhanced micro-nutrient content. *Plant and Soil* **245**, 205–214.

Willett, W.C., Manson, J.E. and Liu, S. (2002) Glycemic index, glycemic load, and risk of type 2 diabetes. *American Journal of Clinical Nutrition* **76**, 274S–280S.

World Health Organization (2004) The world health report 2004—changing history.

World Health Organization (1998)Vitamin A deficiency (VAD) prevalence by level of public health significance (map). WHO Global Database on vitamin A deficiency.

Wu, G., Truksa, M., Datla, N., Vrinten, P., Bauer, J., Zank, T., Cirpus, P., Heinz, E. and Qiu, X. (2005) Stepwise engineering to produce high yields of very long-chain polyunsaturated fatty acids in plants. *Nature Biotechnology* **23**, 1013–1017.

Zimmermann, M.B. and Hurrell, R.F. (2002) Improving iron, zinc and vitamin A nutrition through plant biotechnology. *Current Opinion in Biotechnology* **13**, 142–145.

Zingg, J.M. and Azzi, A. (2004) Non-antioxidant activities of vitamin E. *Current Medicinal Chemistry* **11**, 1113–1133.

2.3

The Production of Long-Chain Polyunsaturated Fatty Acids in Transgenic Plants

Louise V. Michaelson, Frédéric Beaudoin, Olga Sayanova and Johnathan A. Napier

*Crop Performance and Improvement, Rothamsted Research, Harpenden, Hertfordshire,
AL5 2JQ, United Kingdom*

Introduction

Long-chain polyunsaturated fatty acids (LC-PUFAs) are important in human health and nutrition. In particular, fetal development is dependent on a supply of *n-3* polyunsaturated fatty acids. *n-3* polyunsaturated fatty acids have also been shown to be protective against cardiovascular disease and risks associated with metabolic syndrome. In view of the decline in marine fish stocks, which represent the predominant natural reserves of *n-3* long chain polyunsaturates, alternative sources are urgently required. One approach may be to express the LC-PUFA biosynthetic pathway in transgenic plants. Recent progress in validating this approach has now emerged, demonstrating the feasibility of using transgenic plants to synthesize these important human nutrients.

There is considerable potential in using molecular techniques to produce plants that have been modified to improve or enhance the nutritional composition of their crop (Tucker, 2003). Improving the nutritional composition of the crop may be attempted by increasing the levels of endogenous nutrients (e.g. vitamin E) (Cahoon *et al.*, 2003) or alternatively introducing non-native compounds (e.g. essential fatty acids) into plants (Jaworski and Cahoon, 2003). There are significant economic and ecological drivers for developing transgenic plants as novel sources of some nutrients; many of these

Plant Biotechnology. Edited by Nigel Halford.
© 2006 John Wiley & Sons, Ltd.

compounds are currently obtained from non-sustainable or economically unviable sources (Ohlrogge, 1999; Tucker, 2003).

Transgenic plants engineered to accumulate specific compounds of benefit to human health and nutrition have begun to emerge as a viable alternative source to current production methods (Thelen and Olrogge, 2002). While the concept of the 'green factory' (i.e. a transgenic plant engineered to synthesize a desired product) is not new, it is only in the last few years that this technology has clearly demonstrated its earlier promise (Ohlrogge, 1999; Tucker, 2003). The continued debate over the desirability, or even acceptability, of transgenic plants entering the human food chain has overshadowed the potential benefits of GM-derived nutritional enhancement of plants (Sayanova and Napier, 2004).

The first generation of transgenic plants for which regulatory approval to enter the human food chain was sought were exclusively input traits, meaning they were engineered for traits such as herbicide tolerance or insect resistance (Thelen and Olrogge, 2002). These traits are of benefit to farmers and conventional agricultural practices, but they do not demonstrate obvious benefits to the consumer, not in the face of increased public scepticism regarding GM food and food safety in general. It might be hoped that output traits, in which transgenic plants are engineered to produce compounds that are of value to the consumer, might help to persuade the general public of the benefits of this technology (Sayanova and Napier, 2004; Tucker, 2003). Current examples of output traits engineered into transgenic plants include the synthesis of molecules such as single-chain antibodies, as well as the above-mentioned nutritionally enhanced foodstuffs (Paine *et al.*, 2005; Warzecha and Mason, 2003).

In particular, we are interested in the possibilities of transgenically expressing the biosynthetic pathway of LC-PUFAs, normally found in aquatic microorganisms (Sayanova and Napier, 2004).

LC-PUFAs in Human Health

LC-PUFAs are known to play several discrete roles in human metabolism. These are likely to include biophysical roles in membrane bilayers, as well as those relating to the metabolism of these fatty acids. Perhaps the best known metabolic function of LC-PUFAs is their role as precursors to a class of compounds termed eicosanoids (i.e. metabolites of eicosa [C_{20}] fatty acids) (Funk, 2001). The eicosanoids consist of leukotrienes, prostaglandins and isoprostanes. These molecules have potent biological activities on platelets, blood vessels and most organ systems, exerting their actions via G protein-coupled receptors (GPCRs) or peroxisomal proliferator-activated receptors (PPARs) (Funk, 2001). These compounds perform a number of essential physiological functions including regulation of the immune system, blood clotting, neurotransmission and cholesterol metabolism (Hwang, 2000).

Eicosanoids are formed via physical or chemical insults, which can result in the release (through the action of phospholipases) of LC-PUFAs from the phospholipid membrane and subsequent oxygenation by specific oxygenase enzymes (Funk, 2001). The molecular species of eicosanoid synthesized (and hence the body's responses) are determined by multiple factors, which include the specific cell type receiving and responding to the challenge (e.g. platelets, leukocytes or endothelial cells), the metabolizing oxygenase enzymes (cyclo-oxygenase versus lipoxygenase) and the levels of substrate C_{20} LC-PUFAs in the cell membrane. Eicosanoids derived from *n-6* LC-PUFAs have very distinct metabolic properties to those derived from *n-3* substrates. In general, eicosanoids are classified into three different groups of LC-PUFA metabolites: series-1 and series-3 are anti-inflammatory, whereas series-2 is pro-inflammatory. Eicosanoids derived from *n-6* substrates are generally pro-inflammatory, pro-aggregatory and immuno-active (Hwang, 2000). In contrast, the eicosanoids derived from *n-3* fatty acids such as eicosapentaenoic acid (20:5, *n-3*; EPA) have little or no inflammatory activity and act to modulate platelet aggregation and immuno-reactivity (Funk, 2001). Currently there is increasing interest in the *n-3* fatty acids because of these perceived beneficial properties.

There is also mounting evidence of the importance of *n-3* LC-PUFAs as protective factors in human pathologies such as cardiovascular disease. Dyerberg, Bang and colleagues documented the low incidence of cardiovascular disease in Inuit communities whose diet was rich in oily fish (Bang, Dyerberg and Hjoorne, 1976; Dyerberg and Bang, 1982). Since these fish oils are rich in *n-3* LC-PUFAs, the authors postulated that this dietary component made a very significant contribution to the reduced levels of cardiovascular disease observed in these populations. These detailed studies of nearly 30 years ago formed the basis for many large-scale intervention studies to assess the importance of *n-3* LC-PUFAs in human health. There is now clear evidence to support the assertion that these dietary components can play a major protective role against cardiovascular disease (Burr *et al.*, 1989; GISSI-Prevenzione Investigators, 1999; Hu, Manson and Willett, 2001; von Shacky, 2003).

More recently, it has also emerged that *n-3* LC-PUFAs play a role in reducing the risk for acquisition of metabolic syndrome. Metabolic syndrome is the descriptor for a collection of pathologies, which are indicative of progression toward cardiovascular disease and other diseases such as obesity and type 2 diabetes (Nugent, 2004; Sargent and Tacon, 1999). Thus, it appears that not only can *n-3* LC-PUFAs of the type found in fish oils help in the treatment of chronic conditions such as cardiovascular disease, but also act as positive protective factors by preventing progression toward these diseases. In particular, metabolic syndrome can be treated by dietary intervention, using a diet with reduced carbohydrate intake but with the inclusion of *n-3* LC-PUFA fish oils. Metabolic syndrome is typified by the presence of a number of symptoms (such as increased waistline, hypertension, high plasma triglycerides and abnormal blood sugar levels), which collectively indicate an increased risk of cardiovascular disease, as well as progression toward type 2 diabetes (Nugent, 2004).

There is a growing alarm at the rapidly increasing levels of obesity and diabetes in North American and European populations. Current estimations are that one in five Americans can be classified as suffering from metabolic syndrome and/or obesity. Fish oils are well known to be high in LC-PUFAs. It is estimated that to achieve the recommended levels of very long chain fatty acids in the diet from fish sources a four-fold increase in the consumption of fish would be required in the USA alone (Kris-Etherton *et al.*, 2000). Unfortunately, dietary preferences are strong and aside from the obvious difficulty of persuading people to change their eating habits, fish stocks are declining globally and some species are being driven to extinction (Jackson *et al.*, 2001; James, Ursin and Cleland, 2003). Marine fisheries are in global crisis from overfishing and farmed fish like salmon and tuna are fed with other fish, thus compounding the problem (Pauly *et al.*, 1998; 2003). Finding a sustainable alternative source to marine fish for LC-PUFAs would aid in protecting these ecosystems.

There have also been doubts about the suitability of supplements derived from marine sources because of the accumulation of environmental pollutants such as dioxins, PCBs and heavy metals, which have been found to be concentrated in the liver of fish. This is of particular relevance to the use of supplements in baby foods. Indeed, in the USA the use of fish oils in products for babies and young children is not permitted (Ratledge, 2004).

The Biosynthesis of LC-PUFAs

The human diet must contain LC-PUFAs or their precursors, as we are unable to synthesize these fatty acids *de novo* (Simopoulos, 2000). Mammals, including humans, have an absolute requirement for the dietary ingestion of the two essential fatty acids, linoleic acid (18:2, *n-6*; LA) and α-linolenic acid (18:3, *n-3*; ALA), because they lack the appropriate desaturases to convert monounsaturates into these two essential fatty acids (Wallis *et al.*, 2002). These Δ^{12} and Δ^{15} desaturases would convert oleic acid (Δ^9 18:1) to linoleic acid and α-linolenic acid. Since plants are rich in both LA and ALA, normal diets usually provide sufficient essential fatty acids. These two fatty acids then enter the LC-PUFA biosynthetic pathway, and undergo sequential rounds of aerobic desaturation and chain-length elongation to yield the C_{20-22} PUFAs (Sayanova and Napier, 2004; Wallis, Watts and Browse, 2002) (see Figure 2.3.1 for details).

There are several factors which are important in the biosynthesis of LC-PUFAs in humans. One of the main considerations is that the conversion of essential fatty acids to LC-PUFAs appears to be a relatively inefficient process (Leonard *et al.*, 2004; Simopoulos, 2000), emphasizing the desirability of supplementing our endogenous metabolism with dietary LC-PUFAs. Another important factor is the requirements for various LC-PUFAs change depending on the developmental stage. For example, it is now recognized that fatty acids such as arachidonic acid (20:4, *n-6*; ARA) and docosahex-aenoic acid (22:6, *n-3*; DHA) play essential roles in neonatal and infant health, being of particular importance for acquisition of ocular vision and brain development (Leonard *et al.*, 2004). This is now reflected in the inclusion of these two LC-PUFAs in infant formula milks, which are designed to replace maternal milk (which itself is rich in these two fatty acids). There is also evidence that LC-PUFA metabolism could be impaired during illness and old age, and that some geriatric conditions may be alleviated by treatment with these fatty acids (Leonard *et al.*, 2004).

An important point regarding mammalian LC-PUFA biosynthesis is that there is no capacity to convert *n-6* fatty acids (such as ARA) to *n-3* forms such as EPA or DHA (Leonard *et al.*, 2004; Sayanova and Napier, 2004). The consequence of this has profound implications on the synthesis of LC-PUFAs from the essential fatty acids, and their subsequent conversion to eicosanoids, since LA (*n-6*) can only yield *n-6* LC-PUFAs such as ARA, and ALA (*n-3*) will yield EPA and DHA *n-3* PUFAs. The conversion of these fatty acids into different eicosanoids generates very different bioactive molecules, and illustrates the distinct roles of *n-6* and *n-3* LC-PUFAs.

This situation is exacerbated by the changes in human diet over the last century, which have moved away from a high vegetable and fish composition (i.e. rich in *n-3* fatty acids

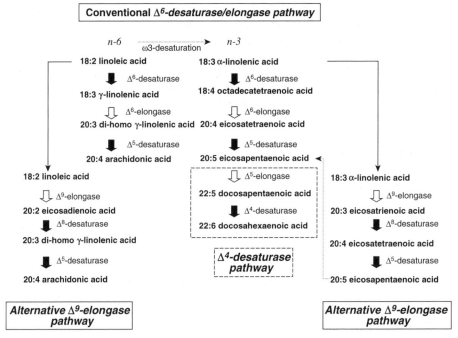

Figure 2.3.1 *Generalized representation of LC-PUFA biosynthesis. The conventional Δ^6-desaturase/elongase pathway for the synthesis of arachidonic acid and eicosapentaenoic acid from the essential fatty acids linoleic and α-linolenic acids is shown, as is the alternative Δ^9-elongase route. The Δ^5-elongase/Δ^4-desaturase route for docosahexaenoic acid synthesis is also indicated (dotted box), as is the potential role of $\omega3$-desaturation in conversion of n-6 substrates into n-3 forms (present in some lower eukaryotes). The 'substrate dichotomy' of LC-PUFA biosynthesis is represented via solid arrows for glycerolipid-linked reactions and open arrows for acyl-CoA reactions.*

in the form of both essential fatty acids and LC-PUFAs), similar to what is known now as the Mediterranean diet, to a much more red meat-rich diet (containing high levels of *n-6* fatty acids) (Simopoulos, 2000). This has led to the assertion that while typical diets of >100 years ago were likely to have an *n-6/n-3* ratio of 2:1, the present ratio found in the modern Western diet is in excess of 10:1 (Simopoulos, 2000). In view of all these factors, it is perhaps unsurprising that human diseases such as metabolic syndrome and obesity, both preventable by dietary ingestion of *n-3* LC-PUFAs, are dramatically increasing in Western societies (Hu *et al.*, 2001).

Given the profound importance and utility of LC-PUFAs in human health and nutrition (GISSI-Prevenzione Investigators, 1999; Hu *et al.*, 2001; Nugent, 2004) and the decline in the marine fish stocks which contain the bulk of consumed LC-PUFAs, alternative sources of these fatty acids are urgently required.

Oil from certain microalgae and fungal species can be used as a source of some *n-3* LC-PUFAs (Ratledge, 2004). The fungus *Mortereilla alpina* is used as a source of ARA, though the cost of maintaining the large facilities required is still relatively high, whereas the output is not high enough. Some aquatic algae, such as *Crypthecodinium cohnii*, which are rich in EPA and/or DHA, are also amenable to cultivation in slightly less controlled conditions. After extraction it can be used in food production or in supplements. Currently, though, oil from microalgae is not being produced in large enough quantities for global impact (Ursin, 2003).

A more reliable and sustainable source of LC-PUFAs could be provided by the engineering of the LC-PUFA biosynthetic pathway into an appropriate (transgenic) oilseed (Abbadi *et al.*, 2001; Graham *et al.*, 2004; Napier, 2004; Opsahl-Ferstad *et al.*, 2003). The identification and characterization of the process in a suitable LC-PUFA synthesizing organism would be followed by the transfer of the genes encoding the primary LC-PUFA biosynthetic enzymes into a heterologous host such as a transgenic plant. The rest of this review covers the current work to satisfy the urgent need for an alternative source of dietary *n-3* LC-PUFAs (German, Roberts and Watkins, 2003).

Characterization of LC-PUFA Biosynthetic Pathways

The linear biosynthetic pathways of LC-PUFA biosynthesis have been subject to much research effort in recent years, and now appear to be fully elucidated at the molecular level (Leonard *et al.*, 2004; Napier, Michaelson and Sayanova, 2003; Sayanova and Napier, 2004; Sperling *et al.*, 2003) (summarized in Figure 2.3.1). The predominant biosynthetic route for the two C_{20} LC-PUFAs, ARA and EPA, is via the Δ^6-desaturation of the precursor essential fatty acids, LA and ALA, to yield γ-linolenic acid (18:3 *n-6*; GLA) and stearadonic acid (18:4, *n-3*; STA), respectively.

The enzyme responsible for this reaction, the microsomal Δ^6 fatty acid desaturase, was first functionally characterized from borage (*Borago officinalis*), one of the very few plant species that carry out this desaturation (Sayanova *et al.*, 1997). Orthologs of the Δ^6-fatty acid desaturase have been isolated from many species, including mammals, invertebrates, fungi and algae, and all examples characterized to date share the common feature of an *N*-terminal cytochrome b_5 domain, distinct from other classes of microsomal desaturases (Napier *et al.*, 2003; Sperling *et al.*, 2003). It appears that most Δ^6-desaturases do not have any particular substrate preference for *n-6* or *n-3* substrates (i.e. the enzyme will desaturate either LA or ALA with equal efficiency). This confirms the importance of the composition of dietary essential fatty acids in their metabolism to LC-PUFAs since, as mentioned above, most higher eukaryotes lack desaturases capable of converting substrates from *n-6* to *n-3* forms (Sayanova and Napier, 2004). It has recently been demonstrated that many lower eukaryotic orthologs of this enzyme utilize as their substrates LA or ALA esterified to the *sn-2* position of phosphatidylcholine (PC) (Domergue *et al.*, 2003). These data confirmed earlier observations that many microsomal plant desaturase reactions (including the Δ^6-desaturation observed in borage) were glycerolipid dependent (Browse and Somerville, 1991). In contrast, Δ^6-desaturases from animals appear to use acyl-CoA substrates (Sperling *et al.*, 2003).

Once the essential fatty acids have undergone Δ^6-desaturation, the products (GLA and STA) are then elongated by the addition of two carbons via the microsomal Δ^6-elongase, which yields di-homo-γ-linolenic acid (20:3, *n-6*; DHGLA) and eicosatetraenoic acid (20:4, *n-3*; ETetA), respectively. The microsomal Δ^6-elongase was first identified from the model lower animal *C. elegans* and the fungus *M. alpina* by heterologous expression of individual candidate genes in yeast (Beaudoin et al., 2000; Parker-Barnes *et al.*, 2000). These experiments indicated that while the elongation process requires four sequential enzyme activities (condensation, ketoreduction, dehydration and enoyl-reduction), only one activity (in the form of the presumptive condensing enzyme) was necessary to reconstitute a heterologous elongase with specific activity toward C_{18} Δ^6-desaturated substrates. The *C. elegans* and *M. alpina* C_{18} Δ^6-elongating activities identified by these studies showed homology to the yeast ELO genes, which are required for the synthesis of the saturated acyl chains found in sphingolipids (Beaudoin *et al.*, 2000; Sayanova and Napier, 2004; Wallis *et al.*, 2002). The demonstration that LC-PUFA elongating activities are paralogous to the yeast ELO genes has facilitated the cloning of many more examples of this activity (see Leonard *et al.*, (2004) for recent review). Microsomal elongation requires acyl-CoA substrates and represents a key step in the biosynthesis of LC-PUFAs.

The final reaction required to synthesize either ARA or EPA is the introduction of an additional double bond into the elongation products DHGLA or ETetA. This reaction is catalyzed by the microsomal Δ^5-desaturase, which was first identified from *M. alpina* by heterologous expression in yeast (Knutzon et al., 1998). The fatty acid Δ^5-desaturase also displays the same substrate specificities as the Δ^6-desaturase, with mammalian forms of the enzyme acting on acyl-CoA substrates, in contrast to the glycerolipid-dependent requirements of orthologs from lower organisms such as fungi (Domergue *et al.*, 2003).

Recently, an additional so-called alternative route for the synthesis of ARA and EPA has been characterized at the molecular level. Previously, a number of examples of organisms in which ARA or EPA were synthesized in the absence of Δ^6-desaturation had been described (Leonard *et al.*, 2004; Napier *et al.*, 2003; Sayanova and Napier, 2004; Sperling *et al.*, 2003). This was found to occur via an alternative route for LC-PUFA biosynthesis, in which the C_2 elongation step precedes desaturation. In this pathway, the substrates LA and/or ALA are elongated to eicosadienoic acid (20:2, *n-6*; EDA) and eicostrienoic acid (20:3, *n-3*; ETriA), which are then desaturated by the C_{20} Δ^8-desaturase. This yields DHGLA and ETetA, which are then Δ^5-desaturated in the same way as the conventional pathway to yield ARA and EPA respectively. This alternative pathway is therefore also known as the Δ^9-elongase/Δ^8-desaturase pathway, due to the distinct activities that comprise this system (Domergue *et al.*, 2003; Napier *et al.*, 2003).

While the Δ^9-elongation step (which utilizes the Δ^9-desaturated substrates LA and ALA) requires acyl-CoAs, the substrate requirements of the C_{20} Δ^8-desaturase are currently undefined. The molecular identification and functional characterization of the C_{20} Δ^8-desaturase from the aquatic microorganism *Euglena gracilis* have revealed similarity to the other LC-PUFA desaturases such as the microsomal Δ^6 and Δ^5 desaturases (Wallis and Browse, 1999). In that respect, it is tempting to speculate that the Δ^8-desaturase might also display similar substrate requirements (i.e. glycerolipid-linked in the case of lower organisms); this however remains to be demonstrated. More recently, the Δ^9-elongating activity that commences the alternative pathway was serendipitously isolated from the marine microorganism *Isochrysis galbana*. Functional characterization of an ELO-like elongating activity from *Isochrysis* revealed this to utilize only LA and ALA as substrates (Qi *et al.*, 2002). Although *Isochrysis* was previously known to synthesize LC-PUFAs, the presence of the alternative biosynthetic pathway was unexpected. Importantly, the Δ^9-elongating activity from *Isochrysis* did not differ significantly at the molecular level from the Δ^6-elongating activities of the conventional pathway (Qi *et al.*, 2002), perhaps indicating a shared ancestry.

As a result of the initial molecular identifications of the activities required for ARA and EPA synthesis (by both the conventional and alternative pathways), many additional orthologs have been cloned from PUFA-synthesizing organisms (for a review, see Napier *et al.*, (2003)). Perhaps more importantly, the identification of cDNA or genomic sequences which encoded PUFA-biosynthetic enzymes provided for the first time the opportunity to introduce the LC-PUFA trait into organisms which lack this pathway and yet contain high levels of substrates such as LA and ALA. It is for this reason that considerable effort (and progress) has been made in the heterologous reconstitution of C_{20} LC-PUFA biosynthesis in transgenic plants. There have been two recent reports that demonstrate for the first time the possibility of synthesizing ARA and EPA in transgenic plants. These two complementary studies utilized different approaches to the heterologous reconstitution of C_{20}-LC-PUFA biosynthesis, although their outcomes were similar in terms of the levels of ARA and EPA that accumulated in the transgenic plants. Besides serving as important 'proof-of-concepts' for the prospect of engineering LC-PUFA biosynthetic pathways into transgenic plants, both these studies provided better insights into the regulation of fatty acid and lipid biosynthesis in plants.

The Successful Synthesis of C_{20} PUFAs in Transgenic Plants

The identification of an alternative pathway of Δ^9-elongating activity from *Isochrysis* provides an ideal approach for attempting to reconstitute LC-PUFA synthesis in transgenic plants because, unlike the conventional Δ^6-desaturase/elongase pathway, the appropriate substrates (in the form of LA-CoA and ALA-CoA) are already present as endogenous components of higher plant lipid metabolism. Expression (under the control of a constitutive promoter) of the *Isochrysis* C_{18} Δ^9-elongating activity in transgenic Arabidopsis (*Arabidopsis thaliana*) resulted in the synthesis of the C_{20} elongation products EDA and ETriA to significant levels ($\sim$15 % of total fatty acids) in all vegetative tissues (Qi *et al.*, 2004). This demonstrated for the first time the efficient reconstitution of an LC-PUFA elongase in transgenic plants and confirmed the feasibility of engineering transgenic plants to accumulate C_{20} polyunsaturated fatty acids. It is worthy of note that although the C_{20} di- and tri-enoic fatty acids accumulated to relatively high levels in membrane lipids of vegetative tissues, this did not result in any disruption to Arabidopsis morphology or development (Qi *et al.*, 2004). This was in contrast to previous studies on the constitutive expression of the Arabidopsis *FAE1* gene, which encodes a condensing enzyme responsible for the synthesis of C_{20-22} monounsaturated fatty acids in seed lipids. Expression of *FAE1* resulted in profound disruptions to plant morphology when levels of $>$10 % C_{20+} monounsaturates were present in vegetative tissue (Millar, Wrischer and Kunst, 1998).

The distribution among different lipid classes of EDA and ETriA in the transgenic Arabidopsis expressing the *Isochrysis* Δ^9-elongase was analyzed and revealed a number of insights into the channeling of non-native fatty acids in plants (Fraser *et al.*, 2004). For example, the levels of EDA (*n-6*) versus ETriA (*n-3*) varied among different tissues, but did not necessarily reflect the ratios of the Δ^9-elongase substrates (LA and ALA). The accumulation and ratios of the two novel C_{20} fatty acids differed dramatically in their accumulation in the different lipid classes present in Arabidopsis. For example, *n-3* ETriA was particularly abundant in plastidial galactolipids, accumulating to almost 30% of the total fatty acids at the *sn-1* position (Browse and Somerville, 1991). Conversely, *n-6* EDA was the predominant C_{20} fatty acid in phospholipids, and accumulated to ~20% of total fatty acids present at the *sn-2* position of either PA or PC. It is of particular interest that high levels of EDA were detected at the *sn-2* position of PC, consistent with the re-acylation of elongated LA (Browse and Somerville, 1991). Such a process is likely to be central to efficient reconstitution of C_{20} LC-PUFA biosynthesis, since desaturation in higher plants usually occurs on glycerolipid-linked substrates, in contrast to the cytosolic acyl-CoA-dependent elongation reaction. The ability to exchange substrates and products between the glycerolipid and acyl-CoA pools is likely to be a major consideration in attempts to produce LC-PUFAs in transgenic plants.

The observation that the *Isochrysis* Δ^9-elongase is capable of directing the synthesis of significant levels of EDA and ETriA facilitated an attempt to fully reconstitute the alternative LC-PUFA biosynthetic pathway for ARA and EPA (Figure 2.3.1). The *Euglena* Δ^8 desaturase and the *M. alpina* Δ^5-desaturase were co-expressed with the *Isochrysis* Δ^9-elongase. All three transgenes were placed under the control of the same constitutive promoter and the different constructs were introduced into Arabidopsis by sequential transformation using different selectable markers (Qi *et al.*, 2004). The resulting (triple) transgenic plants were morphologically indistinguishable from wild-type Arabidopsis. However, analysis of the fatty acid composition of these transgenics revealed the presence of several C_{20} LC-PUFAs including ARA and EPA (Sayanova *et al.*, 1997). These two LC-PUFAs accumulated in leaf tissues of transgenic Arabidopsis plants to a combined level of ~10% of total fatty acids, the majority being ARA (*n-6*). Again, this did not reflect the proportions of *n-6* or *n-3* substrate, which is predominantly ALA (*n-3*) in vegetative tissues.

Two other C_{20} PUFAs, sciadonic acid ($20:3\Delta^{5,11,14}$) and juniperonic acid ($20:4\Delta^{5,11,14,17}$), were identified in the transgenic Arabidopsis, in addition to ARA and EPA (Qi *et al.*, 2004). These two non-methylene-interupted PUFAs are likely to have resulted from the promiscuous activity of the Δ^5-desaturase, acting on substrates which might usually be expected to undergo Δ^8-desaturation. Whether this represents some aspect of perturbation to substrate channeling remains unclear, although both desaturases are assumed to utilize similar substrates (C_{20} acyl chains at the *sn-2* position of PC). Although sciadonic and juniperonic acids were not primary targets for synthesis and accumulation in the transgenic plants, recent evidence suggests that these LC-PUFAs may also be beneficial to health and play a role in modulating some aspects of human metabolism. Moreover, both sciadonic and juniperonic acids are found in a number of species of pine seeds, and as such have been previously consumed by humans without demonstrating any anti-nutritional effects.

The *M. alpina* Δ^5-desaturase that was used in the reconstitution of the alternative LC-PUFA biosynthetic pathway was previously observed to utilize unexpected substrates when individually expressed in transgenic canola, resulting in the accumulation of the unusual Δ^5-desaturated C_{18} fatty acids, taxoleic and pinolenic acid (Knutzon *et al.*,

1998). This may indicate that the determinants affecting the substrate specificity of this desaturase are not fully understood at present.

The reconstitution of the alternative Δ^9-elongase/Δ^8-desaturase LC-PUFA biosynthetic pathway in transgenic plants has been recognized as a major milestone in the production of these nutritional compounds in a sustainable manner (Green, 2004). However, while current data represent a significant demonstration of 'proof-of-concept' in vegetative tissues, it will be crucial to demonstrate that a similar efficient reconstitution of the alternative LC-PUFA biosynthetic pathway is possible in developing seeds, with the concomitant accumulation of ARA and (more preferably) EPA in storage lipids such as triacylglycerol.

A second major study on the accumulation of LC-PUFAs has recently been reported. Components of the conventional Δ^6-desaturase/elongase pathway were expressed in transgenic plants (Abbadi *et al.*, 2004). This study provides some additional insights into heterologous LC-PUFA synthesis during seed development of transgenic oilseeds. Using genes encoding enzymes from a number of different LC-PUFA-accumulating species, transgenic linseed and tobacco lines were engineered to express the three primary activities of the conventional pathway, the Δ^6-desaturase, the Δ^6-elongase and the Δ^5-desaturase (Abbadi *et al.*, 2004) (Figure 2.3.1). In contrast to the above study on the alternative LC-PUFA pathway, these three activities were placed under the transcriptional regulation of a seed-specific promoter. Additionally, these three heterologous genes were introduced into transgenic tobacco or linseed as a single integration event, rather than via sequential transformation. Analysis of homozygous seeds of resultant transgenic tobacco and linseed confirmed the presence of very high levels of Δ^6-desaturated fatty acids ($\sim$30 % of total fatty acids), yet only relatively low amounts of ARA and EPA. These data clearly demonstrated the seed-specific reconstitution of the conventional Δ^6-desaturase LC-PUFA biosynthetic pathway in transgenic oilseeds, and they also paralleled earlier observations in yeast on the inefficient synthesis of C_{20} PUFAs. These earlier studies had revealed a potential bottleneck at the elongation step in the pathway, which had been ascribed to the inefficient (acyl) exchange between the glycerolipid and acyl-CoA pools.

Further detailed analysis of transgenic linseed expressing these activities revealed a number of important observations. First, although the Δ^6-desaturase and the Δ^6-elongase appeared to function at very different efficiencies (as estimated by the accumulation of their biosynthetic products), the two transgenes encoding these activities were transcribed at similar levels. Second, *in vitro* elongation assays carried out on microsomal fractions isolated from transgenic developing seeds demonstrated the activity of the heterologous Δ^6-elongase when supplied with exogenous acyl-CoA substrates. Third, although high levels of Δ^6-desaturated fatty acids accumulated in the microsomal membranes, particularly at the *sn-2* position of PC, this was not reflected in a concomitant increase in the Δ^6-desaturated acyl-CoAs (as determined by profiling of the acyl-CoA pool). While it is clear that a lack of Δ^6-desaturated fatty acids in the acyl-CoA pool will prevent the Δ^6-elongase from functioning efficiently, it is less clear why these substrates remain in the microsomal membrane lipids. This may reflect inefficient exchange from PC into the acyl-CoA pool.

A further subtlety was identified by analysis of the acyl composition of TAGs from these transgenic seeds, revealing the presence of high levels of the *n-3* Δ^6-desaturated fatty acid STA when compared with its distribution in other lipid classes such as PC, PE and DAG (Abbadi *et al.*, 2004). In contrast, although the *n-6* Δ^6-desaturated fatty acid GLA was abundant in PC, STA was present at a very much lower level, even though the relevant substrates (LA, ALA) were present at similar levels. Full positional analysis of transgenic TAGs was used to determine the precise distribution of these novel Δ^6-desaturated fatty acids (Abbadi *et al.*, 2004). This found STA predominantly at the *sn-3* position, whereas GLA was found at both *sn-2* and *sn-3* positions.

These data pose a number of possibilities regarding the channeling of fatty acids into different lipid classes. In particular, the absence of Δ^6-desaturated fatty acids in the acyl-CoA pool could reflect a number of scenarios including: inefficient exchange between the CoA and PC pools; rapid channeling into lipids of any Δ^6-desaturated acyl-CoAs such that their presence is not detected; channeling into lipids via an acyl-CoA-independent process, such as the enzyme phospholipid: diacylglycerol acyltransferase (PDAT) (Beaudoin and Napier, 2004). In that respect, it seems most likely that the *n-3* Δ^6-desaturated fatty acid STA is channeled from PC directly into TAG by the PDAT enzyme, precluding it from further elongation and desaturation. In addition, it may be that exchange of any Δ^6-desaturated fatty acids (*n-3* or *n-6*) from PC into the acyl-CoA pool is inefficiently catalyzed by the endogenous *lyso*-phosphatidylcholine: acyltransferase (LPCAT) enzyme and so also substrate-limits the activity of the heterologous LC-PUFA elongase (Abbadi *et al.*, 2004; Beaudoin and Napier, 2004). The combination of (at least) these two channeling activities is therefore likely to contribute to the low levels of C_{20} LC-PUFAs in the transgenic oilseeds (less than 10% of the novel C_{18} Δ^6-PUFAs, and skewed toward the accumulation of *n-6*) (Abbadi *et al.*, 2004).

Taking these observations together, it seems likely that a major constraint for the efficient reconstitution of C_{20} LC-PUFAs via the conventional Δ^6-desaturase/elongase route is the dichotomy of substrate requirements for glycerolipid desaturation and acyl-CoA elongation. However, the levels of ARA and EPA obtained in the seed lipids of transgenic linseed are still significant, even allowing for the clearly suboptimal exchange and channeling of acyl-substrates. Thus, these results can be taken as a highly encouraging 'proof-of-concept' for the seed-specific synthesis of LC-PUFAs via this pathway.

Future Directions and Prospects

It is clear from the two studies described above that heterologous reconstitution of C_{20} LC-PUFA synthesis in transgenic plants has now been demonstrated (Abbadi *et al.*, 2004; Qi *et al.*, 2004). This has been achieved by the 'reverse engineering' of the primary biosynthetic enzymes, and has yielded significant levels of the nutritionally important C_{20} LC-PUFAs such as ARA and EPA. Perhaps of equal significance, these data have indicated that our understanding of the biochemical processes that underpin the synthesis and accumulation of these fatty acids is still incomplete. In particular, the role of acyl-channeling, either in terms of substrate presentation or compartmentation of lipids, is still an emerging topic (Beaudoin and Napier, 2004).

Another consideration arises directly from the initial rationale to synthesize LC-PUFAs in transgenic plants, that is the need to replace the diminishing stocks of fish oils with a sustainable alternative (Napier, 2004; Opsahl-Ferstad *et al.*, 2003). While transgenic plants can clearly synthesize and accumulate C_{20} LC-PUFAs, this results in a mixture of *n-6* and *n-3* fatty acids. This is in contrast to the aquatic food web (i.e. the LC-PUFA-synthesizing algae and the fish that consume them), which is predominantly rich in *n-3* fatty acids such as EPA and DHA (Opsahl-Ferstad *et al.*, 2003; Sayanova and Napier, 2004). Seen from the perspective of human health and nutrition the *n-3* LC-PUFAs are beneficial via the derived protection from metabolic syndrome and cardiovascular disease, whereas *n-6* LC-PUFAs such as ARA may give rise to pro-inflammatory responses through their metabolism via the eicosanoid pathway (Nugent, 2004; Sargent and Tacon, 1999). In that respect, the channeling of *n-3* fatty acids into storage lipids (i.e. TAG) observed in linseed may represent another potential bottleneck in the efficient synthesis of *n-3* LC-PUFAs, since it precludes *n-3* substrates from the heterologous LC-PUFA biosynthetic pathway (Abbadi *et al.*, 2004). It remains to be seen if other oilseeds display the same strong channeling of *n-3* fatty acids into TAGs.

The demonstration that EPA can be synthesized in transgenic plants and accumulated specifically in seed TAGs is a major step toward providing a sustainable source of LC-PUFAs, but an additional goal must also be the production of the C_{22} fatty acid DHA. To that end, the very recent identification of the C_{20} Δ^5-elongase (which elongates EPA to 22:5) (Meyer *et al.*, 2004; Pereira *et al.*, 2004), together with the earlier functional characterization of the C_{22} Δ^4-desaturase (Napier *et al.*, 2003; Sperling *et al.*, 2003), will facilitate the heterologous reconstitution of DHA synthesis. Initial 'proof-of-concept' experiments have been carried out in yeast and revealed low but significant levels of DHA in strains that have been engineered to contain activities of the conventional LC-PUFA biosynthetic pathway (i.e. the C_{18} Δ^6-elongase, C_{20} Δ^5-desaturase, C_{20} Δ^5-elongase and the C_{22} Δ^4-desaturase) (Meyer *et al.*, 2004). A very high proportion of the *n-3* C_{18} STA supplied to the transgenic yeast was elongated to ETetA, probably due to the high availability of the substrate as an acyl-CoA. EPA is efficiently elongated to DPA by the newly identified Δ^5-elongase and DPA is correctly Δ^4-desaturated to DHA, but the resultant levels of DHA are low ($\sim$1 % of total fatty acids). This appears to be due to the very poor conversion of ETetA to EPA by the microsomal Δ^5-desaturase (Meyer *et al.*, 2004).

As discussed above, the microsomal desaturation reactions, which underpin LC-PUFA biosynthesis utilize substrates at the *sn-2* position of PC, and the inefficiency of the Δ^5-desaturase may simply reflect the lack of glycerolipid-linked substrate (even though total levels of ETetA are high). In that respect, the data on the heterologous reconstitution of the $C_{20} > C_{22}$ LC-PUFA biosynthetic pathway in yeast confirm the bottlenecks observed for the $C_{18} > C_{20}$ pathway in both yeast and transgenic plants (Abbadi *et al.*, 2004; Beaudoin *et al.*, 2000; Sperling *et al.*, 2003; Qi *et al.*, 2004). In particular, the dichotomy of substrates required for elongation and desaturation indicates the need for additional factors (such as acyltransferases) to improve the efficiency of this process. Attempts to accumulate DHA in transgenic plants are currently determining additional constraints on heterologous LC-PUFA biosynthesis in these organisms.

Conclusions

The efficient biosynthesis of C_{20} LC-PUFAs in transgenic plants has now been demonstrated conclusively, using two different approaches. Not only do these data clearly indicate the potential to use transgenic plants as an alternative sustainable source of these important fatty acids, but they also provide new insights into our understanding of lipid biochemistry, in particular the channeling of fatty acids into various different lipids. In that respect, these heterologous expression systems have not only realized the possibility of producing these important nutritional compounds in transgenic plants, but also provided a new experimental tool with which to better investigate plant lipid metabolism.

Acknowledgments

Rothamsted Research receives grant-aided support from the Biotechnology and Biological Sciences Research Council (BBSRC), UK. The authors thank BASF Plant Sciences for financial support.

References

Abbadi, A., Domergue, F., Meyer, A., Riedel, K., Sperling, P., Zank, T. and Heinz, E. (2001) Transgenic oilseeds as sustainable source of nutritionally relevant C_{20} and C_{22} polyunsaturated fatty acids? *European Journal of Lipid Science and Technology* **103**, 106–113.

Abbadi, A., Domergue, F., Fahl, A., Ott, C., Bauer, J., Napier, J.A., Welti, R., Cirpus, P. and Heinz, E. (2004) Biosynthesis of very long chain polyunsaturated fatty acids in transgenic oilseeds: constraints on their accumulation. *Plant Cell* **16**, 2734–2748.

Bang, H.O., Dyerberg, J. and Hjoorne N. (1976) The composition of food consumed by Greenland Eskimos. *Acta Medica Scandinavica* **200**, 69–73.

Beaudoin, F., Michaelson, L.V., Hey, S.J., Lewis, M.J., Shewry, P.R., Sayanova, O. and Napier J.A. (2000) Heterologous reconstitution in yeast of the polyunsaturated fatty acid biosynthetic pathway. *Proceedings of the National Academy of Sciences USA* **97**, 6421–6426.

Beaudoin, F. and Napier, J.A. (2004) Biosynthesis and compartmentalisation of triacyglycerol in higher plants. *Topics in Current Genetics* **6**, 267–287.

Browse, J. and Somerville, C. (1991) Glycerolipid synthesis—biochemistry and regulation. *Annual Review of Plant Physiology and Plant Molecular Biology* **63**, 467–506.

Burr, M.L., Fehily, A.M., Gilbert, J.F., Rogers, S., Holliday, R.M., Sweetnam, P.M., Elwood, P.C. and Deadman, N.M. (1989) Effects of changes in fat, fish, and fibre intakes on death and myocardial reinfarction: diet and reinfarction trial (DART). *Lancet* **2**, 757–761.

Cahoon, E.B., Hall, S.E., Ripp, K.G., Ganzke, T.S., Hit, W.D. and Coughlan, S.J. (2003) Metabolic redesign of vitamin E biosynthesis in plants for tocotrienol production and increased antioxidant content. *Nature Biotechnology* **21**, 1082–1087.

Domergue, F., Abbadi, A., Ott, C., Zank, T.K., Zahringer, U. and Heinz, E. (2003) Acyl carriers used as substrates by the desaturases and elongases involved in very long-chain polyunsaturated fatty acids biosynthesis reconstituted in yeast. *Journal of Biological Chemistry* **278**, 35115–35126.

Dyerberg, J. and Bang, H.O. (1982) A hypothesis on the development of acute myocardial infarction in Greenlanders. *Scandinavian Journal of Clinical and Laboratory Investigation Supplement* **161**, 7–13.

Fraser, T.C.M., Qi, B., Elhussein, S., Chatrattanakunchai, S., Stobart, A.K. and Lazarus, C.M. (2004) Expression of the *Isochrysis* C18-Δ9 polyunsaturated fatty acid specific elongase component alters *Arabidopsis* glycerolipid profiles. *Plant Physiology* **135**, 859–866.

Funk, C.D. (2001) Prostaglandins and leukotrienes: advances in eicosanoids bioilogy. *Science* **294**, 1871–1875.

German, J.B., Roberts, M.A. and Watkins, S.M. (2003) Genomics and metabolomics as markers for the interaction of diet and health: lessons from lipids. *Journal of Nutrition* **133**, 2078S–2083S.

GISSI-Prevenzione Investigators (1999) Dietary supplementation with *n-3* polyunsaturated fatty acids and vitamin E after myocardial infarction: results of the GISSI-Prevenzione trial. Gruppo Italiano per lo Studio della Sopravvivenza nell'Infarto miocardico. *Lancet* **354**, 447–455.

Graham, I.A., Cirpus, P., Rein, D. and Napier, J.A. (2004) The use of very long chain polyunsaturated fatty acids to ameliorate metabolic syndrome: transgenic plants as an alternative sustainable source to fish oils. *Nutrition Bulletin* **29**, 228–233.

Green, A.G. (2004) From alpha to omega—producing essential fatty acids in plants. *Nature Biotechnology* **22**, 680–682.

Hu, F.B., Manson, J.E. and Willett, W.C. (2001) Types of dietary fat and risk of coronary heart disease: a critical review. *Journal of the American College of Nutrition* **20**, 5–19.

Hwang, D. (2000) Fatty acids and immune responses—a new perspective in searching for clues to mechanism. *Annual Review of Nutrition* **20**, 431–456.

Jackson, J.B., Kirby, M.X., Berger, W.H., Bjorndal, K.A., Botsford, L.W., Bourque, B.J., Bradbury, R.H., Cooke, R., Erlandson, J., Estes, J.A., Hughes, T.P., Kidwell, S., Lange, C.B., Lenihan, H.S., Pandolfi, J.M., Peterson, C.H., Steneck, R.S., Tegner, M.J. and Warner, R.R. (2001) Historical overfishing and the recent collapse of coastal ecosystems. *Science* **293**, 629–637.

James, M.J., Ursin, V.M. and Cleland, L.G. (2003) Metabolism of stearidonic acid in human subjects: comparison with the metabolism of other *n-3* fatty acids. *American Journal of Clinical Nutrition* **77**, 1140–1145.

Jaworski, J. and Cahoon, E.B. (2003) Industrial oils from transgenic plants. *Current Opinion in Plant Biology* **6**, 178–184.

Knutzon, D.S., Thurmond, J.M., Huang, Y.S., Chaudhary, S., Bobik, Jr. E.G., Chan, G.M., Kirchner, S.M. and Mukerji, P. (1998) Identification of Δ5-desaturase from *Mortierella alpina* by heterologous expression in Bakers, yeast and canola. *Journal of Biological Chemistry* **273**, 29360–29366.

Kris-Etherton, P.M., Taylor, D.S., Yu-Poth, S., Huth, P., Moriarty, K., Fishell, V., Hargrove, R.L., Zhao, G. and Etherton, T.D. (2000) Polyunsaturated fatty acids in the food chain in the United States. *American Journal of Clinical Nutrition* **71**, 179–188.

Leonard, A.E., Pereira, S.L., Sprecher, H. and Huang, Y.S. (2004) Elongation of long-chain fatty acids. *Progress in Lipid Research* **43**, 36–54.

Meyer, A., Kirsch, H., Domergue, F., Abbadi, A., Sperling, P., Bauer, J., Cirpus, P., Zank, T.K., Moreau, H., Roscoe, T.J., Zahringer, U. and Heinz, E. (2004) Novel fatty acid elongases and their use for the reconstitution of docosahexaenoic acid biosynthesis. *Journal of Lipid Research* **45**, 1899–1909.

Millar, A.A., Wrischer, M. and Kunst, L. (1998) Accumulation of very-long-chain fatty acids in membrane glycerolipids is associated with dramatic alterations in plant morphology. *Plant Cell* **10**, 1889–1902.

Napier, J.A., Michaelson, L.V. and Sayanova, O. (2003) The role of cytochrome b_5 fusion desaturases in the synthesis of polyunsaturated fatty acids. *Prostaglandins, Leukotrienes and Essential Fatty Acids* **68**, 135–143.

Napier, J.A. (2004) The production of long chain polyunsaturated fatty in transgenic plants: a sustainable source for human health and nutrition. *Lipid Technology* **16**, 103–107.

Nugent, A.P. (2004) The metabolic syndrome. *Nutrition Bulletin* **29**, 36–43.

Ohlrogge, J. (1999) Plant metabolic engineering: are we ready for phase two? *Current Opinion in Plant Biology* **2**, 121–122.

Opsahl-Ferstad, H-G., Rudi, H., Ruyter, B. and Refstie, S. (2003) Biotechnological approaches to modify rapeseed oil composition for applications in aquaculture. *Plant Science* **165**, 349–357.

Paine, J.A., Shipton, C.A., Chaggar, S., Howells, R.M., Kennedy, M.J., Vernon, G., Wright, S.Y., Hinchliffe, E., Adams, J.L., Silverstone, A.L. and Drake, R. (2005) Improving the nutritional value of golden rice through increased pro-vitamin A content. *Nature Biotechnology* **23**, 482–487.

Parker-Barnes, J.M., Das, T., Bobik, E., Leonard, A.E., Thurmond, J.M., Chaung, L.T., Huang, Y.S. and Mukerji, P. (2000) Identification and characterization of an enzyme involved in the elongation of *n-6* and *n-3* polyunsaturated fatty acids. *Proceedings of the National Academy of Sciences USA* **97**, 8284–8289.

Pauly, D., Alder, J., Bennett, E., Christensen, V., Tyedmers, P. and Watson, R. (2003) The future for fisheries. *Science* **302**, 1359–1361.

Pauly, D., Christensen, V., Dalsgaard, J., Froese, R. and Torres, F., Jr. (1998) Fishing down marine food webs. *Science* **279**, 860–863.

Pereira, S.L., Leonard, A.E., Huang, Y.S., Chuang, L.T. and Mukerji, P. (2004) Identification of two novel microalgal enzymes involved in the conversion of the omega-3 fatty acid, eicosapentaenoic acid, into docosahexaenoic acid. *Biochemistry Journal* **384**, 357–366.

Qi, B., Beaudoin, F., Fraser, T., Stobart, A.K., Napier, J.A. and Lazarus, C.M. (2002) Identification of a cDNA encoding a novel C_{18}-$\Delta 9$ polyunsaturated fatty acid-specific elongating activity from the docosahexaenoic acid (DHA)-producing microalga, *Isochrysis galbana*. *FEBS Letters* **510**, 159–165.

Qi, B., Fraser, T., Mugford, S., Dobson, G., Sayanova, O., Butler, J., Napier, J.A., Stobart, A.K. and Lazarus, C.M. (2004) The production of very long chain polyunsaturated omega-3 and omega-6 fatty acids in transgenic plants. *Nature Biotechnology* **22**, 739–745.

Ratledge, C. (2004) Fatty acid biosynthesis in microorganisms being used for single cell oil production. *Biochimie* **86**, 807–815.

Sargent, J.R. and Tacon, A.G. (1999) Development of farmed fish: a nutritionally necessary alternative to meat. *Proceedings of the Nutrition Society* **58**, 377–383.

Sayanova, O. and Napier, J.A. (2004) Eicosapentaenoic acid: biosynthetic routes and the potential for synthesis in transgenic plants. *Phytochemistry* **65**, 147–158.

Sayanova, O., Smith, M.A., Lapinskas, P., Stobart, A.K., Dobson, G., Christie, W.W., Shewry, P.R. and Napier, J.A. (1997) Expression of a borage desaturase cDNA containing an N-terminal cytochrome b5 domain results in the accumulation of high levels of Δ^6-desaturated fatty acids in transgenic tobacco. *Proceedings of the National Academy of Sciences USA* **94**, 4211–4216.

Simopoulos, A.P. (2000) Human requirement for *n-3* polyunsaturated fatty acids. *Poultry Science* **79**, 961–970.

Sperling, P., Ternes, P., Zank, T.K. and Heinz, E. (2003) The evolution of desaturases. *Prostaglandins, Leukotrienes and Essential Fatty Acids* **68**, 73–95.

Thelen, J.J. and Ohlrogge, J.B. (2002) Metabolic engineering of fatty acid biosynthesis in plants. *Metabolic Engineering* **4**, 12–21.

Tucker, G. (2003) Nutritional enhancement of plants. *Current Opinion in Biotechnology* **14**, 221–225.

Ursin, V.M. (2003) Modification of plant lipids for human health: development of functional land-based omega-3 fatty acids. *Journal of Nutrition* **133**, 4271–4274.

von Schacky, C. (2003) The role of omega-3 fatty acids in cardiovascular disease. *Current Atherosclerosis Reports* **5**, 139–145.

Wallis, J.G. and Browse, J. (1999) The $\Delta 8$-desaturase of *Euglena gracilis*: an alternate pathway for synthesis of 20-carbon polyunsaturated fatty acids. *Archives of Biochemistry and Biophysiology* **365**, 307–316.

Wallis, J.G., Watts, J.L. and Browse, J. (2002) Polyunsaturated fatty acid synthesis: what will they think of next? *Trends in Biochemical Sciences* **27**, 467–470.

Warzecha, H. and Mason, H.S. (2003) Benefits and risks of antibody and vaccine production in transgenic plants. *Journal of Plant Physiology* **160**, 755–764.

2.4

The Application of Genetic Engineering to the Improvement of Cereal Grain Quality

Peter R. Shewry

Crop Performance and Improvement, Rothamsted Research, Harpenden,
Hertfordshire, AL5 2JQ, United Kingdom

Introduction

Cereals are the most important crops in the world, with total annual yields in excess of 2000 million tonnes compared with less than 700 million tonnes for root and tuber crops, and about 380 million tonnes for legumes and oilseeds (FAO, 2003). Three cereal species, wheat, maize and rice, are particularly dominant with total annual production of about 600 million tonnes of each.

The major component of all cereal grain is starch, which accounts for about 70–80 % of the total dry weight. Hence, cereals are traditionally regarded as sources of energy. However, they also contain between about 8 % and 15 % protein, meaning that they are significant sources of protein for humans and livestock. In fact, it can be estimated that the total amount of protein harvested in cereal grains is approximately fourfold greater than that in soybeans (which contain up to 40 % protein) and over twice that harvested in all other seed crops. In addition, cereal grains are important dietary sources of fiber and some vitamins and minerals, although these are all concentrated in the aleurone layer and consequently depleted on milling (wheat) or polishing (rice). Furthermore, it is important to note that wheat is the major cause of one of the most widespread forms of food intolerance (coeliac disease) as well as causing several important dietary and respiratory

Plant Biotechnology. Edited by Nigel Halford.
© 2006 John Wiley & Sons, Ltd.

allergies. All of these aspects will be covered in the present chapter with the exception of starch composition and quality, which is discussed in Chapter 2.5.

Nutritional Quality for Food and Feed

Amino Acid Composition

The nutritional quality of the grain proteins is important when cereals are used as feed for non-ruminant livestock or consumed by humans as a high proportion of the diet. Protein nutritional quality is determined by the amounts of essential amino acids. These are the amino acids that cannot be synthesized by animals and hence must be provided in the diet. If only one of these amino acids is limiting, the others will be broken down and excreted, resulting in loss of nitrogen to the environment. Nine of the 20 protein amino acids are essential (lysine, isoleucine, leucine, phenylalanine, tyrosine, threonine, tryptophan, valine and methionine) but cysteine is also often added as it can only be produced in animals from methionine, which is essential.

Because storage proteins account for half or more of the total grain proteins in cereals their compositions are the major determinants of the nutritional quality of the whole grain. In wheat, barley, maize, sorghum and most other cereals the major storage proteins are alcohol-soluble prolamins. These contain low proportions of lysine, which results in this being the first limiting amino acid (approximately 2.0, 3.1 and 3.5 g % in wheat, barley and maize compared with a WHO recommended level of 5.5 g %) (Shewry, 2000). The levels of methionine and threonine are also low and, in maize, the level of tryptophan. Oats and rice differ from other major cereals in that their major storage proteins belong to the 11S/12S globulin family and have adequate contents of essential amino acids.

Two complementary approaches have been taken to increase the lysine content of cereal grain. First, although most of the amino acids in seeds are present in proteins, small proportions (typically about 1 % or less of the total) are present as free amino acids. Lysine, threonine and methionine are all synthesized in plants from aspartic acid, with the entry into the pathway and the branch point to lysine being catalyzed by two feedback-regulated enzymes called aspartate kinase and dihydrodipicolinate synthase (DHPS), respectively. In order to increase the accumulation of these free amino acids, it is necessary to eliminate the feedback regulation of the enzymes, which has been achieved for lysine by transformation with feedback-insensitive forms of DHPS from bacteria or from maize. Mazur, Krebbers and Tingey (1999) showed that the expression of a *Corynebacterium* enzyme in the maize embryo and aleurone resulted in increases in free lysine that were sufficient to increase the total grain lysine content by up to twofold, but that no increase was observed when the same gene was expressed in the starchy endosperm due to increased lysine breakdown. Similar increased degradation has been observed in other crops (soybean and canola) (Falco *et al.*, 1995; Mazur *et al.*, 1999), but work in Arabidopsis (*Arabidopsis thaliana*) has shown that the effect can be eliminated by knocking out genes encoding enzymes of lysine catabolism (Zhu and Galili, 2003). In contrast, Brinch-Pedersen *et al.* (1996) reported that the expression of a feedback-insensitive form of DHPS from *Escherichia coli* under control of the constitutive 35S promoter resulted in a two-fold increase in free lysine in the grain. Lee *et al.* (2001) also mutated a maize gene encoding DHPS to eliminate the feedback sensitivity of the enzyme and showed that constitutive expression resulted in up to 2.5-fold more free lysine in rice seeds despite increased catabolism.

The second approach to increasing essential amino acids is to increase the amounts of proteins that are enriched in these amino acids. This has been applied particularly to methionine, recognizing the fact that it is the third limiting amino acid in some cereals, and that cereals are frequently mixed with methionine-poor legume seeds for livestock diets. A number of methionine-rich plant proteins have been identified, which fall into two broad groups: 2S storage albumins from dicotyledenous seeds and prolamin storage proteins from maize and related panicoid cereals (sorghum and millets).

The only methionine-rich 2S albumin to be expressed in a cereal is SFA8 from sunflower seeds, which contains 16 methionine and eight cysteine residues of 103 residues in total (Kortt *et al.*, 1991). Expression of this protein in rice seeds resulted in the expected increase in seed methionine of about 27 % but this was accompanied by a decrease in cysteine of about 15 % (Hagan *et al.*, 2003). Hence, there was little impact on the overall nutritional quality of the grain. Similar effects on the redistribution of sulfur within the seed components (rather than an increase in total seed sulfur) were observed when SFA8 or a methionine-rich 2S albumin from Brazil nut was expressed in dicotyledonous seeds, indicating that the supply of sulfur to the seed was limiting (Shewry, Jones and Halford, 2005a). A further drawback to using 2S albumins is that many, including SFA8 and Brazil nut albumin, have been reported to be allergenic (Kelly and Hefle, 2000; Kelly, Hlywka and Hefle, 2000; Melo *et al.*, 1994; Nordlee *et al.*, 1996).

The β-zeins and γ-zeins of maize have methionine contents of up to 11.4 % and 26.9 %, respectively (Pedersen *et al.*, 1986; Swarup *et al.*, 1995), with related methionine-rich prolamins being present in other species such as sorghum (Chamba *et al.*, 2005). Increasing the amount of γ-zein in maize by transformation with additional gene copies (Anthony *et al.*, 1997) or by modifying the stability of its mRNA (Lai and Messing, 2002) has been reported to result in increases in total grain methionine. However, although no data on total grain sulfur were reported, Lai and Messing (2002) showed that the cysteine content of the grain fell slightly. It is therefore probable that increases in the transport of sulfur to the seed will be required to exploit the potential for increasing grain methionine by expression of methionine-rich proteins.

Less work has been carried out on lysine-rich proteins and these also appear to be less widely distributed in plants, possibly due to their high charge. The most well known of them is the barley grain protein, chymotrypsin inhibitor 2 (CI-2), which has been widely studied as a model system for protein folding. Hence, its structure and properties are understood in some detail (see, e.g., Clore *et al.*, 1987a,b; McPhalen and James, 1987; Otzen and Fersht, 1995, 1998). CI-2 is enriched in seeds of the high-lysine barley, Hiproly, where it contributes to the increased lysine content (Hejgaard and Boisen, 1980). Hence, its accumulation within the cereal grain appears to be tolerated without adverse effects on development. Similarly, the accumulation of CI-2 in the seed has no apparent effects on protein digestibility. CI-2 contains eight lysines of 83 residues but mutant forms containing up to 25 mol % lysine have been designed. Furthermore, analysis of the mutant proteins expressed in *E. coli* has shown that some appear to be sufficiently stable for expression in transgenic plants (Forsyth *et al.*, 2005; Roesler and Rao, 1999; 2000).

A similar approach has been used to design high lysine forms of hordothionin, a barley seed protein containing five lysines of 45 residues (Rao *et al.*, 1994). However, this protein would be less acceptable for expression in transgenic plants as thionins are highly toxic *in vitro* to microorganisms (bacteria, fungi and yeasts), invertebrates and animal cells (Florack and Stiekema, 1994).

Elimination of Allergies and Intolerances

Food allergies are considered to affect 1–2% of the population and are even more prevalent in children (up to 8%). Although cereals are not major causes of allergy, they nevertheless have significant effects. For example, dietary allergy to wheat is uncommon but it has been implicated in atopic dermatitis and food-dependent exercise-induced anaphylaxis. However, one class of cereal seed proteins have been studied in some detail in relation to both food and respiratory allergies. These are small sulfur-rich proteins of the cereal trypsin/α-amylase inhibitory family, which are related to the 2S albumins discussed above and to the major prolamins of cereals (Shewry *et al.*, 2004). Proteins of this family have been demonstrated to be the major causes of baker's asthma (respiratory allergy to flours of wheat, barley and rye) and dietary allergy to rice (Salcedo *et al.*, 2004).

These α-amylase/trypsin inhibitors comprise a number of components, which are structurally related and encoded by small multigene families. Hence, it should be possible to use genetic engineering to downregulate either the whole family of proteins or single components. This has already been demonstrated in rice where antisense expression of a single sequence resulted in reduction of the amount of a group of M_r 14 000–M_r 16 000 proteins responsible for dietary allergy to about 25% of the amount present in wild-type plants (Tada *et al.*, 1996). Although this transgenic rice is hypoallergenic, it still contains significant amounts of α-amylase/trypsin inhibitors and hence is not suitable for consumption by individuals suffering from rice allergy.

The most widespread food intolerance is coeliac disease, which is considered to affect up to 1% of the population in some countries. Coeliac disease is a T cell-mediated autoimmune response, which is triggered by gluten proteins of wheat and a range of related proteins in barley and rye (Kasarda, 2001). Although the response is triggered by sequences present in many gluten proteins, there is evidence that the proteins vary in their relative toxicities, and that the proportions of these proteins also varies between cultivars. Thus, it could be possible to use a combination of classical plant breeding and genetic engineering to produce lines with reduced coeliac toxicity, even if it could not be eliminated completely. However, it must be borne in mind that wheat is consumed after extensive processing, and that any changes to the gluten protein composition should not compromise the functional properties.

Enhancing the Amounts of Minerals and Vitamins

Although cereal seeds are potentially significant sources of minerals for nutrition, a high proportion of these are unavailable as they are in the form of phytin, mixed salts of phytic acid (*myo*-inositol-(1,2,3,4,5,6)-hexa*kis* phosphate). These salts (phytates) account for over 70% of the total phosphate in cereal grains as well as significant amounts of other minerals such as Mg^{2+}, K^+, Fe^{3+}, Zn^{2+}, Ca^{2+} and Cu^{2+}. Phytin is present as granules in the aleurone and embryo cells of cereals, and is digested to release the minerals during germination. However, phytates cannot be digested by humans or livestock and are therefore excreted. This can lead to eutrophication of natural waters by phosphate in areas of intensive livestock production and to mineral deficiency in humans. The latter is particularly significant for women and children in developing countries.

Genetic engineering has been used to reduce the amount of phytin in a number of crops, including cereals, by the expression of genes encoding phytase, notably from the fungus *Aspergillus niger*, which secretes an extracellular phytase that has been produced commercially as an additive for animal feed. Feeding studies with transgenic soybean and canola have shown benefits in terms of reduced phosphate excretion or increased growth of pigs and chickens (Denbow *et al.*, 1998; Zhang *et al.*, 2000a,b). Expression of fungal phytase has also been reported in wheat and rice (Brinch-Pedersen *et al.*, 2000; 2003; Lucca, Hurrell and Potrykus, 2001). The current interest in these two cereals focuses on the expression of heat-stable forms of phytase (from *A. fumigatus*) to reduce the loss of activity, which occurs when the grain is heated during the preparation of food or feed (Brinch-Pedersen, Sorensen and Holm, 2002; Lucca *et al.*, 2001).

Iron deficiency affects up to 30 % of the total world population (WHO, 1992) and is the most widespread human mineral deficiency. A significant proportion of the iron in cereal grain is present in phytates and consequently can be released by digestion with phytase. An alternative, and complementary, approach is to increase the amounts of other components which bind iron in the seed. The most well-studied of these is ferritin, a protein which binds iron to form a storage reserve in plants, animals and bacteria (Theil, 1987). Lucca *et al.* (2001) have shown that the expression of ferritin genes from soybean in developing seeds of rice results in two- to three fold increases in the iron content.

Cereals are also a significant source of dietary selenium, accounting for about 10 % of the intake in the UK. However, in this case the content in the grain appears to be determined primarily, if not solely, by the availability of selenium in the soil (Adams *et al.*, 2002) with little or no opportunity for improvement by GM.

Cereal grains contain a range of fat-soluble and water-soluble vitamins but these tend to be concentrated in the embryo and aleurone, and hence are depleted by milling (wheat) or polishing (rice). Consequently, consumption of a high proportion of refined cereal products in the diet can be associated with vitamin deficiency. The best known example of this is vitamin A (retinol) and rice consumption. It has been estimated that a quarter of a million children in South East Asia alone go blind every year as a result of vitamin A deficiency related to consumption of white rice. This has led to the development of the widely publicized 'Golden Rice', in which two genes from daffodil and one gene from the bacterium *Erwinia uredova* have been transferred into rice, leading to the accumulation of β-carotene, which can be converted into vitamin A by humans (Ye *et al.*, 2000). Golden Rice also expresses a transgene for phytase to increase mineral availability (see above), but these traits have not yet been introduced into commercial lines for human consumption (see Chapter 2.2).

A further important target for GM in cereals is folic acid, with deficiency leading to defects in neural tube development during pregnancy and to a range of detrimental effects in adults. Consequently, all grain products are now fortified with folic acid in the USA but not in the UK. A recent study has shown that the expression of a single gene from *E. coli* (expressing GTP cyclohydrolase) in Arabidopsis resulted in an increase in folates of up to 3.3-fold (Hossain *et al.*, 2004); this strategy could also be applied to cereals.

Special Requirements for Animal Feed

Massive volumes of cereal grain are used for animal feed, particularly maize, wheat, barley and oats for cattle, pigs and poultry. Although the broad nutritional requirements are similar across species, there are notable differences.

In the case of protein quality, the precise composition of amino acids is only important in monogastric (i.e. non-ruminant) animals as the bacteria present in the rumen of ruminants are capable of synthesizing the whole range of essential amino acids irrespective of the composition of the feedstuff. In contrast, reducing the levels of phytate in order to increase the availability of phosphate and other minerals is important for all species of livestock. Cell wall composition and content are particularly important when cereals are fed to chickens and other poultry as high levels of β-glucan or arabinoxylan can lead to sticky feces. Increasing the hardness of wheat and barley grains could lead to improved feed quality for both ruminant and monogastric livestock, but for different reasons. Whole grain are usually used to feed ruminants and in this case the stronger adhesion of matrix proteins to the starch granules could result in reduced loss of starch by digestion in the rumen. In contrast, grain for monogastric animals is frequently milled and the stronger protein/starch interactions in hard grain leads to greater starch damage during milling and hence increased digestibility. Approaches to manipulating grain texture are discussed below.

Food Processing Quality

Rice is largely consumed after boiling the whole grain, and maize after cooking of whole or pearled grain. In contrast, wheat is only consumed after processing into a range of food products (notably bread, cakes, pastries, biscuits, pasta and noodles) and hence the processing properties are of paramount importance in determining the most appropriate products. Similarly, although a small proportion of barley is consumed in food products, including whole pearled grains and malted grain, the main consumption by humankind is after malting and distilling to give beer and whisky. Thus, in this case, the quality for malting and distilling is crucial.

Barley Quality for Malting

The malting process involves the controlled germination of the grain followed by drying and gentle cooking (kilning). During the first part of this process, the enzymes secreted by the aleurone cells and scutellum digest the walls of the starchy endosperm cells and some of the proteins present in the cells, rendering the grain friable and easy to mill. It also allows access of enzymes to the starch granules but limited digestion of the starch occurs. The drying and kilning reduce the water level to stabilize the grain and prevent further digestion, but the temperature regime is carefully selected to prevent excessive loss of enzyme activity. The kilning also develops the characteristic malt flavors with higher temperatures resulting in smoky or burnt flavors (Bamforth, 1998).

It is clear from this brief description that malting is a complex process which is affected by a range of parameters. Also, these include factors operating during grain development which determine the structure and composition of the mature grain and processes occurring during the malting itself.

Factors operating during grain development determine the relative thickness of the starchy endosperm cell walls, the grain protein content, the grain texture and the synthesis and accumulation of two hydrolases, β-amylase and α-glucosidase (maltase). The major factors operating during germination determine the production, activity and stability of a range of hydrolytic enzymes, including β-glucanase and other cell wall degrading enzymes, proteases and amylolytic enzymes (α-amylase, limit dextrinase and α-glucosidase). These present a wide range of targets for improvement, but only a limited number of them have been or can readily be targeted using a GM approach.

β-amylase is one of the major components of diastatic power, which is particularly important if high levels of starch adjunct (e.g. other cereal starches) are used for brewing. β-amylase is a hydrolytic enzyme, which is only synthesized in the developing grain where it is concentrated in the aleurone layer and starchy endosperm. It appears to act as a storage protein in that the amount is regulated by nitrogen availability (Giese and Hejgaard, 1984; Yin *et al.*, 2002). There is also genetic variation in β-amylase activity, which allows the selection for high β-amylase lines (Yin *et al.*, 2002). However, a more important target is to improve thermostability as enzymic activity is lost during kilning and the subsequent mashing at 65 °C (Bamforth and Quain, 1988). Natural variation in the thermostability of β-amylase occurs between cultivars of cultivated barley (Kihara *et al.*, 1998, 1999; Malysheva, Ganal and Roder, 2004; Zhang, Kaneko and Takeda, 2004). It is also possible to increase the thermostability of barley β-amylase by protein engineering, with mutations in seven individual residues resulting in an increase in the T_{50} (i.e. the temperature at which half of the enzyme activity is lost in over 30 min) by 11.6 °C, from 57.4 °C to 69 °C (Yoshigi *et al.*, 1995). This improved enzyme would be ideal for expression in barley by genetic engineering. Other sources of heat stable β-amylases have been sought in thermophilic microorganisms such as *Clostridium thermosulfurogens* (Kitamoto *et al.*, 1998).

Kihara *et al.* (1997; 2000) have reported the expression of their engineered thermostable form of barley β-amylase (Yoshigi *et al.*, 1995) in seeds of transgenic barley under the control of its own promoter. The transgene was expressed in six out of nine lines, but there was little or no effect on the specific activity of the enzyme in mature seeds. However, significant increases in the stability of the enzyme at 65 °C were observed in five of the lines, with between 20 % and 60 % of the initial activity remaining after 30 min compared with none in the control seeds.

During the malting process, the cell walls are digested to allow the solubilization of the grain proteins and to release the starch granules for fermentation. In addition to limiting the fermentation of starch, the failure to digest cell walls can also lead to high wort viscosity and poor filtration in breweries (Fox *et al.*, 2003). It is therefore not surprising that there are strong correlations between malting quality and either the amount of β-glucans (the major cell wall components) in the mature seed or the amount of β-glucanase produced during germination.

There is also a requirement to increase the heat stability of β-glucanase to prevent denaturation during kilning and mashing. von Wettstein and colleagues (Jensen *et al.*, 1996; Horvath *et al.*, 2000) have therefore expressed an engineered heat-stable form of (1,3-1,4)-β-glucanase derived from *Bacillus* spp. in transgenic barley using two strategies: expression in the developing starchy endosperm using the D hordein (*Hor 3*) promoter and in the aleurone cells of the germinating grain using an α-amylase gene promoter and signal peptide. Both strategies resulted in synthesis and accumulation of the enzyme. Furthermore, subsequent studies showed that incorporation of malted grain from a line expressing the enzyme under control of the α-amylase promoter into feed for chickens resulted in increased weight gain and reduced incidence of sticky feces (resulting from viscosity) (von Wettstein *et al.*, 2000). However, the malting performance of these lines has not yet been reported.

In considering the potential impact of GM crops it should be noted that the malting, brewing and distilling industries are understandably conservative regarding the introduction of new materials and hence are unlikely to accept the use of GM raw materials in the absence of widespread public acceptance.

Grain Texture and Dough Strength of Wheat

Two main characters are used to classify wheats into groups for different end uses: grain texture (hardness) and gluten strength. Hardness has been described as 'the most important single characteristic that affects the functionality of a common wheat' (Pomeranz and Williams, 1990) and affects the properties for milling (tempering, yield, size shape and density of the flour particles) and processing (for making bread, noodles, cakes and biscuits). An important difference between hard and soft wheats is in their water absorption. Flours milled from hard wheats have a higher baking absorbance, giving better quality and ultimately increased profit. This difference is assumed to result from a higher degree of starch damage during milling of hard wheats, due to the starch granules being more tightly bound into the protein matrix. However, it is important to distinguish hardness from vitreousness, in which the endosperm appears to be glassy. Vitreousness is not under strong genetic control but can occur in all varieties, particularly when grown with high nitrogen fertilization or at high temperature (Pomeranz and Williams, 1990). Similarly, hardness is not associated with dough strength although concurrent selection by breeders has resulted in breadmaking varieties generally being hard and strong.

It is generally accepted that differences in grain hardness are largely determined by the strength of adhesion between the surface of the starch granule and the matrix (i.e. gluten) proteins, although adhesion between cell wall components and matrix proteins may also contribute.

The main breakthrough in our understanding of grain hardness was the demonstration that an M_r 15 000 protein was present on the surface of water-washed starch from soft wheats but was reduced in amount or absent from hard bread wheats and completely absent from durum wheat, which is ultra-hard (Greenwell and Schofield, 1986). This protein was subsequently called 'friabilin' as it appears to determine grain friability and has been proposed to act as a 'non-stick' agent preventing adhesion between starch and proteins. Subsequent studies have shown that friabilin is a mixture of proteins, with the major components being two related proteins called puroindolines (Pins) a and b (Gautier *et al.*, 1994). Genetic studies have shown that Pin a and Pin b are encoded by genes present at the *Ha* (hardness) locus on chromosome 5D of bread wheat. This locus actually confers softness and its absence from durum wheat accounts for the ultra-hard texture.

Friabilin also comprises several other minor proteins (Morris *et al.*, 1994; Oda and Schofield, 1997), with one component (called grain softness protein or GSP) also mapping to the *Ha* locus (Chantret *et al.*, 2004; Jolly, Glen and Rahman, 1996). However, the functional significance of this protein has not been demonstrated.

In contrast, there is clear genetic and molecular evidence that the Pins do play a direct functional role in determining hardness. Detailed studies carried out mainly by Morris, Giroux and colleagues have shown that softness is associated with mutations in the *Pin a* and/or *Pin b* genes of bread wheat (Giroux and Morris, 1997; 1998; Morris, 2002). These mutations are of two types, either null mutations in either gene, which result in the absence of the corresponding protein, or point mutations in the coding sequence of the *Pin b* gene, which affect the amino acid sequence of the protein. The most widespread *Pin b* mutation results in the substitution of a glycine residue for a serine adjacent to a cluster of tryptophan residues, which have been suggested to act as a lipid-binding site (Kooijman *et al.*, 1997).

Giroux and co-workers have provided direct evidence for the role of puroindolines in conferring grain softness by functional complementation using genetic transformation. This was initially carried out using the wheat *Pin a* and *Pin b* genes in rice (Krishnamurthy and Giroux, 2001) and subsequently with the wild-type *Pin b* gene in hard bread wheat (Beecher *et al.*, 2002).

We have also obtained similar results, by expressing *Pin a* in durum wheat and *Pin b* in a hard genotype of bread wheat (unpublished results of Guang He, Paul Wiley, Paola Tosi, Sue Steele, Huw Jones and Peter Shewry).

These results clearly demonstrate that it is possible to manipulate the milling texture of cereals by expression of genes encoding puroindolines. This has been demonstrated for bread and durum wheats and presumably also applies to barley and oats in which proteins related to puroindolines are present (Beecher *et al.*, 2001; Darlington *et al.*, 2001; Gautier *et al.*, 2000). In addition, it also applies to rice, a species which appears to lack puroindoline homologs, although a putative GSP gene has been reported (Chantret *et al.*, 2004). This provides opportunities to develop new cereal varieties with optimized textural characteristics for animal feed and processing.

Breadmaking is one of humankind's oldest forms of 'biotechnology', dating back to ancient Egypt and Mesopotamia. It is also the most widespread form of food processing in the world today, with a vast array of breads being baked and consumed in different cultures. These range from the leavened pan and hearth breads most familiar to readers in Europe and N. America to steamed breads in China and the Far East and a wide spectrum of flat and pocket breads, including leavened and unleavened forms, in the Middle East, North Africa and the Indian subcontinent. It is also important to note that comparable products cannot be produced from other cereals, even related species such as barley and rye, although breads produced from these species, or from mixtures with wheat, are consumed to a limited extent, notably in Northern and Eastern Europe. This unique ability to make bread from wheat flour is determined largely by the gluten proteins which confer cohesiveness and viscoelasticity to dough.

The gluten proteins of wheat correspond to the storage proteins, which account for over half of the total nitrogen in the mature grain. These proteins are deposited in the cells of the starchy endosperm in discrete protein bodies, which fuse to form a continuous matrix during the later stages of grain maturation. They are therefore concentrated in white flour derived from the starchy endosperm cells. When flour is mixed with water to form dough, the gluten proteins present in the individual flour particles come together to form a continuous network which is responsible for the viscoelastic properties.

Gluten is a highly complex and polymorphic mixture of proteins which can be divided into groups based on their functions or their chemical and molecular characteristics. The traditional classification, which has been used for over a century, is into two broad groups which are termed 'gliadins' and 'glutenins'. The longevity of this classification reflects its usefulness as the gliadins and glutenins appear to play different roles in determining dough properties. The gliadins are monomeric proteins which interact with other gluten proteins by non-covalent interactions, principally hydrogen bonds. Consequently, they appear to be largely responsible for gluten and dough viscosity. In contrast, the glutenins comprise high-molecular-mass polymers (probably up to M_r of 10 million) which are stabilized by inter-chain disulfide bonds. These polymers also interact with each other and with the gliadins by non-covalent forces and are largely responsible for gluten elasticity, also called strength.

Samples of wheat flour vary widely in strength, and this is exploited in selecting grain for specific end uses. For example, weak doughs are preferred for making cakes and biscuits and strong doughs for breadmaking. Low dough strength is the most widespread problem in sourcing wheat for breadmaking, and hence developing cultivars which give reliably strong doughs is a major target for plant breeders.

Although the environment in which the wheat is grown has a major impact on dough strength, there is also strong genetic control. In particular, a range of studies have shown a clear relationship between allelic variation in the amount and composition of one group of gluten proteins and dough strength. These are the high-molecular-weight (HMW) subunits of glutenin.

These proteins account for up to about 12 % of the total grain proteins, or 1–1.7 % of the flour dry weight (Halford *et al.*, 1992; Nicolas, 1997; Seilmeier *et al.*, 1991). However, they have been calculated to account between 45 % and 70 % of the genetic variation in breadmaking performance within European wheats (Branlard and Darderet, 1985; Payne *et al.*, 1987; 1988). Consequently, the HMW subunits have been the most intensively studied group of wheat grain proteins over the past two decades (see reviews by Payne, 1987; Shewry *et al.*, 2003).

It is clear from these studies that the HMW subunits play a major role in determining the amounts of high-molecular-mass glutenin polymers and their properties. However, it appears that the allelic variation has two separate effects: quantitative and qualitative. The quantitative effects relate primarily to differences in the numbers of expressed genes. Cultivars of bread wheat all have six HMW subunit genes but only three, four or five of these are expressed. Since each HMW subunit accounts, on average, for about 2 % of the total grain protein (Halford *et al.*, 1992; Seilmeier *et al.*, 1991) this results in quantitative effects on the total amount of HMW subunit protein. In contrast, the qualitative effects relate to differences in the structures and properties of allelic subunits, with some allelic forms appearing to promote the formation of larger and more elastic polymers (Popineau *et al.*, 1994).

The clear correlation between the numbers of expressed genes and dough elasticity has resulted in several attempts to improve dough strength by transformation with additional HMW subunit genes (Altpeter *et al.*, 1996; Alvarez *et al.*, 2000; 2001; Anderson and

Blechl, 2000; Barro *et al.*, 1997; Blechl and Anderson, 1996; Darlington *et al.*, 2003; Popineau *et al.*, 2001; Rooke *et al.*, 1999; Vasil *et al.*, 2001; Zhang *et al.*, 2003). In general, the levels of transgene expression reported are comparable to those of the endogenous HMW subunits. However, exceptionally high levels of gene expression have been reported which may result from multiple transgene copies (Rooke, *et al.*, 1999).

Several groups have also reported the functional properties of the transgenic lines grown either in glasshouse or field trials, although the range of genetic backgrounds used make the results difficult to compare. In particular, the cultivar Bobwhite is frequently used for transformation due to its high regeneration frequency but is a poor background to explore gluten functionality due to the presence of the chromosome 1BL/1RS translocation which results in 'sticky dough'. Nevertheless, some conclusions can be drawn from these studies, the most important of which are that it is possible to increase dough strength and breadmaking quality by transformation with HMW subunit genes but that the effects observed vary between genes.

This is illustrated in Figure 2.4.1, which compares the expression of two HMW subunit genes in a line of wheat (L88-31) selected because it expresses only two HMW subunits (subunits 1Bx17 and 1By18, which migrate as the single band indicated by the arrow in Part a). The mixograph shows a relatively low peak resistance and short mixing time, and the loaf volume is only moderate. Expression of the 1Ax1 HMW subunit gene in this

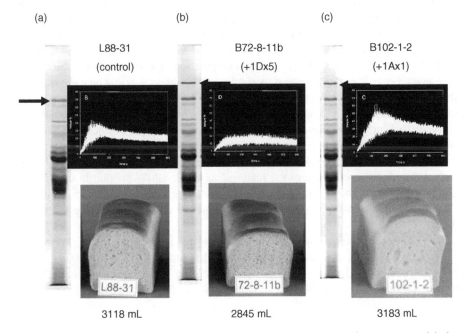

Figure 2.4.1 *Effect of expression of HMW subunit transgenes on the mixing and baking properties of wheat. Line L88-31 expressing the endogenous HMW subunits 1Bx17 + 1By18 (co-migrating and indicated by the arrow in Part a) was transformed to express subunit 1Dx5 (line B72-8-11b) (arrow in Part b) or subunit 1Ax1 (line B102-1-2) (arrow in Part c). Parts a–c show the proteins separated by SDS-PAGE, mixograph curves and loaves baked from field-grown grain of the three lines. Taken in part from Popineau et al. (2001), with permission Journal of Agricultural Food Chemistry (2001) 49: pp 399, Figure 4 (part).*

background (Part c) results in an improved mixing curve (i.e. longer mixing time and higher peak resistance) and greater loaf volume, the impact being similar to that which would be observed if the gene was transferred by crossing. In contrast, expression of the 1Dx5 transgene at a similar level (Part b) resulted in detrimental effects on the mixing properties, and on the loaf volume and texture. These differences have been ascribed to differences in the amino acid sequences of the two proteins with the subunit 1Dx5 protein containing an additional cysteine residue which could lead to the formation of highly cross-linked polymers, which do not hydrate properly during dough mixing (Popineau *et al.*, 2001).

The experiment shown in Figure 2.4.1 was carried out using a 'model' wheat line in order to compare the functional properties of the two subunits. We have since shown that expression of the 1Ax1 transgene can also lead to increased dough strength in commercial group 1 and 2 cultivars grown in the UK, such as Canon and Cadenza (Pastori *et al.*, 2000; Shewry *et al.*, 2005b).

Dough strength is also an important quality characteristic for making pasta from durum wheats. It is possible to use a similar transgenic approach to that described above, with the expression of the 1Ax1 transgene resulting in increased dough strength measured using the Mixograph (He *et al.*, 1999). However, studies of durum wheat have shown that the low-molecular-weight (LMW) subunits of glutenin have greater impacts than the HMW subunits (Pogna *et al.*, 1990; Ruiz and Carillo, 1995). In addition, these impacts may be associated with differences in the expression levels of the subunit proteins with one group (called LMW-2) being synthesized in larger amounts than the commonly occurring allele (LMW-1) (Masci *et al.*, 1995; Payne *et al.*, 1984). Masci *et al.* (2003) therefore determined the effect of overexpression of a single LMW subunit on the functional properties of durum wheat. A single line was identified in which the transgenic subunit was expressed at about 12 times that of the endogenous proteins. In view of this level of overexpression it is not surprising that negative effects on dough strength were observed. In contrast, we have generated a number of transgenic lines of commercial wheat cultivars expressing LMW subunit transgenes (Tosi *et al.*, 2004). Although two of the lines showed similar levels of transgene expression, they differed in that this was associated with co-suppression of the major endogenous LMW subunit in one line. The lines also differed in their functional properties, with the dough strength being increased in the line without co-suppression but reduced in the co-suppressed line (Tosi *et al.*, 2005).

These studies not only demonstrate that it is feasible to manipulate the gluten protein composition of bread and durum wheats but also indicate that high levels of gene expression can be used to introduce new properties (e.g. increased cross-linking), which may be of value in developing lines suitable for novel food or non-food uses.

Conclusions

The technology for the transformation of all the major cereals is now established, although in some cases, such as sorghum, it is still far from facile and improvements are still required even for wheat. Similarly, a number of quality traits are now understood at the biochemical and molecular levels, and the key genes have been isolated and characterized. The main limitation to the widespread use of this technology remains public acceptance, particularly when applied to the manipulation of food crops in the UK and Western Europe.

I consider that the key to improving public acceptance is to develop crops with positive health benefits and, in particular, to target the major health risks faced by consumers in the 21st century: type 2 diabetes, obesity, diet-related cancers and cardiovascular disease. These targets are challenging, but plant biotechnology has the potential to make important impacts.

Acknowledgments

Rothamsted Research receives grant-aided support from the Biotechnology and Biological Sciences Research Council of the United Kingdom.

References

Adams, M.L., Lombi, E., Zhao, F-J. and McGrath, S.P. (2002) Evidence of low selenium concentrations in UK bread-making wheat grain. *Journal of the Science of Food and Agriculture* **82**, 1160–1165.

Altpeter, F., Vasil, V., Srivastava, V. and Vasil, I.K. (1996) Integration and expression of the high-molecular-weight glutenin subunit 1Ax1 gene into wheat. *Nature Biotechnology* **14**, 1155–1159.

Alvarez, M.L., Guelman, S., Halford, N.G., Lustig, S., Reggiardo, M.I., Ryabushkina, N., Shewry, P.R., Stein, J. and Vallejos, R.H. (2000) Silencing of HMW glutenins in transgenic wheat expressing extra HMW subunits. *Theoretical and Applied Genetics* **100**, 319–327.

Alvarez, M.L., Gomez, M., Carrillo, J.M. and Vallejos, R.H. (2001) Analysis of dough functionality of flours from transgenic wheat. *Molecular Breeding* **8**, 103–108.

Anderson, O.D. and Blechl, A.E. (2000) Transgenic wheat—challenges and opportunities, in, *Transgenic Cereals* (eds L. O'Brien and R. Henry), American Association of Cereal Chemists, St. Paul, MN, USA pp. 1–27.

Anthony, A., Brown, W., Buhr, D., Ronhovde, G., Genovesi, D., Lane, T., Yingling, R., Rosato, M. and Anderson, P. (1997) Transgenic maize with elevated 10 kDa zein and methionine, in, *Sulphur Metabolism in Higher Plants* (eds W. G. Cram *et al.*), Backhuys Publishers, Leiden, p. 295.

Bamforth, C.W. and Quain, D.E. (1988) Enzymes in brewing and distilling, in *Cereal Science and Technology* (ed G.H. Palmer), Aberdeen University Press, Aberdeen, UK, pp. 326–366.

Bamforth, C. (1998) *Beer—Tap Into the Art and Science of Brewing*, Insight Books, New York, USA.

Barro, F., Rooke, L., Békés, F., Gras, P., Tatham, A.S., Fido, R.J., Lazzeri, P., Shewry, P.R. and Barcelo, P. (1997) Transformation of wheat with HMW subunit genes results in improved functional properties. *Nature Biotechnology* **15**, 1295–1299.

Beecher, B., Smidansky, E.D., See, D., Blake, T.K. and Giroux, M.J. (2001) Mapping and sequence analysis of barley hordoindolines. *Theoretical and Applied Genetics* **102**, 833–840.

Beecher, B., Bettge, A., Smidansky, E. and Giroux, M.J. (2002) Expression of wild-type *pinB* sequence in transgenic wheat complements a hard phenotype. *Theoretical and Applied Genetics* **105**, 870–877.

Blechl, A.E. and Anderson, O.D. (1996) Expression of a novel high-molecular-weight glutenin subunit gene in transgenic wheat. *Nature Biotechnology* **14**, 875–879.

Branlard, G. and Dardevet, M. (1985) Diversity of grain protein and bread wheat quality. II. Correlation between high molecular weight subunits of glutenin and flour quality characteristics. *Journal of Cereal Science* **3**, 345–354.

Brinch-Pedersen, H., Galili, G., Knudsen, S. and Holm, P.B. (1996) Engineering of the aspartate family biosynthetic pathway in barley (*Hordeum vulgare* L.) by transformation with heterologous genes encoding feed-back-insensitive aspartate kinase and dihydrodipicolinate synthase. *Plant Molecular Biology* **32**, 611–620.

Brinch-Pedersen, H., Olesen, A., Rasmussen, S.K. and Holm, P.B. (2000) Generation of transgenic wheat (*Triticum aestivum* L.) for constitutive accumulation of an *Aspergillus* phytase. *Molecular Breeding* **6**, 195–206.

Brinch-Pedersen, H., Sorensen, L.D. and Holm, P.B. (2002) Engineering crop plants: getting a handle on phosphate. *Trends in Plant Science* **7**, 118–125.

Brinch-Pedersen, H., Hatzack, F., Sorensen, L.D. and Holm, P.B. (2003) Concerted action of endogenous and heterologouse phytase on phytic acid degredation in seeds of transgenic wheat (*Triticum aestivum* L.). *Transgenic Research* **12**, 649–659.

Chamba, E.B., Halford, N.G., Wilkinson, M. and Shewry, P.R. (2005) Molecular cloning of β-kafirin, a methionine-rich protein of sorghum grain. Journal of Cereal Science 41, 381–383.

Chantret, N., Cenci, A., Sabot, F., Anderson, O. and Dubcovsky, J. (2004) Sequencing of the *Triticum monococcum* Hardness locus reveals good microlinearity with rice. *Molecular and General Genomics* **271**, 377–386.

Clore, G.M., Gronenborn, A.M., Kjaer, M. and Poulsen, F.M. (1987a) The determination of the three-dimensional structure of barley serine protease inhibitor 2 by nuclear magnetic resonance, distance geometry and restrained molecular dynamics. *Protein Engineering* **1**, 305–311.

Clore, M.G., Gronenborn, A.M., James, M.N.G., Kjaer, M., McPhalen, C.A. and Poulsen, F.M. (1987b) Comparison of the solution and X-ray structures of barley serine protease inhibitor 2. *Protein Engineering* **1**, 313–318.

Darlington, H.F., Rouster, J., Hoffmann, L., Halford, N.G., Shewry, P.R. and Simpson, D. (2001) Identification and molecular characterisation of hordoindolines from barley grain. *Plant Molecular Biology* **47**, 785–794.

Darlington, H., Fido, R., Tatham, A.S., Jones, H., Salmon, S.E. and Shewry, P.R. (2003) Milling and baking properties of field grown wheat expressing HMW subunit transgenes. *Journal of Cereal Science* **38**, 301–306.

Denbow, D.M., Grabau, E.A., Lacy, G.H., Kornegay, E.T., Russell, D.R. and Umbeck, P.F. (1998) Soybeans transformed with a fungal phytase gene improve phosphorous availability for broilers. *Poultry Science* **77**, 878–881.

FAO (2003) Production Vol. 56 2002, FAO, Rome.

Falco, S.C., Guida, T., Locke, M., Mauvais, J., Sandres, C., Ward, R.T. and Webber, P. (1995) Transgenic canola and soybean seeds with increased lysine. *Bio/Technology* **13**, 577–582.

Florack, D.E.A. and Stiekema, W.J. (1994) Thionins: properties, possible biological roles and mechanisms of action. *Plant Molecular Biology* **26**, 25–37.

Forsyth, J.L., Beaudoin, F., Halford, N.G., Sessions, R., Clark, A.R. and Shewry, P.R. (2005) Design, expression and characterisation of lysine-rich forms of the barley seed protein CI-2. *Biochimica et Biophysica Acta*, 1747, 221–227.

Fox, G.P., Panozzo, J.F., Li, C.D., Lance, R.C.M., Inkerman, P.A. and Henry, R.J. (2003) Molecular basis of barley quality. *Australian Journal of Agricultural Research* **54**, 1081–1101.

Gautier, M.F., Aleman, M.E., Guirao, A., Marion, D. and Joudrier P. (1994) *Triticum aestivum* puroindolines, two basic cystine-rich seed proteins: cDNA sequence analysis and developmental gene expression. *Plant Molecular Biology* **25**, 43–57.

Gautier, M.F., Cosson, P., Guirao, A., Alary, R. and Joudrier, P. (2000) Puroindoline genes are highly conserved in diploid ancestor wheats and related species but absent in tetraploid *Triticum* species. *Plant Science* **153**, 81–91.

Giese, H. and Hejgaard, J. (1984) Synthesis of salt-soluble proteins in barley. Pulse-labelling study of grain filling in liquid-cultured detached spikes. *Planta* **161**, 172–177.

Giroux, M.J. and Morris C.F. (1997) A glycine to serine change in puroindoline b is associated with wheat grain hardness and low levels of starch-surface friabilin. *Theoretical and Applied Genetics* **95**, 857–864.

Giroux, M.J. and Morris, C.F. (1998) Wheat grain hardness results from highly conserved mutations in the friabilin components puroindoline a and b. *Proceedings of the National Academy of Sciences USA* **95**, 6262–6266.

Greenwell, P. and Schofield, J.D. (1986) A starch granule protein associated with endosperm softness in wheat. *Cereal Chemistry* **63**, 378–380.

Hagan, N.D., Upadhyaya, N., Tabe, L.M. and Higgins, T.J.V. (2003) The redistribution of protein sulphur in transgenic rice expressing a gene for a foreign, sulphur-rich protein. *Plant Journal* **34**, 1–11.

Halford, N.G., Field, J.M., Blair, H., Urwin, P., Moore, K., Robert, L., Thompson, R., Flavell, R.B., Tatham, A.S. and Shewry, P.R. (1992) Analysis of HMW glutenin subunits encoded by chromosome 1A of bread wheat (*Triticum aestivum* L.) indicates quantitative effects on grain quality. *Theoretical and Applied Genetics* **83**, 373–378.

He, G.Y., Rooke, L., Steele, S., Békés, F., Gras, P., Tatham, A.S., Fido, R., Barcelo, P., Shewry, P.R. and Lazzeri, P. (1999) Transformation of pasta wheat (*Triticum durum* L.) with HMW glutenin subunit genes and modification of dough functionality. *Molecular Breeding* **5**, 377–386.

Hejgaard, J. and Boisen, S. (1980) High-lysine proteins in Hiproly barley breeding: identification, nutritional significance and new screening methods. *Hereditas* **93**, 311–320.

Horvath, H., Huang, J., Wong, Oi., Kohl, E., Okita, T., Kannangara, C.G. and von Wettstein, D. (2000) The production of recombinant proteins in transgenic barley grains. *Proceedings of the National Academy of Sciences USA* **97**, 1914–1919.

Hossain, T., Rosenberg, I., Selhub, J., Kishore, G., Beachy, R. and Schubert, K. (2004). Enhancement of folates in plants through metabolic engineering. *Proceedings of the National Academy of Sciences USA* **101**, 5158–5163.

Jensen, L.G., Olsen, O., Kops, O., Wolf, N., Thomsen, K.K. and von Wettstein, D. (1996) Transgenic barley expressing a protein-engineered, thermostable (1,3–1,4)- β-glucanase during germination. *Proceedings of the National Academy of Sciences USA* **93**, 3487–3491.

Jolly, C.G., Glen, G.M. and Rahman, S. (1996) *GSP-1* genes are linked to the grain hardness locus (*Ha*) on wheat chromosome 5D. *Proceedings of the National Academy of Sciences USA* **93**, 2408–2413.

Kasarda, D.D. (2001) Grains in relation to celiac disease. *Cereal Foods World* **46**, 209–210.

Kelly, J.D. and Hefle, S.L. (2000) 2S methionine-rich protein (SSA) from sunflower seed is an IgE-binding protein. *Allergy* **55**, 556–560.

Kelly, J.D., Hlywka, J.J. and Hefle, S.L. (2000) Identification of sunflower seed IgE-binding proteins. *International Archives of Allergy and Applied Immunology* **121**, 19–24.

Kihara, M., Okada, Y., Kuroda, H., Saeki, K. and Ito, K. (1997) Generation of fertile transgenic barley synthesizing thermostable β-amylase, in *European Brewing Convention Proceedings, 26th Congress, Maastricht*, Oxford University Press, Oxford, UK, pp. 83–90.

Kihara, M., Kaneko, T. and Ito, K. (1998) Genetic variation of β-amylase thermostability among varieties of barley, *Hordeum vulgare* L. and relation to malting quality. *Plant Breeding* **117**, 425–428.

Kihara, M., Kaneko, T., Ito, K., Aida, Y. and Takeda, K. (1999) Geographical variation of beta-amylase thermostability among varieties of barley (*Hordeum vulgare*) and beta-amylase deficiency. *Plant Breeding* **118**, 453–455.

Kihara, M., Okada, Y., Kuroda, H., Saeki, K., Yoshigi, N. and Ito, K. (2000) Improvement of β-amylase thermostability in transgenic barley seeds and transgene stability in progeny. *Molecular Breeding* **6**, 511–517.

Kitamoto, N., Yamagata, H., Kato, T., Tsukagoshi, N. and Udaka, S. (1998) Cloning and sequencing of the gene encoding thermophilic β-amylase of *Clostridium thermosulfurogenes*. *Journal of Bacteriology* **170**, 5848–5854.

Kooijman, M., Orsel, R., Hessing, M., Hamer, R.J. and Bekkers, A.C.A.P.A. (1997) Spectroscopic characterisation of the lipid-binding properties of wheat puroindolines. *Journal of Cereal Science* **26**, 145–159.

Kortt, A.A., Caldwell, J.B., Lilley, G.G. and Higgins, T.J.V. (1991) Amino acid and cDNA sequences of a methionine-rich 2S protein from sunflower seed (*Helianthus annus* L.). *European Journal of Biochemistry* **195**, 329–334.

Krishnamurthy, K. and Giroux, J. (2001) Expression of wheat puroindolines genes in transgenic rice enhances grain softness. *Nature Biotechnology* **19**, 162–166.

Lai, J. and Messing, J. (2002) Increasing maize seed methionine by mRNA stability. *Plant Journal* **30**, 395–402.

Lee, S.I., Kim, H.U., Lee, Y.H., Suh, S.C., Lim, Y.P., Lee, H.Y. and Kim, H.I. (2001) Constitutive and seed-specific expression of a maize lysine-feedback-insensitive-dihydrodipicolinate synthase gene leads to increased free lysine levels in rice seeds. *Molecular Breeding* **8**, 75–84.

Lucca, P., Hurrell, P. and Potrykus, I. (2001) Genetic engineering approaches to improve the bioavailability and the level of iron in the rice grains. *Theoretical and Applied Genetics* **102**, 392–397.

Malysheva, L., Ganal, M.W. and Roder, M.S. (2004) Evaluation of cultivated barley (*Hordeum vulgare*) germplasm for the presence of thermostable alleles of β-amylase. *Plant Breeding* **123**, 128–121.

Masci, S., Lew, E.J-L., Lafiandra, D., Porceddu, E. and Kasarda, D.D. (1995) Characterisation of low-molecular-weight glutenin subunits in durum wheat by RP-HPLC and N-terminal sequencing. *Cereal Chemistry* **72**, 100–104.

Masci, S., D'Ovidio, R., Scossa, F., Patacchini, C., Lafiandra, D., Anderson, O.D. and Blechl, A.E. (2003) Production and characterisation of a transgenic bread wheat line over-expressing a low-molecular-weight glutenin subunit gene. *Molecular Breeding* **12**, 209–222.

Mazur, B., Krebbers, E. and Tingey, S. (1999) Gene discovery and product development for grain quality traits. *Science* **285**, 372–375.

McPhalen, C.A. and James, M.N.G. (1987) Crystal and molecular structure of the serine protease inhibitor CI2 from barley grains. *Biochemistry* **26**, 261–269.

Melo, V.M.M., Xavier-Filho, J., Lima, M.S. and Prouvost-Danon, A. (1994) Allergenicity and tolerance to proteins from Brazil nut (*Bertholletia excelsa* HBK). *Food and Agricultural Immunology* **6**, 185–195.

Morris, C.F., Greenblatt, G.A., Bettge, A.D. and Malkawi, H.I. (1994) Isolation and characterisation of multiple forms of friabilin. *Journal of Cereal Science* **21**, 167–174.

Morris, C.F. (2002) Puroindolines: the molecular genetic basis of wheat grain hardness. *Plant Molecular Biology* **48**, 633–647.

Nicolas, Y. (1997) Les prolamines de blé: extraction exhaustive et développement de dosages chromatographiques en phase inverse et de dosages immunochimiques à l'aide d'anticorps anti-peptides. PhD thesis, University of Nantes.

Nordlee, J.A., Taylor, S.L., Townsend, J.A., Thomas, L.A. and Bush, R.K. (1996) Identification of a Brazil-nut allergen in transgenic soybeans. *New England Journal of Medicine* **334**, 688–692.

Oda, S. and Schofield, J.D. (1997) Characterisation of friabilin polypeptides. *Journal of Cereal Science* **26**, 29–36.

Otzen, D.E. and Fersht, A.R. (1995) Side-chain determinants of (β-sheet stability. *Biochemistry* **34**, 5718–5724.

Otzen, D.E. and Fersht, A.R. (1998) Folding of circular and permuted chymotrypsin inhibitor 2: retention of the folding nucleus. *Biochemistry* **37**, 8139–8146.

Pastori, G.M., Steele, S.H., Jones, H.D. and Shewry, P.R. (2000) Transformation of commercial wheat varieties with high molecular weight glutenin subunit genes, in *Wheat Gluten* (eds P.R. Shewry and A.S. Tatham), Royal Society of Chemistry, Cambridge, UK, pp. 88–92.

Payne, P.I., Jackson, E.A. and Holt, L.M. (1984) The association between γ-gliadin 45 and gluten strength in durum wheat varieties. A direct causal effect or the result of genetic linkage. *Journal of Cereal Science* **2**, 73–81.

Payne, P.I. (1987) Genetics of wheat storage proteins and the effect of allelic variation on breadmaking quality. *Annual Review of Plant Physiology* **38**, 141–153.

Payne, P.I., Nightingale, M.A., Krattiger, A.F. and Holt, L.M. (1987) The relationship between HMW glutenin subunit composition and the breakmaking quality of British grown wheat varieties. *Journal of the Science of Food and Agriculture* **40**, 51–65.

Payne, P.I., Holt, L.M., Krattiger, A.F. and Carrillo, J.M. (1988) Relationship between seed quality characteristics and HMW glutenin subunit composition determined using wheats grown in Spain. *Journal of Cereal Science* **7**, 229–235.

Pedersen, K., Argos, P., Naravana, S.V.L. and Larkins, B.A. (1986) Sequence analysis and characterization of a maize gene encoding a high-sulphur zein protein of M_r 15 000. *Journal of Biological Chemistry* **261**, 6279–6284.

Pogna, N.E., Autran, J.C., Mellini, F., Lafiandra, D. and Feillet, P. (1990) Chromosome 1B-encoded gliadins and glutenin subunits in durum wheat: genetics and relationship to gluten strength. *Journal of Cereal Science* **11**, 15–34.

Pomeranz, Y. and Williams, P.C. (1990) Wheat hardness: its genetic, structural and biochemical background, measurement and significance, in *Advances in Cereal Science and Technology*, Volume 10 (ed Y. Pomeranz), American Association of Cereal Chemists, St. Paul, MN, USA, pp. 471–544.

Popineau, Y., Cornec, M., Lefebvre, J. and Marchylo, B. (1994) Influence of high M_r glutenin subunits on glutenin polymers and rheological properties of gluten and gluten subfractions of near-isogenic lines of wheat Sicco. *Journal of Cereal Science* **19**, 231–241.

Popineau, Y., Deshayes, G., Lefebvre, J., Fido, R., Tatham, A.S. and Shewry, P.R. (2001) Prolamin aggregation, gluten viscoelasticity and mixing properties of transgenic wheat lines expressing 1Ax and 1Dx high molecular weight glutenin subunit transgenes. *Journal of Agriculture and Food Chemistry* **49**, 395–401

Rao, A.G., Hassan, M. and Hempel, J.C. (1994) Structure-function validation of high lysine analogs of α-hordothionin designed by protein modelling. *Protein Engineering* **7**, 1485–1493.

Roesler, K.R. and Rao, A.G. (1999) Conformation and stability of barley chymotrypsin inhibitor-2 (CI-2) mutants containing multiple lysine substitutions. *Protein Engineering* **12**, 967–973.

Roesler, K.R. and Rao, A.G. (2000) A single disulfide bond restores thermodynamic and proteolytic stability to an extensively mutated protein. *Protein Science* **9**, 1642–1650.

Rooke, L., Békés, F., Fido, R., Barro, F., Gras, P., Tatham, A.S., Barcelo, P., Lazzeri, P. and Shewry, P.R. (1999) Over-expression of a gluten protein in transgenic wheat results in greatly increased dough strength. *Journal of Cereal Science* **30**, 115–120.

Ruiz, M. and Carrillo, J.M. (1995) Relationships between different prolamin proteins and some quality properties in durum wheat. *Plant Breeding* **114**, 40–44.

Salcedo, G., Sanchez-Monge, R., Garcia-Casado, G., Armentia, A., Gomez, L. and Barber, D. (2004) The cereal α-amylase/trypsin inhibitor family associated with Bakers' asthma and food allergy, in *Plant Food Allergens* (eds E.N.C. Mills and P.R. Shewry), Blackwell Science, Oxford, UK.

Seilmeier, W., Belitz, H.D. and Wieser, H. (1991) Separation and quantitative determination of high-molecular-weight subunits of glutenin from different wheat varieties and genetic variants of the variety Sicco. *Z Lebensm Unters Forsch* **192**, 124–129.

Shewry, P.R. (2000) Seed proteins, in *Seed Technology and its Biological Basis* (eds M. Black and J.D. Bewley), Sheffield Academic Press, Sheffield, UK, pp. 42–84.

Shewry, P.R., Halford, N.G., Tatham, A.S., Popineau, Y., Lafiandra, D. and Belton, P.S. (2003) The high molecular weight subunits of wheat glutenin and their role in determining wheat processing properties. *Advances in Food and Nutrition Research* **45**, 219–302.

Shewry, P.R., Jenkins, J., Beaudoin, F. and Mills, E.N.C. (2004) The classification, functions and evolutionary relationships of plant proteins in relation to food allergens, in *Plant Food Allergens* (eds E.N.C. Mills and P.R. Shewry), Blackwell Science, Oxford, UK, pp. 24–41.

Shewry, P.R., Jones, H.D. and Halford, N.G. (2005a) Plant biotechnology: transgenic crops, in *Food Biotechnology* (ed Prof. U. Stahl), Springer-Verlag Heidelberg, Berlin, Germany. In press.

Shewry, P.R., Pastori, G.M., Bekes, F. and Jones, H.D. (2005b) Wheat transgenics: where are we now? in Cereals 2004, Royal Australian Chemical Institute Conference, Canberra, Australia, 21–24 September 2004.

Swarup, S., Timmermans, M.C.P., Chaudhuri, S. and Messing, J. (1995) Determinants of the high-methionine trait in wild and exotic germplasm may have escaped selection during early cultivation of maize. *Plant Journal* **8**, 359–368.

Tada, Y., Nakase, M., Adachi, T., Nakamura, R., Shimada, H., Takahashi, M., Fujimura, T. and Matsuda, T. (1996) Reduction of 14–16 kDa allergenic proteins in transgenic rice plants by antisense gene. *FEBS Letters* **391**, 341–345.

Theil, E.C. (1987) Ferritin: structure, gene regulation and cellular function in animals, plants and microorganisms. *Annual Review of Biochemistry* **56**, 289–315.

Tosi, P., D'Ovidio, R., Napier, J.A., Békés, F. and Shewry, P.R. (2004) Expression of epitope tagging LMW glutenin subunits in the starchy endosperm of wheat and their incorporation into the glutenin polymers. *Theoretical and Applied Genetics* **108**, 468–476.

Tosi, P., Masci, S., Giovangrossi, A., D'Ovidio, R., Békés, F., Larroque, O., Napier, J.A. and Shewry, P.R. (2005) Functional analysis of a LMW subunit of glutenin using transgenic pasta wheat. Molecular Breeding, in press.

Vasil, I.K., Bean, S., Zhao, J., McCluskey, P., Lookhart, G., Zhao, H-P., Altpeter, F. and Vasil, V. (2001) Evaluation of baking properties and gluten protein composition of field grown transgenic wheat lines expressing high molecular weight glutenin gene 1Ax1. *Journal of Plant Physiology* **158**, 521–528.

von Wettstein, D., Mikhaylenko, G., Froseth, J.A. and Kannangara, C.G. (2000) Improved barley broiler feed with transgenic malt containing heat-stable (1,3–1,4)-β-glucanase. *Proceedings of the National Academy of Sciences USA* **97**, 13512–13517.

WHO (1992) National strategies for overcoming micro-nutrient malnutrition. Document A45/3, WHO, Geneva, Switzerland.

Ye, X., Al-Babili, S., Klöti, A., Zhang, J., Lucca, P., Beyer, P. and Potrykus, I. (2000) Engineering pro-vitamin A (β-carotene) biosynthetic pathway into (carotenoid-free) rice endosperm. *Science* **287**, 303–305.

Yin, C., Zhang, G.P., Wang, J.M. and Chen, J.X. (2002) Variation of β-amylase activity in barley as affected by cultivar and environment and its relation to protein content and grain weight. *Journal of Cereal Science* **36**, 307–312.

Yoshigi, N., Okada, Y., Maeba, H., Sahara, H. and Tamaki, T. (1995) Construction of a plasmid used for the expression of a sevenfold-mutant barley β-amylase with increased thermostability in *Escherichia coli* and properties of the sevenfold-mutant β-amylase. *Journal of Biochemistry* **118**, 562–567.

Zhang, W.S., Kaneko, T. and Takeda, K. (2004) β-amylase variation in wild barley accessions. *Breeding Science* **54**, 41.

Zhang, X., Liang, R., Chen, X., Yang, F. and Zhang, L. (2003) Transgene inheritance and quality improvement by expressing novel HMW glutenin subunit (HMW-GS) genes in winter wheat. *Chinese Science Bulletin* **48**, 771–776.

Zhang, Z.B., Kornegay, E.T., Radcliffe, J.S., Denbow, D.M., Veit, H.P. and Larsen, C.T. (2000a) Comparison of genetically engineered microbial and plant phytase for young broilers. *Poultry Science* **79**, 709–717.

Zhang, Z.B., Kornegay, E.T., Radcliffe, J.S., Wilson, J.H. and Veit, H.P. (2000b) Comparison of phytase from genetically engineered *Aspergillus* and canola in weanling pig diets. *Journal of Animal Science* **78**, 2868–2878.

Zhu, X. and Galili, G. (2003) Increased lysine synthesis coupled with a knockout of its catabolism synergistically boosts lysine content and also transregulates the metabolism of other amino acids in *Arabidopsis* seeds. *Plant Cell* **15**, 845–853.

2.5

Improvements in Starch Quality

Michael M. Burrell

Department of Animal and Plant Sciences, University of Sheffield,
Sheffield, S10 2TN, United Kingdom

Introduction

Starch is the main storage product of the major agricultural crops important to man. In 2003 worldwide cereal and potato production reached 2386 million metric tons (FAO). The starch in these crops provides, in some parts of the world, up to 90 % of man's calorific intake (Burrell 2003; Hels *et al.*, 2003) whereas in the UK it provides directly about 30 % (UK National Statistics 2004). Thus, increasing the proportion of photo-synthetically fixed carbon that is stored as a harvestable crop could provide a means of reducing malnutrition in the world. However, the focus of this chapter is the starch that is purified from these plants and how biotechnology might be used to improve the suitability of the starch for different applications. The aim is to provide an overview allowing the reader to delve more deeply into aspects of interest rather than to cover certain aspects of the topic comprehensively.

Starch has many diverse uses both food and non-food. The Egyptians used it as a binder in paper ($\sim$4000 BC), the Romans used it as an adhesive and the Greeks used it in medicine (Tester and Karkalas, 2001; 2002). Maize is the main source of extracted starch and much of it is produced in the US. In 2002, over 15 % of the US maize crop was processed by corn refiners and of this over 60 % was converted into various corn syrups while 10 % was shipped as starch products (Corn Refiners Association Inc, 2003). The US farmers and corn industry have embraced biotechnology with some 34 % of the crop planted in 2003 as biotech varieties. Maltodextrin produced from starch and starch itself are important energy sources, in particular in baby foods because they have a low

Plant Biotechnology. Edited by Nigel Halford.
© 2006 John Wiley & Sons, Ltd.

fermentability. However, starch not only is important as an energy source but also has a wide range of uses in processed food to provide different types of texture and preservation (Burrell, 2003). Modified waxy maize starch is important in processed meat products where its gelling properties are useful as a binder to maintain the texture and stability of the product. It is used as a thickener in sauces, a stabilizing agent in canning and a coating on pre-prepared meals.

While starch is used in food its industrial uses are also very important. Starch is used as an additive in cement to improve the setting time. It is also used to improve the viscosity of drilling muds in oil wells and hence seal the walls of bore holes and prevent fluid loss (Burrell, 2003). In paper making, starch is used for several purposes but different types of starch are required for these. As a filler, it binds the cellulose fibers together and improves the strength of the product. It is used as an adhesive in paper bags. It is also used as a coating to improve writing and erasing properties and improve the finish quality. One particular use is in carbonless copy papers (Nachtergaele and Vannuffel, 1989; White, 1998). Here starch granules of a uniform size are mixed with encapsulated ink particles of a slightly smaller size. The starch keeps the sheets of paper apart, but when compressed by the tip of a pen, the ink capsules break and ink is released onto the surface of the paper. For this the starch granules need to be rounded and approximately 10 μm in diameter. One particular fraction of wheat starch best meets these criteria. For a more detailed listing of current uses of starch the reader is referred to the International Starch Institute web pages (www.starch.dk). For all these uses it is the structure of the starch and the size of the granule that are important.

Starch Structure

Starch is produced as granules in plastids. Granules vary greatly in size and shape, both between species and between the same variety grown under different conditions (Banks and Muir, 1980; Boyer *et al.*, 1976; Chojecki *et al.*, 1986; Haase and Plate, 1996; Jobling, 2004) (Table 2.5.1). Potato (*Solanum tuberosum*) has some of the largest granules found in nature. In tubers they may be as large as 100 μm and oval/spherical in shape. On the

Table 2.5.1 *Characteristics of starch from different species. Compiled from Banks and Muir (1980), Davydova et al., (1995) and Whistler et al., (1984).*

Species	Property				
	Type	Crystallinity	Shape	Diameter	Other
Cassava	Root	B	Oval	4–35	
Barley	Cereal	A	Round/elliptical	2–25	Bimodal
Maize	Cereal	A	Round/polygonal	5–30	
Pea	Legume	C	Round/oval	10–45	
Potato	Tuber	B	Oval/spherical	5–100	
Rice	Cereal	A	Lenticular/round	3–8	Compound
Wheat	Cereal	A	Lenticular/round	1–45	Bimodal

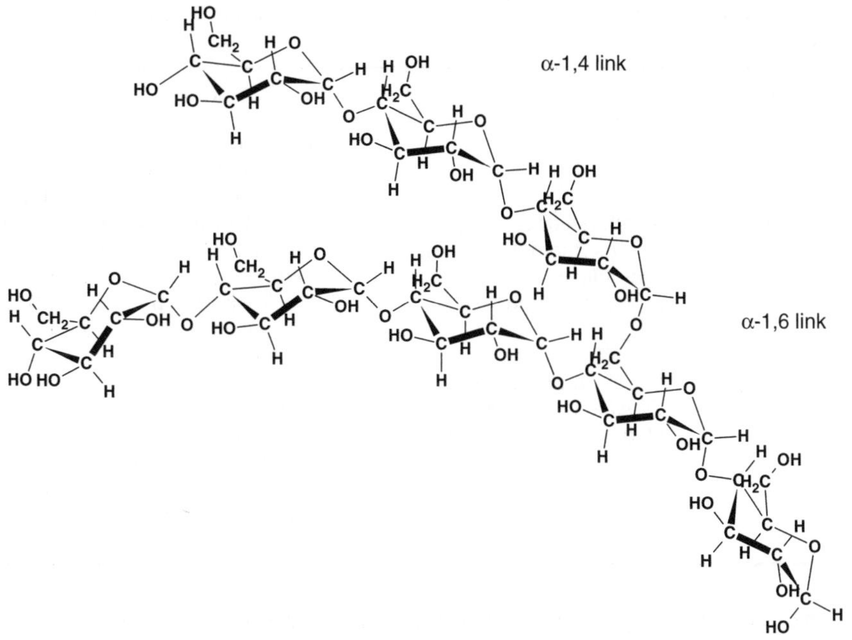

Figure 2.5.1 *The structure of the branch in an amylopectin chain.*

other hand, in barley and wheat there are usually two sizes of granule, with the smaller ones only being 1–2 μm and the larger ones being round and 10–15 μm. The shape of the A granules (the larger ones) in wheat makes them ideal for carbonless copy papers, where the granule provides the spacer between the sheets of paper protecting the ink capsules until they are broken by the pressure of a pen.

The granules consist of mainly two carbohydrate polymers, amylose and amylopectin (Banks and Muir, 1980). It is the behavior of these two molecules on heating in water that gives starch the properties that lead to its multitude of uses. Both polymers consist of α-1, 4-linked glucan chains, which have α-1,6 branches (Figure 2.5.1). Amylopectin has shorter chains of up to 50 glucan units, depending on the source, and many branches, whereas amylose has much longer linear chains with fewer branches. In potato starch the amylose chains can be 100 glucan units or more. These two molecules are assembled together to form a semi-crystalline starch granule. The granule also contains small amounts of lipid and phosphate (Kainuma, 1988), which have significant effects on the properties of the starch.

Our current understanding of the organization of the amylose, amylopectin, lipid and phosphate in the starch granule is quite incomplete. The granule consists of crystalline and amorphous regions and it is this degree of crystallinity, the length of the glucan chains and the quantity of minor components that determine the thermal properties of the starch i.e., the temperature and stability of the gel or paste that is produced.

On heating in water, the starch granules swell as water is absorbed into the amylopectin structure and the helices melt with the formation of a visco-elastic paste.

***Table* 2.5.2** *Common starch modifications (Whistler et al., 1984). Reproduced with permission of Elsevier.*

Modification		Property and uses
Oxidation		Low viscosity, high solid dispersion, resistance to gelling. Used in paper and board manufacture.
Esterification	Starch acetates Carboxymethyl starches Cationic starches Mono-starch phosphates	Reduced gelatinisation temperature, increased clarity and reduced retrogradation. Used in food thickenings, when cross-linked in canned foods, warp sizing of textiles.
+ Cross-linking	Di-starch phosphates Di-starch adipates Hydroxyalkyl starches	Reinforces the hydrogen bonds in the granule and reduces swelling. More stable viscosity at high temperature. Lacks rubbery texture of waxy corn starch. Used in salad dressings, deodorants and, textile printing.

On further heating the granules burst and viscosity decreases as the starch goes into solution. On cooling, a gel is formed as the glucan chains re-associate. A good example of the properties of a natural starch is the clear paste produced by potato starch. The long amylose chains provide a low gelling temperature and a clear gel, which is suitable in the canning industry. However, gels made from natural starches with a high amylose content lose a lot of water from the gel on cooling, a process called retrogradation.

The starch industry has developed many chemical methods to improve and alter the processing properties of starch. One type of modification is to add groups to the glucose residues, which cross-link the glucan chains. Such cross-linking modifications strengthen the granules by reinforcing the hydrogen bonds, which are responsible for holding the granules intact and thus improve the stability of the swollen starch granules under handling and processing conditions. The interaction between the glucan chains and the interaction with other molecules such as clay or cellulose is altered (Table 2.5.2). For a detailed discussion of starch modification the reader is referred to Whistler *et al.* (1984). The efficacy of such modifications depends on the properties of the starch used, which in turn depends not only on the species (Table 2.5.1) but on variety as well. There is therefore a sound basis to justify an understanding of how a plant synthesizes starch and how alterations in structure affect the processing properties. A targeted plant breeding or genetic modification strategy could then be used to tailor the end product to man's needs.

The Synthesis of Starch

The synthesis of the glucan in starch involves at least three enzyme reactions starting with glucose-1-phosphate. However, this hides a very complex set of reactions which involve many isozymes with slightly different properties, and there is still much to learn about how the different enzymes interact.

ADPGlucose Pyrophosphorylase (AGPase) E.C. 2.7.7.27

This enzyme catalyses the first step where carbon is committed to starch synthesis:

$$ATP + \text{glucose 1-P} \longleftrightarrow ADPG + PPi$$

The reaction is reversible with an equilibrium position of approximately 1 (Preiss, 1988). It is generally considered that the direction is synthetic because alkaline pyrophosphatase in the plastid will cleave the pyrophosphate. However, recent evidence has shown that at least in cereal species a considerable proportion of the enzyme is in the cytosol (Burton *et al.*, 2002) and thus other enzymes are required to cleave the pyrophosphate. At present it is quite unclear whether ADPG is synthesized simultaneously in both compartments, how the glucose-1-P in the cytosol is partitioned between glycolysis, the synthesis of UDPG and ADPG, and how the cell regulates this.

AGPase is regulated by several mechanisms. Allosteric regulation occurs via the inhibitor phosphate and activator PGA in many species but the regulation can differ in different tissues of the same species (Ball and Preiss, 1994; Ballicora *et al.*, 1998; Copeland and Preiss, 1981; Dickinson and Preiss, 1969; Preiss 1988; Sowokinos and Preiss, 1982). More recently, it has been shown that a redox-dependent inhibition and activation mechanism is involved in regulating the *in vivo* activity of the enzyme in potato tubers (Gibon *et al.*, 2004; Hendriks *et al.*, 2003; Tiessen *et al.*, 2002). The mechanism appears to involve the formation of a thiol link between the two small subunits of the tetrameric enzyme. Recently, this dimerization has also been found to occur in wheat, suggesting that cereals also regulate the *in vivo* enzyme activity by this mechanism. Thus, the synthesis of ADPG and its regulation varies between species and between tissues (Burrell et al., 2004).

Starch Synthase E.C. 2.4.1.21

All starch synthases use ADPG and add a unit of glucose to a glucan chain, in contrast to mammalian glycogen synthases which use UDPG in essentially the same reaction. Originally, starch synthases were divided into two groups: the granule bound enzyme and the soluble enzymes. However, with the increased understanding of the multiple forms of starch synthases and their expression pattern, this distinction is no longer really valid (Harn *et al.*, 1998). GBSSI, originally described as the granule-bound starch synthase, is the most studied enzyme and is involved in amylose biosynthesis. It is a very abundant enzyme in the amyloplast and its location within the starch granule makes its isolation quite straightforward, but the measurement of maximum catalytic activity difficult. The maize mutation *waxy* leads to a lack of this enzyme, no amylose synthesis and a starch with very different characteristics (as discussed above).

Waxy starches are easily identified because the lack of amylose means that they do not stain blue/black with iodine but stain a red/brown color. They are very important commercially and have been actively sought and discovered in all the starch-producing crops. In some species, plants that are heterozygous for the waxy allele can have intermediate amounts of amylose (see below).

A second GBSS isoform was suggested by Nelson in 1978 (Nelson *et al.*, 1978) as a minor component in maize seed, which was associated with the *su2* locus. This locus is now thought to be a soluble starch synthase zSSIIa (see below). More recently a GBSSII has been identified in the pericarp of waxy wheat seeds and appears to account for amylose synthesis in this tissue (Nakamura et al., 1998).

Multiple forms of soluble starch synthases have been identified in different systems and originally these were identified by their elution pattern from DEAE-cellulose chromatography (Preiss and Levi, 1980). Three soluble starch synthases have been identified in maize, wheat and potato. In describing these enzymes, care must be taken to distinguish between descriptions of enzyme activity and analysis of genes and their function. SSI can be assayed in the absence of glucan primer and the activity is stimulated by the presence of citrate. SSII however requires a glucan primer and is not affected by citrate. Knight et al. (1998) have cloned a cDNA, now designated zSSI, which accounts for the SSI activity in maize endosperm. The gene for SSIIa activity appears to map to the *sugary-2* (*su-2*) locus (Harn et al., 1998). The gene *Dull1* (*Du1*) encodes an SSII activity that, to avoid confusion, has now been designated DU1 and, by sequence comparison, is more similar to other SSIII isoforms than SSII. SSIII appears to be exclusively soluble in maize and wheat but is also found attached to the granule in potato. While GBSSI is only active in the starch granule it is not clear whether the soluble enzymes need to be attached to the starch to function *in vivo*.

All the soluble starch synthases are involved in amylopectin synthesis, but with the complexity of isoforms and expression patterns, it has proved difficult to identify their specific roles. Kinetic evidence has shown that different forms have different affinities for the glucan polymer, suggesting that SSII forms would extend longer chains (Preiss, 1988). *Du1* mutants appear to have as much soluble starch synthase activity as wild-type maize, although the structure of the starch differs, having an increased proportion of branches. However, immuno-depletion experiments indicated that the DU1 protein can account for nearly 30 % of the maize kernel soluble starch synthase activity (Cao et al., 2000). *SSII* mutants in wheat and pea and the production of antisense *SSII* potato tubers revealed starches with an increased proportion of short chains (<10).

Starch Branching Enzyme E.C. 2.4.1.18

Starch branching enzymes introduce α-1,6 linkages into starch. As with the synthases there are multiple isoforms. SBE1 isoforms transfer longer glucan chains and have a higher affinity for amylose while SBEII isoforms have a higher affinity for amylopectin and transfer shorter chains (Burton et al., 1995; Kawasaki et al., 1993; Mizuno et al., 2001; Morell et al., 1997; Takeda et al., 1993). In wheat, maize and pea embryos the *sbe2* gene, encoding SBEII protein, is expressed earlier in development and for longer than its *sbe1* counterpart (Hamada et al., 2002). The maize mutant, amylose extender (*ae*), confirms the role of SBEII isoforms in amylopectin synthesis, since its starch has a higher apparent amylose content and longer amylopectin chains. Recently, it has been shown that a novel sbe1c isoform in wheat can be regulated by protein phosphorylation (Tetlow et al., 2004). This provides a mechanism for regulating the branch frequency and chain length of the glucan polymer and also provides new targets for biotechnology.

In addition to the above enzymes, other enzymes such as debranching enzyme, disproportionating enzyme and isoamylase have been suggested to be involved in granule synthesis (Ball and Morell, 2003). Two models of amylose synthesis have been proposed and are supported with evidence from different species. In one, the amylose chains are extended from the growing starch granule and then cleaved to produce the free amylose chains present in the amorphous region. In the second, small malto-oligosaccharides are used as the primer (Ball *et al.*, 1998; Denyer *et al.*, 1996). The latter was originally proposed based on work with pea and potato and also suggests that GBSSI could be involved in amylopectin biosynthesis.

A current model for the synthesis of amylopectin (Ball *et al.*, 1996; Mouille *et al.*, 1996) involves the discontinuous elongation of glucan chains which, when long enough to accommodate branching enzymes, become highly branched. Debranching enzymes then selectively reduce the complexity of branching allowing further elongation. Clearly, there is still much to learn before we can define the correct roles and sequence of action of all the above enzymes. However, there are plenty of target enzymes which can be modified to further our understanding of the synthesis of the granule and provide modified starches for the marketplace.

Modification of Starch Quality

Starches with Altered Amylose

Amylose and amylopectin can be fractionated by dissolving the starch in water and letting the amylose diffuse out, but these methods are time-consuming and expensive. Thus, there has been a lot of effort to produce plants with native starch of the required properties.

As discussed above, the maize mutant *waxy* is the original low-amylose starch plant used for starch production (Whistler *et al.*, 1984). Pastes made with waxy (high amylopectin) starches have the great advantage of little retrogradation. However, they tend to have a rubbery texture not suitable for food applications and require improved stability. They are therefore usually cross-linked with agents that form covalent bonds between the glucan chains. At low levels of cross-linking this produces a paste with a lower viscosity at high temperatures but better properties for many food applications. At high levels of cross-linking, starches can be produced which do not gelatinize in boiling water.

Waxy forms have now been produced in all the species used for commercial starch production. However, waxy wheat and waxy potatoes were not achieved until the late 1980s and early 1990s (Hovenkamp-Hermelink *et al.*, 1987; Nakaimura *et al.*, 1995). Fully waxy soft wheats have starch with distinctive viscometric profiles, which may not be suitable for current applications but offer new opportunities (Kim *et al.*, 2003). However, partial waxy durum wheats have been found to be useful (Sharma *et al.*, 2002). For example, a wheat with the *wx* alleles on chromosome 4A, containing 5 % amylose, produced better pasta with lower cooking loss. This was correlated with a higher starch peak viscosity, greater swelling power and improved adhesive properties. White salted noodles are produced from wheat flour and a comparison between lines carrying the *waxy* mutation in the A, B and D genomes showed that both single null types, *wx*-B1 or *wx*-D1, had better mechanical and texture properties. Interestingly the double null, *wx*-A1/*wx*-D1, was best (Ishida *et al.*, 2003). Clearly there could be advantages in developing lines with small increments in the amount of amylose for specific purposes.

Visser *et al.* were the first to demonstrate the feasibility of altering starch type by genetic modification. They introduced an antisense clone of GBSSI into potato plants, producing a waxy type of potato starch (Visser *et al.*, 1991). A waxy potato starch has distinct advantages with an improved paste clarity and stability. Commercial lines of GM waxy potatoes are now available and have undergone extensive field tests. The GM approach provides two important benefits: firstly, it provides a means to control the level of expression of a gene and hence potentially the quantitative amount of amylose; secondly, the tetraploid nature of potato makes the introduction of a waxy phenotype into other varieties a complex process. Sense suppression has also been used in sweet potato to produce waxy mutants (Kimura *et al.*, 2001; Noda *et al.*, 2002).

As indicated above, while the improved freeze-thaw stability of waxy starches is an advantage, it is often not enough for many uses. Thus, it would be advantageous to produce more stable starches *in planta*, which do not need chemical modification. Recently, substantial progress has been made on this front by producing potato tubers with antisense constructs for starch synthase II and III as well as GBSSI (Jobling *et al.*, 2002). These triple antisense plants, which illustrate the true power and precision now possible with GM technology, lacked amylose and had shorter glucan chains in the amylopectin. The starch had altered viscometric profiles and excellent freeze-thaw stability. A similar starch can be produced in maize from a double mutant *wx/su-2* (Fergason and Wurzburg, 1984), but such maize starches are difficult to process. Interestingly, the defects that occur in the maize starch granules from double mutants were not present in the potato starch. To produce starches with the freeze-thaw parameters of the triple antisense potato plants normally requires chemical modifications. It is not clear how these novel properties were achieved but the authors hypothesize that the shorter amylopectin chains restricted reassociation of the glucan chains and therefore improved water retention. Thus, these new starches offer considerable environmental and consumer benefits.

High-Amylose Starches

High-amylose starches are also of considerable economic importance. The high gel strength of these starches makes them important in sweets and the film properties make them excellent coatings, especially for fried products. A high amylose content also produces 'resistant starch', which is now recognized to have some health benefits because it is not digested in the small intestine but fermented in the large intestine, producing compounds which reduce the risk of cancer (Brouns *et al.*, 2002).

In maize, high amylose starch was recognized in the *ae* mutants (Deatherage *et al.*, 1954). This gene encodes SBE-IIb (Hedman and Boyer, 1982). Antisense experiments in potato with an *SBE-A* gene, which on the basis of sequence analysis would encode an SBEII-type enzyme, showed that although it was a very minor component of branching enzyme activity it had a large effect on the starch, increasing the amylose content from 30% to 38.5%, increasing the phosphorus content and changing the amylopectin structure (Jobling *et al.*, 1999).

Previously, a clone for a potato SBEI had been introduced in both sense and antisense constructs into the *waxy* mutant of potato and, although it had no apparent effect on the starch structure, the physico-chemical properties were changed (Flipse *et al.*, 1996). To obtain very high amylose potato starches both SBEI and SBEII activities had to be inhibited (Schwall *et al.*, 2000). In this case, the branching enzyme activity was reduced to almost none. The starch in these plants had altered granule morphology, changed glucan molecular weight, altered branch length and increased phosphorus content. Since the chain length and the phosphate content of normal potato starches are greater than those of cereal starches, one can expect that these new starches will have some interesting functional properties. It is clear that even with our present limited knowledge, a wide range of novel starches could now be produced by systematically altering the amounts and proportions of the different branching enzymes and starch synthases.

Amylopectin Structure

Changes in amylopectin structure have been achieved in potato by introducing antisense constructs for SSII and SSIII isoforms separately and in combination (Edwards *et al.*, 1999). Reduction of either or both activities caused a change in granule morphology, although this was minor in the case of reduced SSII, and an increase in shorter glucan chains. In general, there was an increase in the number of short chains below about DP 12 and a decrease in the longer chains, although the individual transgenic lines had distinct patterns. The starch had a very low gelatinization temperature, which suggests some valuable properties for use in the food industry.

Recently, changes in amylopectin structure have been achieved in rice by alterations in SBEI (Satoh *et al.*, 2003) and possibly SSIIa (Umemoto *et al.*, 2002), although the latter needs confirmation. The introduction of a glycogen branching enzyme into the *amf* mutant of potato increased the number of branches and the number of short glucan chains in the amylopectin. This was associated with a lower peak viscosity and weaker gels (Kortstee *et al.*, 1996; 1998).

Covalent Modification

The only known natural covalent modification of starch is the attachment of phosphate to amylopectin chains. As already indicated above, the amount of phosphate can be altered by altering amylopectin and amylose contents and structure. The phosphate group is added to the amylopectin molecule by a water glucan dikinase (Ritte *et al.*, 2002). It is not understood what function the presence of phosphate serves in storage starches, such as in tubers, but in leaves, where transitory starch occurs, it appears to be involved in starch degradation (Ritte *et al.*, 2004). However, we know very little about how this dikinase enzyme works. Phosphate is attached at the C6, C2 and C3 positions of the glucose molecule (Nielsen *et al.*, 1994), which might suggest that more than one protein is involved. Also, the best substrate for the dikinase has a longer glucan chain length than the reported position of the phosphate in potato starch (Blennow *et al.*, 1998).

Commercially, the addition of more phosphate or other acidic groups to the starch molecule *in planta* would find many uses. Many enzymes have now been described from microorganisms, which could be used to modify starch *in vivo* and we have the knowledge to target them to the amyloplast. Such a biotechnological approach would avoid many of the chemical processes used to create the starches shown in Table 2.5.2.

Granule Size

As indicated above, granule size is an important factor for many applications, including thin films, paper coatings, cosmetic products and carbonless copy papers. However, we have very little knowledge on how this is controlled. The timing of expression of the cereal starch branching enzymes and starch synthases might affect it. Alternatively, granule size may be related to amyloplast size (Burrell and Coates, 2003), although our understanding of plastid division is still in its infancy. Recent observations on isoamylase mutants of barley suggest that the timing and expression of isoamylase genes may control granule initiation or size (Burton *et al.*, 2002), but in neither of these cases is it clear how the cell balances granule initiation and growth to control the final size and shape of the granule.

Conclusion

Henry Nix stated at a meeting of the Australian Urban and Regional Information Systems Association in 1990: 'Data does not equal information, information does not equal knowledge and, most importantly of all, knowledge does not equal wisdom'. We certainly have plenty of data on starch synthesis and some information. Clearly, we have a little knowledge on how to apply it to improve starch quality but we are lacking in wisdom on how to do this predictively.

References

Ball, K. and Preiss, J. (1994) Allosteric sites of the large subunit of the spinach leaf ADPglucose pyrophosphorylase. *Journal of Biological Chemistry* **269**, 24706–24711.

Ball, S., Guan, H.-P., James, M., Myers, A., Keeling, P., Mouille, G., Buleon, A., Colonna, P. and Preiss, J. (1996) From glycogen to amylopectin: a model for the biogenesis of the plant starch granule. *Cell* **86**, 349–352.

Ball, S.G. and Morell, M.K. (2003) From bacterial glycogen to starch: understanding the biogenesis of the plant starch granule. *Annual Review of Plant Biology* **54**, 207–233.

Ball, S.G., van de Wal, M.H.B.J. and Visser, R.G.F. (1998) Progress in understanding the biosynthesis of amylose. *Trends in Plant Science* **3**, 462–467.

Ballicora, M.A., Fu, Y., Nesbitt, N.M. and Preiss, J. (1998) ADP-glucose pyrophosphorylase from potato tubers. Site directed mutagenesis studies of the regulatory sites. *Plant Physiology* **118**, 265–274.

Banks, W. and Muir, D.D. (1980) Structure and chemistry of the starch granule, in *The Biochemistry of Plants* (ed J. Preiss), Academic Press, New York, pp. 321–369.

Blennow, A., Bay-Smidt, A.M., Wischmann, B., Olsen, C.E. and Moller B.L. (1998) The degree of starch phosphorylation is related to the chain length distribution of the neutral and the phosphorylated chains of amylopectin. *Carbohydrate Research* **307**, 45–54.

Boyer, C.D., Shannon, J.C, Garwood, D.L. and Creech, R.G. (1976) Changes in starch granule size and amylose percentage during kernel development in several *Zea mays* L. genotypes. *Cereal Chemistry* **53**, 327–337.

Brouns, F., Kettlitz, B. and Arrigoni, E. (2002) Resistant starch and "the butyrate revolution". *Trends in Food Science and Technology* **13**, 251–261.

Burrell, M.M. (2003) Starch: the need for improved quality or quantity—an overview. *Journal of Experimental Botany* **54**, 451–456.

Burrell, M.M. and Coates, S.A. (2003) Manipulation of starch granule size and number, in Patent Application WO2002GB04806 20021024, Gemstar Cambridge Ltd (GB), Europe.

Burrell, M.M., Tetlow, I.J., Burrell, M.M. and Emes, M.J. (2004) Post translational regulation of AGPase activity in wheat endosperm. *Comparative Biochemistry and Physiology* 134: S215

Burton, R.A., Bewley, J.D., Smith, A.M., Bhattacharyya, M.K., Tatge, H., Ring, S., Bull, V., Hamilton, W.D.O. and Martin, C. (1995) Starch branching enzymes belonging to distinct enzyme families are differentially expressed during pea embryo development. *The Plant Journal* **7**, 3–15.

Burton, R.A., Jenner, H., Carrangis, L., Fahy, B., Fincher, G.B., Hylton, C., Laurie, D.A., Parker, M., Waite, D., van Wegen, S., Verhoeven, T. and Denyer, K. (2002a) Starch granule initiation and growth are altered in barley mutants that lack isoamylase activity. *The Plant Journal* **31**, 97–112.

Burton, R.A., Johnson, P.E., Beckles, D.M., Fincher, G.B., Jenner, H.L., Naldrett, M.J. and Denyer, K. (2002b) Characterization of the genes encoding the cytosolic and plastidial forms of ADP-glucose pyrophosphorylase in wheat endosperm. *Plant Physiology* **130**, 1464–1475.

Cao, H.P., James, M.G. and Myers, A.M. (2000) Purification and characterization of soluble starch synthases from maize endosperm. *Archives of Biochemistry and Biophysics* **373**, 135–146.

Chojecki, A.J.S., Gale, M.D. and Bayliss, M.W. (1986) The number and sizes of starch granules in the wheat endosperm, and their association with grain weight. *Annals of Botany* **58**, 819–831.

Copeland, L. and Preiss, J. (1981) Purification of spinach leaf ADPglucose pyrophosphorylase. *Plant Physiology* **68**, 996–1001.

Corn Refiners Association Inc (2003) Corn Annual 2003. In.

Davydova, N.I., Leont'ev, S.P., Genin, Y.V., Sasov, A.Y. and Bogracheva, T.Y. (1995) Some physico-chemical properties of smooth pea starches. *Carbohydrate Polymers* **27**, 109–115.

Deatherage, W.L., MacMasters, M.M., Vineyard, M.L. and Bear, R.P. (1954) High amylose content from corn with high starch content. *Cereal Chemistry* **31**, 50.

Denyer, K., Clarke, B., Hylton, C., Tatge, H. and Smith, A.M. (1996) The elongation of amylose and amylopectin chains in isolated starch granules. *The Plant Journal* **10**, 1135–1143.

Dickinson, D.B. and Preiss, J. (1969) ADP glucose pyrophosphorylase from maize endosperm. *Archives of Biochemistry and Biophysics* **130**, 119–128.

Edwards, A., Fulton, D.C., Hylton, C.M., Jobling, S.A., Gidley, M., Rossner, U., Martin, C. and Smith, A.M. (1999) A combined reduction in activity of starch synthases II and III of potato has novel effects on the starch of tubers. *The Plant Journal* **17**, 251–261.

Fergason, V.L.U. and Wurzburg, O.B.U. (1984) Starch thickener characterized by improved low-temperature stability in National Starch Chemical Corporation (US), US, p. 9.

Flipse, E., Suurs, L., Keetels, C.J.A.M., Kossman, J., Jacobsen, E. and Visser, R.G.F. (1996) Introduction of sense and antisense cDNA for branching enzyme in the amylose-free potato mutant leads to physico-chemical changes in the starch. *Planta* **198**, 340–347.

Gibon, Y., Blasing, O.E., Palacios-Rojas, N., Pankovic, D., Hendriks, J.H.M., Fisahn, J., Hohne, M., Gunther, M. and Stitt, M. (2004) Adjustment of diurnal starch turnover to short days: depletion of sugar during the night leads to a temporary inhibition of carbohydrate utilization, accumulation of sugars and post-translational activation of ADP-glucose pyrophosphorylase in the following light period. *The Plant Journal* **39**, 847–862.

Haase, N.U. and Plate, J. (1996) Properties of potato starch in relation to varieties and environmental factors. *Starch-Starke* **48**, 167–171.

Hamada, S., Ito, H., Hiraga, S., Inagaki, K., Nozaki, K., Isono, N., Yoshimoto, Y., Takeda, Y. and Matsui, H. (2002) Differential characteristics and subcellular localization of two starch-branching enzyme isoforms encoded by a single gene in *Phaseolus vulgaris* L. *Journal of Biological Chemistry* **277**, 16538–16546.

Harn, C., Knight, M., Ramakrishnan, A., Guan, H.P., Keeling, P.L. and Wasserman, B.P. (1998) Isolation and characterization of the zSSIIa and zSSIIb starch synthase cDNA clones from maize endosperm. *Plant Molecular Biology* **37**, 639–649.

Hedman, K.D. and Boyer, C.D. (1982) Gene dosage at the amylose-extender locus of maize—effects on the levels of starch branching enzymes. *Biochemical Genetics* **20**, 483–492.

Hels, O., Hassan, N., Tetens, I. and Thilsted, S.H. (2003) Food consumption, energy and nutrient intake and nutritional status in rural Bangladesh: changes from 1981–1982 to 1995–1996. *European Journal of Clinical Nutrition* **57**, 586–594.

Hendriks, J.H.M., Kolbe, A., Gibon, Y., Stitt, M. and Geigenberger, P. (2003) ADP-glucose pyrophosphorylase is activated by post-translational redox-modification in response to light and to sugars in leaves of *Arabidopsis* and other plant species. *Plant Physiology* **133**, 838–849.

Hovenkamp-Hermelink, J.H.M., Jacobsen, E., Ponstein, A.S., Visser, R.G.F., Vos-Scheperkeuter, G.H., Bijmolt, E.W., de Vries, J.N., Witholt, B. and Feenstra, W.J. (1987) Isolation of an amylose-free starch mutant of the potato (*Solanum tuberosum* L.). *Theoretical and Applied Genetics* **75**, 217–221.

Ishida, N., Miura, H., Noda, T. and Yamauchi, H. (2003) Mechanical properties of white salted noodles from near-isogenic wheat lines with different wx protein-deficiency. *Starch-Starke* **55**, 390–396.

Jobling, S. (2004) Improving starch for food and industrial applications. *Current Opinion in Plant Biology* **7**, 210–218.

Jobling, S.A., Schwall, G.P., Westcott, R.J., Sidebottom, C.M., Debet, M., Gidley, M.J., Jeffcoat, R. and Safford, R. (1999) A minor form of starch branching enzyme in potato (*Solanum tuberosum* L.) tubers has a major effect on starch structure: cloning and characterisation of multiple forms of SBE A. *The Plant Journal* **18**, 163–171.

Jobling, S.A., Westcott, R.J., Tayal, A., Jeffcoat, R. and Schwall, G.P. (2002) Production of a freeze-thaw-stable potato starch by antisense inhibition of three starch synthase genes. *Nature Biotechnology* **20**, 295–299.

Kainuma, K. (1988) Structure and chemistry of the starch granule, in The Biochemistry of Plants (ed J. Preiss), Academic Press, New York, pp. 141–180.

Kawasaki, T., Mizuno, K., Baba, T. and Shimada, H. (1993) Molecular analysis of the gene encoding a rice starch branching enzyme. *Molecular and General Genetics* **237**, 10–16.

Kim, W., Johnson, J.W., Graybosch, R.A. and Gaines, C.S. (2003) Physicochemical properties and end-use quality of wheat starch as a function of waxy protein alleles. *Journal of Cereal Science* **37**, 195–204.

Kimura, T., Otani, M., Noda, T., Ideta, O., Shimada, T. and Saito, A. (2001) Absence of amylose in sweet potato *Ipomoea batatas* (L.) Lam. following the introduction of granule-bound starch synthase I cDNA. *Plant Cell Reports* **20**, 663–666.

Knight, M.E., Harn, C., Lilley, C.E.R., Guan, H.P., Singletary, G.W., Mu-Forster, C.M., Wasserman, B.P. and Keeling, P.L. (1998) Molecular cloning of starch synthase I from maize (W64) endosperm and expression in *Escherichia coli*. *The Plant Journal* **14**, 613–622.

Kortstee, A.J., Suurs, L., Vermeesch, A.M.G., Keetels, C., Jacobsen, E. and Visser, R.G.F. (1998) The influence of an increased degree of branching on the physicochemical properties of starch from genetically modified potato. *Carbohydrate Polymers* **37**, 173–184.

Kortstee, A.J., Vermeesch, A.M.S., deVries, B.J., Jacobsen, E. and Visser, R.G.F. (1996) Expression of *Escherichia coli* branching enzyme in tubers of amylose-free transgenic potato leads to an increased branching degree of the amylopectin. *The Plant Journal* **10**, 83–90.

Mizuno, K., Kobayashi, E., Tachibana, M., Kawasaki, T., Fujimura, T., Funane, K., Kobayashi, M. and Baba, T. (2001) Characterization of an isoform of rice starch branching enzyme, RBE4, in developing seeds. *Plant and Cell Physiology* **42**, 349–357.

Morell, M.K., Blennow, A., Kosar-Hashemi, B. and Samuel, M.S. (1997) Differential expression and properties of starch branching enzyme isoforms in developing wheat endosperm. *Plant Physiology* **113**, 201–208.

Mouille, G., Maddelein, M-L., Libessart, N., Talaga, P., Decq, A., Delrue, B. and Ball, S. (1996) Preamylopectin processing: a mandatory step for starch biosynthesis in plants. *The Plant Cell* **8**, 1353–1366.

Nachtergaele, W. and Vannuffel, J. (1989) Starch as a stilt material in carbonless copy paper—new developments. *Starch* **41**, 386–392.

Nakaimura, T., Yamamori, M., Hirano, H., Hidaka, S. and Nagamine, T. (1995) Production of waxy (amylose free) wheat. *Molecular and General Genetics* **248**, 253–259.

Nakamura, T., Vrinten, P., Hayakawa, K. and Ikeda, J. (1998) Characterization of a granule-bound starch synthase isoform found in the pericarp of wheat. *Plant Physiology* **118**, 451–459.

Nelson, O.E., Chourey, P.S. and Chang, M.T. (1978) Nucleoside diphosphate sugar–starch glucosyl transferase activity of wx starch granules. *Plant Physiology* **62**, 383–386.

Nielsen, T.H., Wischmann, B., Enevoldsen, K. and Moller, B.L. (1994) Starch phosphorylation in potato tubers proceeds concurrently with de novo biosynthesis of starch. *Plant Physiology* **105**, 111–117.

Noda, T., Kimura, T., Otani, M., Ideta, O., Shimada, T., Saito, A. and Suda, I. (2002) Physicochemical properties of amylose-free starch from transgenic sweet potato. *Carbohydrate Polymers* **49**, 253–260.

Preiss, J. (1988) Biosynthesis of starch and its regulation, in *The Biochemistry of Plants* (ed J. Preiss), Academic Press, New York, London, Toronto, pp. 181–254.

Preiss, J. and Levi, C. (1980) Starch biosynthesis and its degradation, in *The Biochemistry of Plants* (ed J. Preiss), Academic Press, New York.

Ritte, G., Lloyd, J.R., Eckermann, N., Rottmann, A., Kossmann, J. and Steup, M. (2002) The starch-related R1 protein is an alpha-glucan, water dikinase. *Proceedings of the National Academy of Sciences USA* **99**, 7166–7171.

Ritte, G., Scharf, A., Eckermann, N., Haebel, S. and Steup, M. (2004) Phosphorylation of transitory starch is increased during degradation. *Plant Physiology* **135**, 2068–2077.

Satoh, H., Nishi, A., Yamashita, K., Takemoto, Y., Tanaka, Y., Hosaka, Y., Sakurai, A., Fujita, N. and Nakamura, Y. (2003) Starch-branching enzyme I-deficient mutation specifically affects the structure and properties of starch in rice endosperm. *Plant Physiology* **133**, 1111–1121.

Schwall, G.P., Safford, R., Westcott, R.J., Jeffcoat, R., Tayal, A., Shi, Y.C., Gidley, M.J. and Jobling, S.A. (2000) Production of very high-amylose potato starch by inhibition of SBE A and B. *Nature Biotechnology* **18**, 551–554.

Sharma, R., Sissons, M.J., Rathjen, A.J. and Jenner, C.F. (2002) The null-4A allele at the wax locus in durum wheat affects pasta cooking quality. *Journal of Cereal Science* **35**, 287–297.

Sowokinos, J.R. and Preiss, J. (1982) Pyrophosphorylases in *Solanum tuberosum* III; purification, physical and catalytic properties of ADPGglucose pyrophosphorylase in potatoes. *Plant Physiology* **69**, 1459–1466.

Takeda, Y., Guan, H-P. and Preiss, J. (1993) Branching of amylose by the branching isoenzymes of maize endosperm. *Carbohydrate Research* **240**, 253–263.

Tester, R.F. and Karkalas, J. (2001) The effects of environmental conditions on the structural features and physico-chemical properties of starches. *Starch* **53**, 513–519.

Tester, R.F. and Karkalas, J. (2002) Starch, in *Biopolymers, volume 6, Polysaccharides II: Polysaccharides* from Eukaryotes (eds E.J. Vandamme, S. De Baets and A. Steinbüchel), Wiley, Weinheim, pp. 381–438.

Tetlow, I.J., Wait, R., Lu, Z.X., Akkasaeng, R., Bowsher, C.G., Esposito, S., Kosar-Hashemi, B., Morell, M.K. and Emes, M.J. (2004) Protein phosphorylation in amyloplasts regulates starch branching enzyme activity and protein–protein interactions. *Plant Cell* **16**, 694–708.

Tiessen, A., Hendriks, J.H.M., Stitt, M., Branscheid, A., Gibon, Y., Farre, E.M. and Geigenberger, P. (2002) Starch synthesis in potato tubers is regulated by post-translational redox modification of ADP-glucose pyrophosphorylase: a novel regulatory mechanism linking starch synthesis to the sucrose supply. *Plant Cell* **14**, 2191–2213.

Umemoto, T., Yano, M., Satoh, H., Shomura, A. and Nakamura, Y. (2002) Mapping of a gene responsible for the difference in amylopectin structure between japonica-type and indica-type rice varieties. *Theoretical and Applied Genetics* **104**, 1–8.

Visser, G.F., Stolte, A. and Jacobsen, E. (1991) Expression of a chimaeric granule bound starch synthase-GUS gene in transgenic potato plants. *Plant Molecular Biology* **17**, 691–699.

Whistler, R.L., Bemiller, J.N. and Paschall, E.F. (1984) *Starch Chemistry and Technology*, Academic Press, San Diego, New York, Boston, London, Sydney.

White, M.A. (1998) The chemistry behind carbonless copy paper. *Journal of Chemical Education* **75**, 1119–1120.

2.6

Production of Vaccines in GM Plants

Liz Nicholson, M. Carmen Cañizares and George P. Lomonossoff

John Innes Centre, Colney Lane, Norwich NR4 7UH, UK

Introduction

In recent years, plants have been proposed as an attractive alternative to the traditional production systems for vaccine production, such as animal cell culture. The production of recombinant proteins in plants has many potential advantages including: (i) it is more economical than traditional expression systems; (ii) proteins are post-translationally modified in a manner similar to that found in mammalian systems; (iii) plants offer the possibility of expressing proteins in different organs and cellular compartments, which may optimize protein accumulation; (iv) the amount of recombinant product that can be produced could be scaled up to industrial levels; (v) absence of risk contamination with animal pathogens and (vi) the possibility of producing edible vaccines.

The expression of foreign proteins in plants can be tackled using stable or transient genetic modification. Among the technologies developed so far, stable genetic modification includes both nuclear genetic transformation and the more recently developed plastid transformation approach. The transient option, on the other hand, is principally represented by *Agrobacterium*-mediated transient expression and the use of viral vectors.

In this chapter we will discuss the main types of plant expression systems considering the advantages and limitations of each system, and will provide an account of the types of antibodies and antigens for passive and active immunization respectively, expressed in plants using these technologies.

Plant Biotechnology. Edited by Nigel Halford.
© 2006 John Wiley & Sons, Ltd.

Main Technologies Used for Vaccine Production in Plants

Nuclear Genetic Modification

The most commonly used approach for the production of vaccines in plants is nuclear genetic modification. This involves the integration of the genes to be expressed into the nuclear genome of the selected plant species. The two principal transformation methods currently used are *Agrobacterium*-mediated gene transfer, mainly to dicotyledonous plants (dicots) such as tobacco and pea (Horsch *et al.*, 1985) and particle bombardment of genes to monocotyledonous plants (monocots), such as wheat, rice or maize (Christou, 1995).

To achieve high yields of the desired protein, the structure of the expression constructs needs to be optimized. To maximize transcription, a strong constitutive promoter is usually chosen. In the case of dicots, the cauliflower mosaic virus 35S (CaMV 35S) promoter is generally the preferred choice while for monocots the maize ubiquitin promoter is favored. The presence of introns in these sequences often increases the rate of transcription. Apart from these constitutive promoters, regulated promoters are increasingly being used. The advantage of using, for example, a tissue-specific promoter is that it would avoid the expression of the protein in vegetative tissues where it might affect plant growth and development. The use of inducible promoters (Padidam, 2003) allows the expression of the recombinant protein when it is required and is a potentially valuable approach in situations when premature protein overexpression would be deleterious to plant growth.

To maximize the translation of transgene-derived mRNAs, translational enhancers, such as the 5′ non-coding region derived from tobacco mosaic virus (TMV) RNA, are usually positioned immediately upstream of the coding region of the inserted gene (Gallie and Walbot, 1992). In addition, the translational start site is generally adjusted to match the Kozak consensus sequence for plants (Kawaguchi and Bailey-Serres, 2002). For some transgenes it has also proven advantageous to modify codon usage of the coding portion of the inserted gene so that it more closely resembles that of the host plant (Kusnadi, Nikolov and Howard, 1997).

To ensure that an expressed protein is appropriately post-translationally modified, and frequently to increase its yield, the sequence of the protein to be expressed can be flanked by signals that can target it to, and retain it within, the plant secretory pathway. The endoplasmic reticulum (ER) provides an environment where correct protein folding and assembly occurs and important post-translational modifications, such as glycosylation, occur only in the endomembrane system. To achieve targeting and retention, an N-terminal signal sequence (which directs proteins to the secretory pathway) and a C-terminal tetrapeptide signal, H/KDEL (which facilitates retrieval from the Golgi apparatus back to the ER; Fiedler *et al.*, 1997; Fischer *et al.*, 1999), are included in the expression construct. In the absence of an H/KDEL signal, the expressed protein is secreted to the apoplast.

Other factors that have an important effect on the level and stability of transgene expression are the position and number of integrations as well as the structure of the transgene locus. To try to minimize variation in transgene expression due to silencing phenomena, such as post-transcriptional gene silencing (PTGS) that are triggered by some of these factors, systems have been developed in which PTGS-impaired mutants or lines of plants overexpressing viral gene silencing suppressors are used for transformation (Butaye *et al.*, 2004; Mallory *et al.*, 2002).

Though the transformation of individual plants cells is reasonably straightforward, the regeneration of lines of fertile transgenic plants can be a time-consuming process which is species dependent. However, once produced, the transgenic lines can be considered to be essentially a plant variety and used in subsequent crossing experiments.

Plastid Genetic Modification

The development of methods for introducing foreign genes into the plastid genome has led to the development of transplastomic plants (Maliga, 2002). There are two major advantages of this approach. First, each cell of a plant contains multiple copies (10 000 in the case of tobacco) of the chloroplast genome per cell, meaning that very high levels of expression can potentially be obtained and levels up to 47 % of the total soluble protein have been reported (DeCosa *et al.*, 2001). Second, because the chloroplast genome is inherited maternally, pollen grains contain no transgenic DNA curtailing the ability of the transgene to spread in the environment and thus providing a level of biocontainment.

Biolistic DNA delivery is the system of choice for most laboratories involved in plastid transformation (Svab, Hajdukiewicz and Maliga, 1990). The integration of the plastid-targeting sequences of the transformation vector into the targeted region of the plastid genome occurs by homologous recombination. To achieve this, plastid vectors have left and right plastid-targeting regions of 1–2 kb that are homologous to the chosen target site.

To obtain high levels of transcription the strong plastid rRNA operon (rrn) promoter (Prrn), which is recognized by the plastid-encoded RNA polymerase, is mostly used. The resultant mRNA can be stabilized by flanking the coding sequence with 5′ and 3′ untranslated regions (5′-UTR and 3′-UTRs) from plastid genes, which typically include stem loop structures. The mRNA 5′-UTR includes sequences that facilitate loading of mRNAs onto ribosomes, facilitating the initiation of translation, whereas the mRNA 3′-UTR functions as an inefficient terminator of transcription (Monde, Schuster and Stern, 2000). The protein levels can be enhanced by including the amino-terminus of the coding region of the source plastid gene, in addition to the 5′-UTR, in the expression cassette (Kuroda and Maliga, 2001). The possible explanation for this is that the sequences around the AUG translation initiation codon have evolved together to ensure the translation of the encoded protein (Zerges, 2000).

Because the foreign gene insertion is at a precise and predetermined location in the plastid genome, the 'position effect' often observed in nuclear transformation is eliminated, resulting in uniform transgene expression among transgenic lines. In addition to this, the phenomenon of gene silencing, frequently observed in nuclear transgenic plants, has not been observed in transplastomic plants. Unfortunately there are also some disadvantages to the transplastomic approach. Apart from being a time-consuming process, as in the case of nuclear transformation, currently plastid transformation is only routine in a few species, though this number is increasing all the time. Perhaps more significant is the fact that glycosylation of proteins does not occur in plastids, so transplastomic technology, at least in its present form, is not suitable for the production of proteins which depend on glycosylation for their activity or stability.

Transient Expression via *Agrobacterium*

This method involves the infiltration of leaves with a suspension of *A. tumefaciens* harboring the construct of interest. The delivery of recombinant *Agrobacterium* into leaf tissue is usually carried out by vacuum infiltration resulting in multiple copies of the construct being delivered to the cells that have been infiltrated. Once *Agrobacterium* is present in the leaf, bacterial proteins transfer the gene of interest into the host cells, a process that is equivalent to the initial stages of nuclear genetic transformation (Section 2.1). However, the transferred T-DNA does not integrate into the host chromosome but is present in the nucleus, where it is transcribed leading to transient expression of the target gene (Kapila *et al.*, 1997). Expression of the introduced gene occurs rapidly, usually in a matter of a few days. The expression vector used for this transient assay can be the same as the ones designed for conventional nuclear transformation, and thus it is also possible to target the sequence of interest to different cellular compartments. The utility of this system for the expression and assembly of multi-component protein complexes has been demonstrated (Vaquero *et al.*, 1999). Though the method is very useful for small-scale experiments aimed at evaluating the properties of gene constructs prior to undertaking lengthy transformation/regeneration studies, scaling up the procedure presents a considerable challenge. However, there has been progress toward this goal by developing systems that either allow higher levels of transient expression to occur (Voinnet *et al.*, 2003) or permit the processing of large amounts of material (Fischer *et al.*, 2004).

Transient Genetic Modification via Viral Vectors

An alternative approach for transient expression involves the use of plant virus-based vectors. One of the main attractions of this approach is that viral genomes multiply within infected cells, potentially leading to very high levels of protein expression. Additional advantages include the fact that viral genomes are very small and therefore relatively easy to manipulate, and that the infection process is simpler than transformation/regeneration. There are, inevitably, disadvantages to the approach: the foreign gene is not heritable, there are limitations on the size and complexity of the sequences which can be expressed in a genetically stable manner and there are concerns about the ability of modified viruses to spread in the environment. Despite these difficulties, a number of different plant virus vectors have been used for protein expression.

Most plant viruses have genomes that consist of one or more strands of positive-sense RNA. As with animal RNA viruses, they use a variety of strategies for gene expression, including the use of subgenomic promoters and polyprotein processing. The availability of infectious cDNA clones was prerequisite for the development of RNA virus-based vectors (Ahlquist *et al.*, 1984). Since then, members of several virus families have been developed as useful vectors (Figure 2.6.1). Initial attempts at vector construction were based on gene replacement strategies in which a sequence encoding a non-essential viral function was replaced by a gene of interest. Though several of these early constructs could replicate in isolated plant cells, they generally could not systemically infect whole plants. A more successful approach has been the development of vectors based on gene addition. In these a foreign sequence is added to the complement of viral genes, rather than substituting for one.

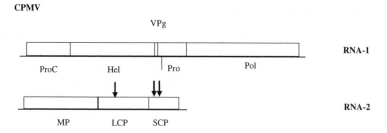

CPMV

VPg

RNA-1

ProC   Hel   Pro   Pol

RNA-2

MP   LCP   SCP

TMV

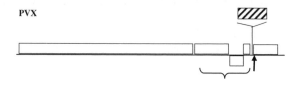

Pol

PVX

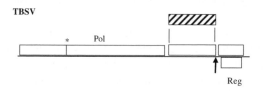

TBSV

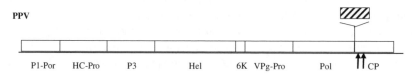

Pol

Reg

PPV

P1-Por   HC-Pro   P3   Hel   6K VPg-Pro   Pol   CP

Figure 2.6.1 *Genome organization of viruses used to express heterologous peptides and proteins in plants. Positions where epitopes have been inserted into the coat proteins of the various viruses are shown by black arrows. The positions where foreign proteins (shown hatched) have been inserted into the viral genomes are also indicated. The functions of various virus genes are shown as: CP, coat protein; HC-Pro, helper component proteinase; Hel, helicase; MP, movement protein; LCP, large coat protein; Pol, RNA-dependent RNA polymerase; Pro, proteinase; ProC, proteinase co-factor; Reg, regulatory protein; TGB, triple gene block; VPg, virus protein genome-linked; P1-Pro, P1-Proteinase; P3, protein P3; 6K, 6 kilodalton protein; SCP, small coat protein; VPg-Pro, VPg-Proteinase. * represents a leaky termination codon.*

Two basic types of systems have been developed for the production of immunogenic peptides and proteins in plants. The first type, often termed epitope presentation systems, involves inserting a sequence encoding an antigenic peptide into the viral coat protein gene in such a manner that the peptide is expressed on the surface of assembled virus particles. The modified virions are often referred to as chimeras or chimeric virus particles (CVPs). Such particles are attractive as potential novel vaccines, since the presentation of multiple copies of an antigenic peptide on the surface of a macromolecular assembly can significantly increase its immunogenicity (Lomonossoff and Johnson, 1996). With these systems it is generally anticipated that the modified particles will be at least partially purified prior to administration to animals. The second type, often referred to as polypeptide expression systems, involves introducing a whole gene into the viral genome in such a manner that it is efficiently expressed in infected cells, usually as an unfused polypeptide. Though purification of the expressed protein may be necessary or desirable, this type of system could be suitable for the production of immunogens that can be supplied orally by direct feeding of plant material to animals.

Choice of Plant Species for Expression

Much of the early work on the expression of foreign proteins was carried out in tobacco (*Nicotiana tabacum*) because of the relative ease with which this plant can be genetically modified and regenerated (Fischer and Emmans, 2000). In addition, there is considerable experience available concerning the large-scale growth of tobacco and a continuing interest in developing new uses for the crop. Although *Nicotiana* species are extremely useful as experimental systems, they suffer the disadvantage that the leaves contain high levels of nicotine and other alkaloids. Thus, any vaccine produced in these plants will have to be extensively purified before use. Consequently, several other leafy plant species have been investigated as production platforms, including lettuce, white clover, soybean, alfalfa and Arabidopsis. Unfortunately, the synthesis of recombinant proteins in the leaf's aqueous environment correlates with low protein stability in the harvested material. As a result, leaves must be frozen or dried for transport, or processed soon after harvest to extract useful amounts of the product.

However, when it is not necessary to synthesize the protein in the leaves of a plant, there are considerable advantages to directing expression to other organs, such as seeds, tubers or fruit. For instance, in contrast to leafy crops, the expression of proteins in seeds enables long-term storage, even at room temperature, because seeds have the appropriate biochemical environment to promote stable protein accumulation. Expression in cereal grains is particularly attractive as methods for the large-scale harvesting and storage of such material are well developed and the expressed proteins are stable over an extended period. Thus, different crops have been investigated for seed-based production, including cereals (maize, rice, wheat and barley) and legumes (pea and soybean). The genetic modification of vegetable crops such as tubers (potato), or roots (carrots) or fruits (tomatoes and bananas) is an attractive prospect. Indeed the possibility of consuming uncooked, unprocessed or partially processed material makes them ideal plant choices for the production of recombinant subunit vaccines and antibodies designed for topical and oral applications.

Along with traditional crop species, a number of more unusual plants with a variety of potential advantages are being investigated for vaccine production. These include the duckweed *Lemna minor* (Gasdaska, Spencer and Dickey, 2003), which is being developed by the US biotechnology company Biolex Inc, the unicellular green alga *Chlamydomonas rheinhardtii* which is amenable to nuclear, plastid and mitochondrial transformation (Franklin and Mayfield, 2004; Mayfield, Franklin and Lerner, 2003) and the moss *Physcomitrella patens* in which knockout mutants may be generated by homologous recombination, allowing efficient targeted gene disruption. Thus, the major drawback of producing human proteins in plants, allergic reactions caused by plant-specific glycosylation, can be diminished by targeted knockout of the responsible genes in moss, ending in the 'humanization' of plant-derived pharmaceuticals (Decker and Reski, 2004).

The successful application of viral vectors for the production of immunogens in plants has provided a valuable alternative approach to stable genetic engineering. Several plant species have been reported in the literature as being useful hosts for vaccine production, including *N. tabacum*, *N. clevelandii*, *Spinacia oleracea*, *Vigna unguiculata* and *N. benthamiana*, with the latter being widely used since it is a host for many different plant viruses.

Recombinant Subunit Vaccines

The term 'subunit vaccine' refers to the use of a pathogen-derived protein, or even just an immunogenic domain of a protein, often called an epitope, rather than the whole pathogen itself (killed or attenuated) to stimulate protective immunity. Subunit vaccines cannot cause disease and obviate the need to culture the pathogen. However, they suffer from the drawback that they may not elicit the entire range of immune responses necessary to provide effective immunity. The original concept behind the use of plants for the expression of subunit vaccines was based on the idea that the plant tissues could be fed to humans or commercially important animals. This method of vaccination would then elicit a mucosal immune response, which represents a first line of defense against most pathogens. Expression of proteins from a variety of microbial and viral pathogens showed the feasibility of transgenic plant expression systems for the production of subunit vaccines. The proteins expressed include those from the Hepatitis B surface antigen (HbsAg) in tobacco and lettuce (Ehsani, Khabiri and Domansky, 1997; Mason, Lam and Arntzen, 1992), a rabies virus antigen in tomato (McGarvey *et al.*, 1995), a cholera antigen in tobacco and potato (Arakawa *et al.*, 1997; Hein *et al.*, 1996), a recombinant bacterial antigen (LT-B) in tobacco and potato (Haq *et al.*, 1995), the Norwalk virus capsid protein (NVCP) in tobacco and potato (Mason *et al.*, 1996) and a human cytomegalovirus antigen in tobacco (Tackaberry *et al.*, 1999).

Interest in the production of subunit vaccines in plants was stimulated when, in 1992, the structural authenticity of HbsAg expressed in tobacco was confirmed, despite low levels of antigen being obtained (0.01 % of soluble leaf protein). Three years later, the immunogenicity of the tobacco-derived HBsAg was demonstrated by immunization of experimental mice (Thanavala *et al.*, 1995). To prove that plant-derived HBsAg can stimulate mucosal immune responses via the oral route, edible plants rather than tobacco were the focus of subsequent research into HBV vaccine production in plants. For instance, clinical trials carried out to test the efficacy of HbsAg expressed in lupin, potato and lettuce, in mice and humans (Kapusta *et al.*, 1999; Ritcher *et al.*, 2000; Thanavala *et al.*, 2005), found that the plant-expressed material was effective at eliciting immune responses to the immunogen supplied orally.

Two other vaccine candidates expressed in tobacco and potato have reached the stage of clinical trials: the heat-labile toxin B subunit (LT-B) of enterotoxigenic *Escherichia coli* (ETEC) and the NVCP (Haq *et al.*, 1995; Mason *et al.*, 1996). These antigens from two enteric pathogens represent ideal oral subunit vaccine candidates, as both are multimeric structures that survive in the extreme conditions of the human gut. Clinical trials in humans showed that consumption of raw potato tubers that contained 0.3–10 mg of LT-B produced high titers of mucosal and systemic antibodies (Tacket *et al.*, 1998). Subsequent studies demonstrated that when LT-B was delivered in transgenic corn meal, it elicited protective immunity (Tacket *et al.*, 2004). Similarly, a mucosal response was also observed in patients receiving potato tubers containing recombinant NVCP (Tacket *et al.*, 2000). In the case of veterinary vaccines, oral and parenteral immunogenicity of the foot and mouth disease virus (FMDV) antigen VP1 has been tested in mice using protein expressed in a variety of transgenic plants including Arabidopsis, potatoes and alfalfa (Carrillo *et al.*, 1998; 2001; Dus Santos *et al.*, 2005), and protection against challenge with virulent FMDV has been shown. In another example, the spike protein of swine transmissible gastroentirits virus (TGEV), expressed in corn seeds and fed to piglets, elicited protective immunity (Lamphear *et al.*, 2002; Streatfield *et al.*, 2001). Other major developments in the expression of subunit vaccines using nuclear or plastid genetic modification are listed in Table 2.6.1.

The viral approach has also been used for the transient expression of subunit vaccines, mainly veterinary ones. Exploiting both the epitope presentation and polypeptide expression approaches, a number of immunogenic peptides and proteins have been expressed in plants (Table 2.6.2). Examples reported so far have made use of vectors based on a number of different plant viruses including cowpea mosaic virus (CPMV), TMV, tomato bushy stunt virus (TBSV), alfalfa mosaic virus (AlMV), plum potyvirus (PPV) and potato virus X (PVX).

A large number of chimeras based on different viruses have been tested for immunogenic properties. In most cases, purified CVPs have been administered parent-erally, although there is increasing evidence that intranasal or oral immunization may also be effective. In a key study, it was found that parenteral admistration of a CPMV chimera (CPMV-PARVO1) that contained a 17-amino acid epitope from the N-terminal region of VP2 capsid protein of canine parvovirus gave protection against canine parvoviruses (CPV) in mink and dogs (Daalsgard *et al.*, 1997; Langeveld *et al.*, 2001). Similarly, Koo *et al.* (1999) were able to produce chimeric TMV particles presenting an epitope from mouse hepatitis virus (MHV). Mice immunized with purified virions produced antibodies against the MHV epitope, and those with high antibody titers were protected against challenge with the virus.

To overcome the limitation on the size of peptide that can be fused directly to the coat proteins of plant viruses without abolishing infectivity, Yusibov *et al.* (1997) developed a system where an appropriately modified version of the AlMV coat protein is expressed from an additional copy of the TMV coat protein subgenomic promoter. Using this approach, a 40-amino acid sequence containing epitopes from rabies virus and a

Table 2.6.1 Proteins with applications for human or animal vaccines and stably expressed by transgenic plants.

Protein	Host plant system	Method of genetic modification	Expression levels	Applications/immunological properties	Reference
Recombinant subunit vaccines					
Hepatitis B virus envelope surface protein (HBsAg)	*N. tabacum* Potato Lupin (*Lupinus* spp.) Lettuce Soybean cell culture	Nuclear	<0.01 % of fresh weight in all plant species. 67 µg/g fresh weight in soybean cell culture	Immunogenic when administered orally in mice and humans.	(Mason *et al.*, 1992) (Kapusta *et al.*, 1999; 2001) (Ritcher *et al.*, 2000) (Thanavala *et al.*, 2005) (Smith *et al.*, 2003)
Norwalk capsid protein	*N. tabacum* Potato	Nuclear	0.23 % of total soluble proteins (TSP) in *N. tabacum* 0.37 % TSP in potato	Virus-like particles are immunogenic when administered orally in mice and humans.	(Mason *et al.*, 1996) (Tacket *et al.*, 2000)
Enterotoxigenic *E. coli* heat-labile toxin B-subunit (LT-B) Non-toxic mutant of LT-B	*N. tabacum* Potato Corn	Nuclear Plastid	<0.01 % TSP (stable) and 3.7 % TSP (plastid) in *N. tabacum* 0.19 % TSP in potato 12 % TSP in corn seeds	Immunogenic and protective when administered orally. Potential adjuvant for oral immunization in humans.	(Haq *et al.*, 1995) (Mason *et al.*, 1998) (Tacket *et al.*, 1998) (Streatfield *et al.*, 2001; 2003) (Kang *et al.*, 2004a)
Cholera toxin (CT-B)	Potato *N. tabacum*	Nuclear Plastid	0.3 % TSP in potato as nuclear 4.1 % TSP in *N. tabacum* chloroplasts	Immunogenic and protective when administered orally. Potential adjuvant for mucosal vaccination.	(Arakawa *et al.*, 1997; 1998a) (Daniell *et al.*, 2001b)
Mouse GAD67	*N. tabacum* Potato	Nuclear	150 µg/g	Diabetes autoantigen for the treatment of an autoimmune disease	(Ma *et al.*, 1997)
Porcine epidemic diarrhea virus (PEDV) synthetic neutralizing epitope	*N. tabacum* *N. tabacum* without nicotine	Nuclear	10 µg/g wet weight in *N. tabacum* 2.1 % TSP in *N. tabacum* without nicotine	Transgenic plants fed to mice, induced both systemic and mucosal immune responses against the antigen.	(Bae *et al.*, 2003) (Kang *et al.*, 2005)
Glycoprotein of the rabies virus	Tomato	Nuclear	1.0 % TSP	Intact protein detected in plants	(McGarvey *et al.*, 1995)

Antigen	Plant	Expression	Expression level	Description	Reference
E2 glycoprotein from classical swine fever virus (CSFV)	Lettuce Alfalfa	Nuclear	160 μg/g dry mass in lettuce 10 μg/g dry mass in alfalfa	Oral route is effective in inducing a specific antibody response against the antigen in mice.	(Legocki et al., 2005)
Cysteine protease from Fasciola hepatica	Lettuce	Nuclear	100 μg/g dry mass	Oral route induces a specific antibody response against the antigen in mice.	(Legocki et al., 2005)
Tetanus toxin fragment C	N. tabacum	Plastid	25 % TSP	It allows the development of a safe, plant-based tetanus vaccine for nasal and oral applications.	(Tregoning et al., 2003)
Protective antigen (PA) of Bacillus anthracis	N. tabacum	Plastid	18.1 % TSP	The antigen assembled into heptamers and retained binding to host cell receptors.	(Watson et al., 2004)
F1-V fusion of Yersinia pestis	N. tabacum	Plastid	14.8 % TSP	Developed to control the spread of plague.	(Daniell et al., 2005)
Hemagglutinin protein from measles virus (MV-H)	N. tabacum	Nuclear	Not reported	Oral immunization of mice with plant-derived MV-H protein resulted in humoral and mucosal immune responses.	(Webster et al., 2005)
CTB-2L21(CPV) fusion	N. tabacum	Plastid	31.1 % TSP	Resulting chimera CTB-2L21 protein retained pentamerization and G_{M1}-ganglioside binding characteristics of the native cholera toxin B (CTB) and induced antibodies able to recognize VP2 protein from canine parvovirus (CPV).	(Molina et al., 2004)

(continued)

Table 2.6.1 (Continued)

Protein	Host plant system	Method of genetic modification	Expression levels	Applications/ immunological properties	Reference
CTB-insulin fusion	Potato	Nuclear	0.1 % TSP	Non-obese diabetic mice fed transformed potato tuber tissues showed a substantial reduction in pancreatic islet inflammation (insulitis) and a delay in the progression of clinical diabetes.	(Arakawa et al., 1998b)
CTB-rotavirus entero-toxin NSP4 fusion	Potato	Nuclear	0.01–0.1 % TSP	It assembles into oligomers that retained enterocyte receptor G_{M1} ganglioside binding affinity.	(Arakawa, Yu and Langridge, 2001)
CTB-rotavirus NSP4$_{175}$ fusion	Potato	Nuclear	0.006–0.026 % TSP	It has the potential to generate more protection against rotavirus infection than first-generation plant-based vaccines containing fewer rotavirus enterotoxin epitopes.	(Kim and Langridge, 2003)
CFA/I-CTA2-CTB-rota-virus enterotoxin NSP4 fusion	Potato	Nuclear	3 µg/g fresh weight	Orally immunized mice generated detectable levels of serum and intestinal antibodies against the pathogen.	(Yu and Langridge, 2001)

Product	Plant	Location	Expression level	Application	Reference
Glycoprotein S from transmissible gastroenteritis coronavirus (TGEV)	A. thaliana N. tabacum Corn	Nuclear	0.06 % TSP in A. thaliana 0.2 % TSP in N. tabacum <0.01 % fresh weight in corn.	Immunogenic when administered to pigs by injection and orally	(Gomez et al., 1998) (Tuboly et al., 2000) (Streatfield et al., 2001) (Lamphear et al., 2002)
Glycoprotein from human cytomegalovirus	N. tabacum	Nuclear	<0.02 % TSP	Immunologically active protein	(Tackaberry et al., 1999)
VP1 from foot and mouth disease virus	A. thaliana Alfalfa Potato	Nuclear	0.01 % TSP in alfalfa	Immunogenic and protective when administered to agricultural domestic animals and mice by injection.	(Carrillo et al., 1998) (Wigdorovitz et al., 1999a) (Carrillo et al., 2001) (Dus santos et al., 2005)
VP60 from rabbit hemorrhagic disease virus (RHDV)	Potato	Nuclear	0.3 % TSP	Immunogenic and protective when administered to rabbits by injection	(Castanon et al., 1999)
Antibodies					
SIgA/G (Guy's 13) Specific to *Streptococcus mutans* adhesin I/II.	N. tabacum Rice	Nuclear	500 µg/g fresh weight tobacco leaves	Therapeutic (topical) for the control of dental caries. Licensed to Planet Biotechnology CA., under the name of CaroRX.	(Ma et al., 1995; 1998; 2003) (Nicholson et al., 2005)
IgG Specific to herpes simplex virus 2 (HSV)	Soybean Rice	Nuclear	Not reported	Therapeutic (topical) for the control of sexually transmitted diseases. Licensed to EPIcyte Pharmaceutical.	(Zeitlin et al., 1998) (Briggs et al., 2001)
SIgA Specific to HSV	Corn	Nuclear	0.3 % TSP	Therapeutic for the control of herpex simplex virus	(Hood, Woodard, and Horn, 2002)
IgG Specific to R9 protein of respiratory syncytial virus	Corn	Nuclear	Not reported	Therapeutic (inhaled). Licensed to EPIcyte Pharmaceutical.	(Cone and Whaley)

(continued)

Table 2.6.1 *(Continued)*

Protein	Host plant system	Method of genetic modification	Expression levels	Applications/ immunological properties	Reference
IgG Specific to rabies virus	*N. tabacum*	Nuclear	0.07 % TSP	Therapeutic. Useful as a post-exposure prophylactic in hamsters.	(Ko et al., 2003)
IgG1 Specific to Rhesus D antigen	*Arabidopsis thaliana*	Nuclear	0.6 % TSP	Therapeutic. Used in the alloimmunization of RhD⁻ mothers carrying a RhD⁺ fetus.	(Bouquin et al., 2002)
Human IgG specific to the protective antigen (PA) of *Bacillus anthracis*	*N. benthamiana*	Transient by agroinfiltration	Not reported	Prophylactic. The resulting antibody was able to neutralize toxin activity *in vitro* and *in vivo*.	(Hull et al., 2005)
scFv-bryodin 1 immunotoxin fusion Specific to CD40.	*N. tabacum* cell culture	Nuclear	30 mg/L	Therapeutic for the control of non-Hodgkin's lymphoma.	(Francisco et al., 1997)
scFv-HER2 Specific to a breast oncogene.	*N. tabacum* *N. benthamiana*	Nuclear and transient by pseudo-viral infection.	Not reported	Therapeutic/diagnostic in the control of breast cancer.	(Galeffi et al., 2005)
scFv84.66, diabody84.66, and chimeric IgG1T84.66. Specific to the carcinoembryonic antigen (CEA)	Rice cell culture and plants Wheat Pea Tomato *N. tabacum*	Nuclear and transient by agroinfiltration	Rice (29 µg/g leaves, 32 µg/g seed, 3.8 µg/g callus) Wheat (0.9 µg/g leaves, 1.5 µg/g seed) *N. tabacum* (1 µg/g)	Therapeutic/diagnostic in the treatment of cancer	(Torres et al., 1999) (Stoger et al., 2000; 2002) (Perrin et al., 2000) (Vaquero et al., 1999; 2002)
Large single-chain Specific to g D of HSV.	*Chlamydomonas reinhardtii*	Plastid	0.5 % TSP	Therapeutic	(Mayfield et al., 2003)

Table 2.6.2 *Proteins with applications for human or animal vaccines and transiently expressed in plants using viral vectors.*

Viral approach	Expressed epitope or polypeptide	Plant used for infection	Species protected	Immunological properties	Reference
Epitope presentation system					
CPMV	VP2 capsid protein of canine parvovirus (CPV)	*Vigna unguiculata*	Mink	Parenteral injection protected mink against challenge with mink enteritis virus (MEV)	(Dalsgaard *et al.*, 1997)
CPMV	Canine parvovirus VP2	*V. unguiculata*	Dog	Ultraviolet light-inactivated form of the chimera protected dogs against challenge with canine parvovirus (CPV)	(Langeveld *et al.*, 2001)
CPMV	*Pseudomonas aeruginosa* outer membrane (OM) protein F	*V. unguiculata*	Mouse	Parenteral injection protected mice against challenge with *P. aeruginosa*	(Brennan *et al.*, 1999)
CPMV	*Staphylococcus aureus* D2 domain of fibronectin-binding protein (FnBP)	*V. unguiculata*	Rat	Parenteral application protected rats against endocarditis	(Rennermalm *et al.*, 2001)
TMV	Glycoprotein ZP3 from the murine zona pellucida	*N. tabacum*	Mouse	The modified virions were capable of eliciting antibodies in parenterally immunized mice	(Fitchen, Beacky and Hein, 1995)
TMV	Mouse hepatitis virus (MHV)	*N. tabacum*	Mouse	Mice parenterally and intranasally immunized were protected against challenge with MHV	(Koo *et al.*, 1999)
TMV	Foot and mouth disease virus (FMDV)	*N. tabacum*	Guinea pigs	Oral immunization was less effective than the parenteral injection against FMDV challenge	(Wu *et al.*, 2003)
TMV	*Pseudomonas aeruginosa* outer membrane (OM) protein F	*N. tabacum*	Mouse	Parenterally immunized mice were protected against challenge with *P. aeruginosa*	Staczek *et al.*, 2000)

(continued)

Table 2.6.2 (Continued)

Viral approach	Expressed epitope or polypeptide	Plant used for infection	Species protected	Immunological properties	Reference
TMV	*Pseudomonas aeruginosa* outer membrane (OM) protein F	N. tabacum	Mouse	Mice immunized with a mixture of a TMV chimera and a chimeric influenza virus, parenterally and nasally administered, respectively, produced antibodies against the two different epitopes expressed and were protected against challenge with *P. aeruginosa*.	(Gilleland et al., 2000)
TMV/AlMV	Rabies virus HIV-1	N. benthamiana Spinacia oleracea	Mouse	Both epitopes produced virus-neutralizing antibodies, being the rabies ones able to protect mice when supplied either intra-peritoneally or orally	(Yusibov et al., 1997) (Modelska et al., 1998)
TMV	Hepatitis C linked to the C-terminus of the CTB subunit	N. benthamiana	Mouse	Nasal administration of crude plant material elicited the production of antibodies against both epitopes.	(Nemchinov et al., 2000)
TBSV	gp 120 of HIV-1	N. benthamiana	Mouse	Parenteral immunization only gave a relatively low synthetic peptide specific primary antibodies	(Joelson et al., 1997)
PPV	VP2 capsid protein of canine parvovirus (CPV)	N. clevelandii	Mouse Rabbit	Parenteral immunization showed the presence of neutralizing antibodies against both CPV and PPV	(Fernandez-Fernandez et al., 1998)
PVX	gp41 of HIV-1	N. benthamiana	Mouse	Mice immunized intra-peritoneally or nasally produced high levels of HIV-1 IgG and IgA antibodies. Immunodeficient mice reconstituted with human peripheral lymphocytes made human primary neutralizing antibodies	(Marusic et al., 2001)

	virus (RSV) G protein	transgenic *N. tabacum*	...nkey	Parenteral injection of recombinant particles generated T- and B-cell responses. *In vitro* human dendritic cells incubated with the chimaera generated strong T-cell response	(Yusibov et al., 2005)
Polypeptide expression system					
TMV	VP1 from the FMDV	*N. benthamiana*	Mouse	Parenteral application in the presence of Freund's complete adjuvant protected all animals in two separate experiments.	(Wigdorovitz et al., 1999b)
TMV	38C13 scFv specific to the 38C13 mouse B cell lymphoma	*N. benthamiana*	Mouse	Mice vaccinated with the affinity purified 38C13 scFv generated >10 μg/mL anti-idiotype immunoglobulins. These mice were protected from challenge by a lethal dose of the syngeneic 38C13 tumor	(McCormick et al., 1999; 2003)
TMV	Betv1 a major birch pollen antigen	*N. benthamiana*	Mouse	Parenteral application with crude leaf extracts generated immunological responses comparable to those induced by the protein expressed in *E. coli* or extracted from birch pollen.	(Krebitz et al., 2000)
TMV	gDc from bovine herpes virus type 1 (BHV-1)	*N. benthamiana*	Mouse Cattle	Parenteral application of oil-based vaccines with crude extracts protected both animals	(Perez-Filgueira et al., 2003)
TMV	Synthetic neutralizing epitope (COE) of porcine epidemic diarrhea virus (PEDV)	*N. tabacum*	Not tested yet	High-level expression reaching 5% TSP, 30-fold higher than that obtained for the native COE gene	(Kang et al., 2004b)
TMV	CO17-1A IgG1 against colon cancer antigen	*N. benthamiana*	Not reported	Therapeutic/diagnostic	(Verch et al., 1998)
TMV	HIV-1 Tat protein	*N. benthamiana* *Spinacia oleracea*	300 μg/g	Mice fed with Tat-producing spinach did not show adverse effects on growth rate, and it primed mice for the subsequent DNA immunization with a Tat-expressing construct	(Karasev et al., 2005)

(continued)

Table 2.6.2 (Continued)

Viral approach	Expressed epitope or polypeptide	Plant used for infection	Species protected	Immunological properties	Reference
PVX	E7 from human papillomavirus 16 (HPV-16)	N. benthamiana	Mouse	Parenteral application with foliar extracts in the presence of Quil A as an adjuvant developed both antibody and cell-mediated immune responses	(Franconi et al., 2002)
PPV	VP60 from RHDV	N. clevelandii	Rabbit	Parenteral application with crude extracts in the presence of adjuvant fully protected rabbits against subsequent challenge with a lethal dose of RHDV	(Fernandez-Fernandez et al., 2001)
CPMV	Hepatitis B virus core antigen (HBcAg)	V. unguiculata	Not reported	Assembled HBcAg particles detected	(Mechtcheriakova et al., 2005)

47-amino acid sequence from HIV-1 were fused to the AlMV CP (Yusibov *et al.*, 1997). Both types of particles elicited the production of appropriate virus-neutralizing antibodies, the ones displaying the rabies virus epitopes being able to protect mice against a normally lethal challenge with the virus (Modelska *et al.*, 1998).

Transient expression of larger polypeptides has been reported using a number of different viral vectors. For example, TMV-based vectors have been used to express a number of antigenic proteins. These include VP1 from FMDV (Wigdorovitz *et al.*, 1999 a,b), which protected mice in two separate experiments, a major birch pollen antigen (Betv1) (Krebitz *et al.*, 2000) that generated immunological responses comparable to those induced by the protein expressed in *E. coli* or extracted from birch pollen, and the cytosolic form of the Bovine herpes virus gD protein (Perez-Filgueira *et al.*, 2003), which protected mice and cattle after challenge with the virulent BHV-1. PVX has been used for the expression of the E7 protein from the human papilloma virus 16. Mice immunized with E7-containing crude foliar extracts developed both humoral and cell-mediated immune responses and were protected from tumor development after challenge with the E7-expressing C3 tumoral cell line (Franconi *et al.*, 2002). Work carried out using a PPV-based vector for the expression of the VP60 structural protein from rabbit hemorrhagic disease virus (RHDV) (Fernandez-Fernandez *et al.*, 2001) found that immunized rabbits were protected against subsequent challenge with a lethal dose of RHDV.

Antibody Production in Plants

Antibodies are complex glycoproteins that recognize and bind to target antigens with great specificity. Because of their highly specific binding properties, antibodies have a number of uses both within and outside medicine. Within medicine, they have applications for both diagnosis and treatment, the latter including their use for vaccination. In most cases, antibodies are used as 'passive' vaccines, where they are used to eliminate a pathogen directly rather than to stimulate the recipient's own immune system. However, there are examples where the administration of antibodies or their fragments are used as active immunogens to stimulate the production of anti-idiotypic antibodies for the treatment of B-cell malignancies.

The production of recombinant antibodies in plants represents a special challenge because the molecules must fold and assemble correctly to recognize their cognate antigens. Full-length mammalian serum antibodies are complex molecules consisting of two identical heavy (H) and two identical light (L) chains held together by disulfide bridges. The variable regions that occur at the N-terminus of each chain determine the specificity of an antibody. Other more complex forms, such as secretory antibodies, are dimers of the typical serum antibody and include two additional polypeptides, the joining (J) chain and the secretory component (SC). Their assembly process requires two different cell types in mammals, but in plants it is achieved in one single cell (Ma *et al.*, 1995; Nicholson *et al.*, 2005). Because antigen recognition is determined entirely by the N-terminal variable domain of each chain, it is possible to produce smaller, less complex, antibody derivatives that still retain their specificity. Among those successfully expressed *in planta* are Fab fragments, single-chain Fv molecules (scFvs), bispecific Fvs, diabodies, minibodies, single variable domains, antibody-fusion proteins, large single-chain antibodies and camelid heavy-chain antibodies (reviewed in Ma, Drake and Christou, 2003). Antibody derivatives such as scFvs have the variable regions of the heavy and light chains of a given antibody fused by a flexible peptide linker. Such molecules are particularly useful for diagnostic and drug targeting purposes as their small size allows efficient penetration of tissue. However, they are less efficient at neutralizing infections as they are monovalent. Major developments in the expression of recombinant antibodies for vaccine production and therapeutic applications in plants are presented in Table 2.6.1.

Recombinant Full-Length Antibodies

The first attempt to produce functional antibodies in plants involved producing separate lines of *N. tabacum* plants transgenic for the light and heavy chains of a monoclonal antibody specific for a low-molecular-weight phosphonate ester, followed by crossing to obtain assembled IgG molecules (Hiatt, Cafferkey and Bowdish, 1989). The levels of assembled IgG that accumulated (up to 1.3 % total leaf protein) considerably exceeded the levels of the individual chains in the initial transformants, indicating that assembly is required for immunoglobulin (Ig) stability. Correct assembly was also obtained when both the heavy and light chains were expressed from a single construct (Düring *et al.*, 1990). Failure of the Ig chains to assemble correctly compromises antigen recognition and binding. In the vast majority of cases, individual light or heavy chains, or misfolded Ig molecules are not functional. In mammalian plasma cells, the mechanism of antibody assembly is partially understood. The Ig light and heavy chains are synthesized as precursor proteins, and signal sequences direct their translocation into the lumen of the ER. Within the ER, there is cleavage of the signal peptides, and stress proteins, such as BiP and GRP94, as well as enzymes such as protein disulfide isomerase (PDI), function as chaperones which bind to unassembled heavy and light chains and direct subsequent folding and assembly. In plants, passage of Ig chains through the ER is also necessary, as in the absence of a signal peptide related either to the light or heavy chain gene, assembly of antibody does not take place (Hiatt *et al.*, 1989). However, both plant and non-plant signal sequences from a variety of sources are sufficient for correct targeting (Düring *et al.*, 1990; Hein *et al.*, 1991). Plant chaperones homologous to mammalian BiP, GRP94 and PDI have been described within the ER (Denecke *et al.*, 1991; Fontes *et al.*, 1991), and expression of Ig chains in plants is indeed associated with increased BiP and PDI expression. Thus, it seems likely that there are broadly similar folding and assembly mechanisms for antibodies in mammals and plants.

Since these pioneering studies a number of full-length antibodies have been expressed in many different plant systems, mainly using the transgenic approach (Daniell *et al.*, 2001a; Fischer *et al.*, 2003; Khoudi *et al.*, 1999; Stoger *et al.*, 2002). One of the most impressive demonstrations of the utility of plants for antibody production concerned the production of functional secretory chimeric IgA/G molecules first in tobacco (Ma *et al.*, 1995) and later in rice (Nicholson *et al.*, 2005). This IgA was based on a monoclonal antibody, Guys 13, which binds to the surface protein of *Streptococcus mutans*, the causative agent of dental caries. Topical application of this monoclonal antibody had been shown to prevent colonization of teeth by *S. mutans*, offering a potential means of combating tooth decay, if the monoclonal could be produced in stable form at low cost.

To assemble the complete secretory antibody, separate tobacco lines individually transgenic for each of the four polypeptides (H, L, J and SC) were created and then crossed (Ma *et al.*, 1995). The quadruple transgenic plants accumulated significant quantities of assembled antibody, which was shown to be active in binding to *S. mutans* (Ma *et al.*, 1998). The Guy's 13 SIgA-G plantibody technology is licensed to Planet Biotechnology Inc (USA) and is currently in Phase II clinical trials under the product name CaroRx™. The trials have shown that topical application, after bacteria have been removed from the mouth, helps to prevent recolonization by *S. mutans* for several months (Ma *et al.*, 2003).

Another example of the use of plants to express functional antibodies concerned the successful expression of a full-length humanized version of IgG1 against herpes simplex virus (HSV)-2 glycoprotein B in transgenic soybean. The plant-expressed antibody had an activity identical to the original monoclonal antibody from which it was derived and could protect mice against vaginal transmission of HSV-2 (Zeitlin *et al.*, 1998). It has now reached advanced stages of development and would represent an inexpensive preventative for sexually transmitted diseases if it worked similarly in humans.

Though the transgenic approach has been the one most widely used for full-length antibody expression, there are some examples of such molecules being expressed transiently or from viral vectors. A full-length antibody, T84.66, which recognizes a carcinoembryonic antigen (CEA), has been transiently expressed in tobacco leaves using agroinfiltration (Vaquero *et al.*, 1999), and it has now been widely tested for cancer imaging and therapy. Likewise, a monoclonal antibody, CO17-1A, specific for a colon cancer antigen, has been produced by infecting *N. benthamiana* leaves with TMV-based constructs expressing the individual H and L chains (Verch, Yusibov and Koprowski, 1998).

Recombinant Antibody Derivatives

Because of their relative simplicity, antibody derivatives such as scFvs have attracted particular attention for plant-based expression using both transgenic and virus-based approaches. For instance, the expression of a scFv 84.66 against CEA has been studied in different plant systems, including rice cell cultures (Torres *et al.*, 1999), rice, wheat, tomato, pea and tobacco plants (Perrin *et al.*, 2000; Stoger *et al.*, 2000; Vaquero *et al.*, 2002), to demonstrate and compare objectively the expression efficiency and product quality. The study also confirmed that the levels of scFvs in seeds did not show a significant decline after storage at room temperature after 6 months, while, in another study, full antigen-binding activity was retained for as long as 3 years at refrigerator temperature.

Numerous laboratories have shown transgenic protein accumulation in seeds including tobacco (Fiedler and Conrad, 1995), corn (Hood and Jilka, 1999; Russell, 1999) or soybean (Zeitlin *et al.*, 1998). A fusion protein that combines scFv84.66 and interleukin-2 has also been produced and could be used to stimulate lymphokine-activated killer cells and tumor-infiltrating lymphocytes near tumor masses (Ma *et al.*, 2003). One of the few examples using a viral approach concerned the expression of scFv molecules corresponding to an antibody expressed on the surface of a mouse lymphoma cell line in *N. benthamiana* using a TMV vector (McCormick *et al.*, 1999). The plant-produced scFv was able to stimulate the production of anti-idiotypic antibodies, which were able to protect mice against challenge with the lymphoma cell line 38C13. This system could be adapted to produce antibodies that recognize unique markers on the surface of any malignant B cell and could therefore be an effective therapy for human diseases such as non-Hodgkin lymphoma. The antibodies were produced using virus-infected plants based on TMV rather than transgenic plants, which is a strategy that is well suited to the rapid and small-scale production that is required to treat individual patients with unique antibodies. Large Scale Biology Corporation (LSBC) has completed phase I trials on such material.

Prospects

As a general expression system for recombinant proteins, plants are highly competitive with other technologies. This includes the production and purification of vaccines for parenteral administration. Though there have been several laboratory-scale experiments which have demonstrated that plant-expressed peptides and proteins can provide protective immunity, there are a number of practical and regulatory issues which will have to be addressed if their use is to become a practical reality. These include the scaling-up of production facilities, addressing any safety concerns about the unwanted spread of transgenic plants or modified viruses and the demonstration of equivalence between plant-expressed proteins and those produced by more conventional means (Miele, 1997).

In the specific area of oral vaccines, the added advantage of using edible plants brings with it its own challenges: the control of dose, the requirement for an adjuvant and the overall control of the immune response. The induction of immune responses following oral immunization is frequently dependent upon the co-administration of appropriate adjuvants that can initiate and support the transition from innate to adaptive immunity. The three bacterial products with the greatest potential to function as mucosal adjuvants are the ADP-ribosylating enterotoxins (cholera toxin (CT) and the heat-labile enterotoxin of *E. coli*), synthetic oligodeoxynucleotides containing unmethylated CpG dinucleotides (CpG ODN) and monophosphoryl lipid A (MPL) (Freytag and Clements, 2005). However, because of their extremely high toxicity, their use in humans is limited. During the past few years, site-directed mutagenesis has permitted the generation of LT and CT mutants which are either fully non-toxic or have dramatically reduced toxicity, but which still retain their strong adjuvanticity at the mucosal level. Two mutants, LTK63 (fully non-toxic) and LTR72 (with some residual enzymatic activity) have been tested as mucosal adjuvants in different animal species using a wide variety of antigens. Mucosal delivery (nasal or oral) of antigens together with LTK63 or LTR72 mutants also conferred protection against challenge in appropriate animal models (e.g. tetanus, *Helicobacter pylori*, pertussis, pneumococci, influenza, etc.) (Pizza *et al.*, 2001).

Making use of plastidial transformation, LTK63 has been expressed in tobacco chloroplasts, reaching up to 3.7 % of total soluble proteins (Kang *et al.*, 2004a). Other developments in plant-derived vaccines include the second generation of edible vaccines, multi-component vaccines, aiming to provide protection against several pathogens. In this context, Yu and Langridge (2001) have expressed the rotavirus enterotoxin and the

ETEC fimbrial antigen fused to both subunits, B and A2, of the CT in potato. In this approach, CT provides a scaffold for presentation of protective epitopes of rotavirus and ETEC, acting as a vaccine candidate in its own right and as a mucosal adjuvant devoid of toxicity. The trivalent edible vaccine, when fed to mice, elicited significant humoral responses, as well as immune memory B cells and T-helper cell responses, important hallmarks of successful immunization. Other recent examples are listed in Table 2.6.1.

A further difficulty however concerns the control of the immune response and in particular the avoidance of oral tolerance, first described by Wells and Osbourne in 1910 (reviewed in Ma, 2001). Oral tolerance has been described as a state of systemic unresponsiveness following mucosal administration of an antigen. This is obviously a rather undesirable adverse reaction to mucosal immunization, but its avoidance is not completely understood (Hershberg and Mayer, 2000). To some extent, controlling antigen dosage and the use of non-toxic derivatives of CT or LT as mucosal adjuvants will address this issue, but the possibility of oral tolerance probably remains one of the major concerns regarding the development of any oral vaccine, whether it be derived from plants or not.

Acknowledgments

The authors acknowledge the support of funding from the BBSRC and from the EC under the Framework 5 Quality of Life Program (contract no. QLK2-CT-2002–01050) and the Framework 6 'PharmaPlanta' Project.

References

Ahlquist, P., French, R., Janda, M. and Loesch-Fries, S. (1984) Multicomponent RNA plant virus infection derived from cloned viral cDNA. *Proceedings of the National Academy of Sciences USA* **81**, 7066–7070.

Arakawa, T., Chong, D.K. and Langridge, W.H. (1998a) Efficacy of a food plant-based oral cholera toxin B subunit vaccine. *Nature Biotechnology* **16**, 292–297.

Arakawa, T., Yu, J., Chong, D.K., Hough, J., Engen, P.C. and Langridge, W.H. (1998b) A plant-based cholera toxin B subunit-insulin fusion protein protects against the development of autoimmune diabetes. *Nature Biotechnology* **16**, 934–938.

Arakawa, T., Chong, D.K.X., Merritt, J.L. and Langridge, W.H.R. (1997) Expression of cholera toxin B subunit oligomers in transgenic potato. *Transgenic Research* **6**, 403–413.

Arakawa, T., Yu, J. and Langridge, W.H.R. (2001) Synthesis of a cholera toxin B subunit-rotavirus NSP4 fusion protein in potato. *Plant Cell Reports* **20**, 343–348.

Bae, J-L., Lee, J-G., Kang, T-J., Jang, H-S., Jang, Y-S. and Yang, M-S. (2003) Induction of antigen-specific systemic and mucosal immune responses by feeding animals transgenic plants expressing the antigen. *Vaccine* **21**, 4052–4058.

Bouquin, T., Thomsen, M., Nielsen, L.K., Green, T.H., Mundy, J. and Dziegiel, M.H. (2002) Human anti-Rhesus D IgG1 antibody produced in transgenic plants. *Transgenic Research* **11**, 115–122.

Brennan, F.R., Jones, T.D., Gilleland, L.B., Bellaby, T., Xu, F., North, P.C., Thompson, A., Staczek, J., Lin, T., Johnson, J.E., Hamilton, W.D. and Gilleland, H.E., Jr. (1999) *Pseudomonas aeruginosa* outer-membrane protein F epitopes are highly immunogenic in mice when expressed on a plant virus. *Microbiology* **145**, 211–220.

Briggs, K., Zeitlin, L., Wang, F., Chen, L., Fitchen, J., Glynn, J., Lee, V., Zhang, S. and Whalley, K. (2001) An anti-HSV antibody produced in transgenic rice plants prevents vaginal HSV-2 infection in mice. *AIDS* **15**, S19–S20.

Butaye, K.M.J., Goderis, I.J.W.M., Wouters, P.F.J., Pues, J.M.T.G., Delaure, S.L., Broekaert, W.F., Depicker, A., Cammue, B.P.A. and De Bolle, M.F.C. (2004) Stable high-level transgene expression in *Arabidopsis thaliana* using gene silencing mutants and matrix attachment regions. *Plant Journal* **39**, 440–449.

Carrillo, C., Wigdorovitz, A., Oliveros, J.C., Zamorano, P.I., Sadir, A.M., Gomez, N., Salinas, J., Escribano, J.M. and Borca, M.V. (1998) Protective immune response to foot-and-mouth disease virus with VP1 expressed in transgenic plants. *Journal of Virology* **72**, 1688–1690.

Carrillo, C., Wigdorovitz, A., Trono, K., Dus Santos, M.J., Castanon, S., Sadir, A.M., Ordas, R., Escribano, J.M. and Borca, M.V. (2001) Induction of a virus specific antibody response to foot and mouth disease virus using the structural protein VP1 expressed in transgenic potato plants. *Viral Immunology* **14**, 49–57.

Castanon, S., Marin, M.S., Martin-Alonso, J.M., Boga, J.A., Casais, R., Humara, J.M., Ordas, R.J. and Parra, F. (1999) Immunization with potato plants expressing VP60 protein protects against rabbit hemorrhagic disease virus. *Journal of Virology* **73**, 4452–4455.

Christou, P. (1995) Particle bombardment. *Methods in Cell Biology* **50**, 375–382.

Cone, R.A. and Whaley, K.J. Topical application of antibodies for contraception and for prophylaxis against sexually transmitted diseases. *US patent* 6; 355, 235.

Dalsgaard, K., Uttenthal, Å., Jones, T.D., Xu, F., Merryweather, A., Hamilton, W.D., Langeveld, J.P., Boshuizen, R.S., Kamstrup, S., Lomonossoff, G.P., Porta, C., Vela, C., Casal, J.I., Meloen, R.H. and Rodgers, P.B. (1997) Plant-derived vaccine protects target animals against a viral disease. *Nature Biotechnology* **15**, 248–252.

Daniell, H., Chebolu, S., Kumar, S., Singleton, M. and Falconer, R. (2005) Chloroplast-derived vaccine antigens and other therapeutic proteins. *Vaccine* **23**, 1779–1783.

Daniell, H., Lee, S.B., Panchal, T., Wiebe and P.O. (2001b) Expression of the native cholera toxin B subunit gene and assembly as functional oligomers in transgenic tobacco chloroplasts. *Journal of Molecular Biology* **311**, 1001–1009.

Daniell, H., Streatfield, S.J. and Wycoff, K. (2001a) Medical molecular farming: production of antibodies, biopharmaceuticals and edible vaccines in plants. *Trends in Plant Science* **6**, 219–226.

De Cosa, B., Moar, W., Lee, S-B., Miller, M. and Daniell, H. (2001) Overexpression of the Bt cry2As2 operon in chloroplasts leads to formation of insecticidal crystals. *Nature Biotechnology* **19**, 71–74.

Decker, E.L. and Reski, R. (2004) The moss bioreactor. *Current Opinion in Plant Biology* **7**, 166–170.

Denecke, J., Goldman, M.H.S., Demolder, J., Seurinck, J. and Botterman, J. (1991) The tobacco luminal binding-protein is encoded by a multigene family. *Plant Cell* **3**, 1025–1035.

Düring, K., Hippe, S., Kreuzaler, F. and Schell, J. (1990) Synthesis and self-assembly of a functional monoclonal antibody in transgenic *Nicotiana tabacum*. *Plant Molecular Biology* **15**, 281–293.

Dus Santos, M.J., Carrillo, C., Ardila, F., Rios, R.D., Franzone, P., Piccone, M.E., Wigdorovitz, A. and Borca, M.V. (2005) Development of transgenic alfalfa plants containing the foot and mouth disease virus structural polyprotein gene *P1* and its utilization as an experimental immunogen. *Vaccine* **23**, 1838–1843.

Ehsani, P., Khabiri, A. and Domansky, N.N. (1997) Polypeptides of hepatitis B surface antigen in transgenic plants. *Proceedings of the National Academy of Sciences USA* **190**, 107–111.

Fernandez-Fernandez, M.R., Martinez-Torrecuadrada, J.L., Casal, J.I. and Garcia, J.A. (1998) Development of an antigen presentation system based on plum pox potyvirus. *FEBS Letters* **427**, 229–235.

Fernandez-Fernandez, M.R., Mourino, M., Rivera, J., Rodriguez, F., Plana-Duran, J. and Garcia, J.A. (2001) Protection of rabbits against rabbit hemorrhagic disease virus by immunization with the VP60 protein expressed in plants with a potyvirus-based vector. *Virology* **280**, 283–291.

Fiedler, U. and Conrad, U. (1995) High-level production and long-term storage of engineered antibodies in transgenic tobacco seeds. *Biotechnology* **13**, 1090–1093.

Fiedler, U., Philips, J., Artsaenko, O. and Conrad, U. (1997) Optimization of scFv antibody production in transgenic plants. *Immunotechnology* **3**, 205–216.

Fischer, R. and Emmans, N. (2000) Molecular pharming of pharmaceutical proteins. *Transgenic Research* **9**, 279–299.

Fischer, R., Liao, Y.C. and Drossard, J. (1999) Affinity-purification of a TMV-specific recombinant full-size antibody from a transgenic tobacco suspension culture. *Journal of Immunological Methods* **226**, 1–10.

Fischer, R., Stoger, E., Schillberg, S., Christou, P. and Twyman, R.M. (2004) Plant-based production of biopharmaceuticals. *Current Opinion in Plant Biology* **7**, 152–158.

Fischer, R., Twyman, R.M. and Schillberg, S. (2003) Production of antibodies in plants and their use for global health. *Vaccine* **21**, 820–825.

Fitchen, J., Beachy, R.N. and Hein, M.B. (1995) Plant virus expressing hybrid coat protein with added murine epitope elicits autoantibody response. *Vaccine* **13**, 1051–1057.

Fontes, E.B.P., Shank, B.B., Wrobel, R.L., Moose, S.P., O'Brian, G.R., Wurtzel, E.T. and Boston, R.S. (1991) Characterization of an immunoglobulin binding-protein homolog in the maize floury-2 endosperm mutant. *Plant Cell* **3**, 483–496.

Francisco, J.A., Gawlak, S.L., Miller, M., Bathe, J., Russell, D., Chace, D., Mixan, B., Zhao, L., Fell, H.P. and Siegall, C.B. (1997) Expression and characterization of bryodin 1 and a bryodin 1-based single-chain immunotoxin from tobacco cell culture. *Bioconjugate Chemistry* **8**, 708–713.

Franconi, R., Di Bonito, P., Dibello, F., Accardi, L., Muller, A., Cirilli, A., Simeone, P., Dona, M.G., Venuti, A. and Giorgi, C. (2002) Plant derived-human papillomavirus 16 E7 oncoprotein induces immune response and specific tumour protection. *Cancer Research* **62**, 3654–3658.

Franklin, S.E. and Mayfield, S.P. (2004) Prospects for molecular farming in the green alga *Chlamydomonas reinhardtii*. *Current Opinion in Plant Biology* **7**, 1–7.

Freytag, L.C. and Clements, J.D. (2005) Mucosal adjuvants. *Vaccine* **23**, 1804–1813.

Galeffi, P., Lombardi, A., Donato, M.D., Latini, A., Sperandei, M., Cantale, C. and Giacomini, P. (2005). Expression of single-chain antibodies in transgenic plants. *Vaccine* **23**, 1823–1827.

Gallie, D.R. and Walbot, V. (1992) Identification of the motifs within the tobacco mosaic virus 5′-leader responsible for enhancing translation. *Nucleic Acids Research* **20**, 4631–4638.

Gasdaska, J.R., Spencer, D. and Dickey, L. (2003) Advantages of therapeutic protein production in the aquatic plant *Lemna*. *Bioprocessing Journal* **2**, 49–56.

Gilleland, H.E., Gilleland, L.B., Straczek, J., Harty, R.N., Garcia-Sastre, A., Palese, P., Brennan, R.F., Hamilton, W.D., Bendahmane, M. and Beacky, R.N. (2000) Chimeric animal and plant viruses expressing epitopes of outer membrane protein F as a combined vaccine against *Pseudomonas aeruginosa* lung infection. *FEMS Immunology and Medical Microbiology* **27**, 291–297.

Gomez, N., Carrillo, C., Salinas, J., Parra, F., Borca, M.V. and Escribano, J.M. (1998) Expression of immunogenic glycoprotein S polypeptides from transmissible gastroenteritis coronavirus in transgenic plants. *Virology* **249**, 352–358.

Haq, T.A., Mason, H.S., Clements, J.D. and Arntzen, C.J. (1995) Oral immunization with a recombinant bacterial antigen produced in transgenic plants. *Science* **268**, 714–716.

Hein, M.B., Tang, Y., McLeod, D.A., Janda, K.D. and Hiatt, A.C. (1991) Evaluation of immunoglobulins from plant cells. *Biotechnology Progress* **7**, 455–461.

Hein, M.B., Yeo, T-C., Wang, F. and Sturtevant, A. (1996) Expression of cholera toxin subunits in plants. *Annals of the New York Academy of Sciences* **792**, 50–56.

Hershberg, R.M. and Mayer, L.F. (2000) Antigen processing and presentation by intestinal epithelial cells—polarity and complexity. *Immunology Today* **21**, 123–128.

Hiatt, A., Cafferkey, R. and Bowdish, K. (1989) Production of antibodies in transgenic plants. *Nature* **342**, 76–78.

Hood E.E., Woodard, S.L. and Horn, M.E. (2002) Monoclonal antibody manufacturing in transgenic plants—myths and realities. *Current Opinion in Biotechnology* **13**, 630–635.

Hood, E.E. and Jilka, J.M. (1999) Plant-based production of xenogenic proteins. *Current Opinion in Biotechnology* **10**, 382–386.

Horsch, R.B., Fry, J.E., Hoffmann, N.L., Eichholtz, D., Rogers, S.G. and Fraley, R.T. (1985) A simple and general method for transferring genes into plants. *Science* **227**, 1229–1231.

Hull, A.K., Criscuolo, C.J., Mett, V., Groen, H., Steeman, W., Westra, H., Chapman, G., Legutki, B., Baillie, L. and Yusibov, V. (2005) Human-derived, plant-produced monoclonal antibody for the treatment of anthrax. *Vaccine* **23**, 2082–2086.

Joelson, T., Åkerblom, L., Oxelfelt, P., Strandberg, B., Tomenius, K. and Morris, T.J. (1997) Presentation of a foreign peptide on the surface of tomato bushy stunt virus. *Journal of General Virology* **78**, 1213–1217.

Kang, T-J., Han, S-C., Kim, M-Y., Kim, Y-S. and Yang M-S. (2004a) Expression of non-toxic mutant of *Escherichia coli* heat-labile enterotoxin in tobacco chloroplasts. *Protein Expression and Purification* **38**, 123–128.

Kang, T-J., Kang, K-H., Kim, J-A., Kwon, T-H., Jang, Y-S. and Yang, M-S. (2004b) High-level expression of the neutralizing epitope of porcine epidemic diarrhea virus by a tobacco mosaic virus-based vector. *Protein Expression and Purification* **38**, 129–135.

Kang, T-J., Kim, Y-S., Jang, Y-S. and Yang, M-S. (2005) Expression of the synthetic neutralizing epitope gene of porcine epidemic diarrhea virus in tobacco plants without nicotine. *Vaccine* **23**, 2294–2297.

Kapila, J., De Rycke, R., Van Montagu, M. and Angenon, G. (1997) An *Agrobacterium*-mediated transient gene expression system for intact leaves. *Plant Science* **122**, 101–108.

Kapusta, A., Modelska, A., Figlerowicz, M., Pniewski, T., Letellier, M., Lisowa, O., Yusibov, V., Koprowski, H., Plucienniczak, A. and Legocki, A.B. (1999) A plant-derived edible vaccine against hepatitis B virus. *FASEB Journal* **13**, 1796–1799.

Kapusta, J., Modelska, A., Pniewski, T., Figlerowicz, M., Jankowski, K., Lisowa, O., Plucienniczak, A., Koprowski, H. and Legocki, A.B. (2001) Oral immunization of human with transgenic lettuce expressing hepatitis B surface antigen. *Advances in Experimental Medicine and Biology* **495**, 299–303.

Karasev, A.V., Foulke, S., Wellens, C., Rich, A., Shon, K.J., Zwierzynski, I., Hone, D., Koprowski, H. and Reitz, M. (2005) Plant based HIV-1 vaccine candidate: Tat protein produced in spinach. *Vaccine* **23**, 1875–1880.

Kawaguchi, R. and Bailey-Serres, J. (2002) Regulation of translational initiation in plants. *Current Opinion in Plant Biology* **5**, 460–465.

Khoudi, H., Laberge, S., Ferullo, J.M., Bazin, R., Darveau, A., Castonguay, Y., Allard, G., Lemieux, R. and Vezina, L.P. (1999) Production of a diagnostic monoclonal antibody in perennial alfalfa plants. *Biotechnology and Bioengineering* **64**, 135–143.

Kim, T.G. and Langridge, W.H.R. (2003) Assembly of cholera toxin B subunit full-length rotavirus NSP4 fusion protein oligomers in transgenic potato. *Plant Cell Reports* **21**, 884–890.

Ko, K., Tekoah, Y., Rudd, P.M., Harvey, D.J., Dwek, R.A., Spitsin, S., Hanlon, C.A., Rupprecht, C., Dietzschold, B., Golovkin, M. and Koprowski, H. (2003) Function and glycosylation of plant-derived antiviral monoclonal antibody. *Proceedings of the National Academy of Sciences USA* **100**, 8013–8018.

Koo, M., Bendahmane, M., Lettieri, G.A., Paoletti, A.D., Lane, T.E., Fitchen, J.H., Buchmeier, M.J. and Beachy, R.N. (1999) Protective immunity against murine hepatitis virus (MHV) induced by intranasal or subcutaneous administration of hybrids of tobacco mosaic virus that carries an MHV epitope. *Proceedings of the National Academy of Sciences USA* **96**, 7774–7779.

Krebitz, M., Wiedermann, U., Essl, D., Steinkellner, H., Wagner, B., Turpen, T.H., Ebner, C., Scheiner, O. and Breiteneder, H. (2000) Rapid production of the major birch pollen allergen Bet v 1 in *Nicotiana benthamiana* plants and its immunological *in vitro* and *in vivo* characterization. *FASEB Journal* **14**, 1279–1288.

Kuroda, H. and Maliga, P. (2001) Sequences downstream of the translation initiation codon are important determinants of translation efficiency in chloroplasts. *Plant Physiology* **125**, 430–436.

Kusnadi, A.R., Nikolov, Z.L. and Howard, J.A. (1997) Production of recombinant proteins in transgenic plants: practical considerations. *Biotechnology and Bioengineering* **56**, 473–484.

Lamphear, B.J., Streatfield, S.J., Jilka, J.A., Brooks, C.A., Barker, D.K., Turner, D.D., Delaney, D.E., Garcia, M., Wiggins, B., Woodard, S.L., Hood, E.E., Tizard, I.R., Lawhorn, B. and Howard, J.A. (2002) Delivery of subunit vaccines in maize seed. *Journal of Controlled Release* **83**, 169–180.

Langeveld, J.P., Brennan, F.R., Martinez-Torrecuadrada, J.L., Jones, T.D., Boshuizen, R.S., Vela, C., Casal, J.I., Kamstrup, S., Dalsgaard, K., Meloen, R.H., Bendig, M.M. and Hamilton, W.D. (2001) Inactivated recombinant plant virus protects dogs from a lethal challenge with canine parovovirus. *Vaccine* **19**, 3661–3670.

Legocki, A.B., Miedzinska, K., Czaplinska, M., Plucieniczak, A. and Wedrychowicz, H. (2005) Immunoprotective properties of transgenic plants expressing E2 glycoprotein from CSFV and cysteine protease from *Fasciola hepatica*. *Vaccine* **23**, 1844–1846.

Lomonossoff, G.P. and Johnson, J.E. (1996) Use of macromolecular assemblies as expression systems for peptides and synthetic vaccines. *Current Opinion in Structural Biology* **6**, 176–182.

Ma, JK-C. (2001) Genes, greens and vaccines. *Nature Biotechnology* **18**, 1141–1142.

Ma, JK-C., Drake, P.M.W. and Christou, P. (2003) The production of recombinant pharmaceutical proteins in plants. *Nature Reviews Genetics* **4**, 794–805.

Ma, JK-C., Hiatt, A., Hein, M., Vine, N.D., Wang, F., Stabila, P., Vandolleweerd, C., Mostov, K. and Lehner, T. (1995) Generation and assembly of secretory antibodies in plants. *Science* **268**, 716–719.

Ma, JK-C., Hikmat, B.Y., Wycoff, K., Vine, N.D., Chargelegue, D., Yu, L., Hein, M.B. and Lehner, T. (1998) Characterization of a recombinant plant monoclonal secretory antibody and preventive immunotherapy in humans. *Nature Medicine* **4**, 601–606.

Ma, S-W., Zhao, D.L., Yin, Z.Q., Mukherjee, R., Singh, B., Qin, H.Y., Stiller, C.R. and Jevnikar, A.M. (1997) Transgenic plants expressing autoantigens fed to mice to induce oral immune tolerance. *Nature Medicine* **3**, 793–796.

Maliga, P. (2002) Engineering the plastid genome of higher plants. *Current Opinion in Plant Biology* **5**, 164–172.

Mallory, A.C., Parks, G., Endres, M.W., Baulcombe, D., Bowman, L.H., Pruss, G.J. and Vance, V.B. (2002) The amplicon-plus system for high-level expression of transgenes in plants. *Nature Biotechnology* **20**, 622–625.

Marusic, C., Rizza, P., Lattanzi, L., Mancini, C., Spada, M., Belardelli, F., Benvenuto, E. and Capone, I. (2001) Chimeric plant virus particles as immunogens for inducing murine and human immune responses against human immunodeficiency virus type 1. *Journal of Virology* **75**, 8434–8439.

Mason, H.S., Ball, J.M., Shi, J-J., Jiang, X., Estes, M.K. and Arntzen, C.J. (1996) Expression of Norwalk virus capsid protein in transgenic tobacco and potato and its oral immunogenicity in mice. *Proceedings of the National Academy of Sciences USA* **93**, 5340–5353.

Mason, H.S., Haq, T.A., Clements, J.D. and Arntzen, C.J. (1998) Edible vaccine protects mice against *Escherichia coli* heat-labile enterotoxin (LT): potatoes expressing a synthetic *LT-B* gene. *Vaccine* **16**, 1336–1343.

Mason, H.S., Lam, DM-K. and Arntzen, C.J. (1992) Expression of hepatitis B surface antigen in transgenic plants. *Proceedings of the National Academy of Sciences USA* **89**, 11745–11749.

Mayfield, S.P., Franklin, S.E. and Lerner, R.A. (2003) Expression and assembly of a fully active antibody in algae. *Proceedings of the National Academy of Sciences USA* **100**, 438–442.

McCormick, A.A., Kumagai, M.H., Hanley, K., Turpen, T.H., Hakim, I., Grill, L.K., Tuse, D., Levy, S. and Levy, R. (1999) Rapid production of specific vaccines for lymphoma by expression of the tumor-derived single-chain Fv epitopes in tobacco plants. *Proceedings of the National Academy of Sciences USA* **96**, 703–708.

McCormick, A.A., Reinl, S.J., Cameron, T.I., Vojdani, F., Fronefield, M., Levy, R. and Tuse, D. (2003) Individualized human scFv vaccines produced in plants: humoral anti-idiotype responses in vaccinated mice confirm relevance to the tumor IgG. *Journal of Immunological Methods* **278**, 95–104.

McGarvey, P.B., Hammond, J., Dienelt, M.M., Hooper, D.C., Fu, Z.F., Dietzschold, B., Kiprowski, H. and Michaels, F.H. (1995) Expression of the rabies virus glycoprotein in transgenic tomatoes. *Biotechnology* **13**, 1484–1487.

Mechtcheriakova, I.A., Eldarov, M.A., Nicholson, L., Shanks, M., Skryabin, K.G. and Lomonossoff, G.P. (2005). The use of viral vectors to produce hepatitis B virus core particles in plants. *J. Virol. Meth.*, **in press**.

190 Plant Biotechnology

Miele, L. (1997) Plants as bioreactors for biopharmaceuticals: regulatory considerations. *Trends in Biotechnology* **15**, 45–50.

Modelska, A., Dietzschold, B., Sleysh, N., Fu, Z.F., Steplewski, K., Hooper, D.C., Koprowski, H. and Yusibov, V. (1998) Immunization against rabies with plant-derived antigen. *Proceedings of the National Academy of Sciences USA* **95**, 2481–2485.

Molina, A., Hervas-Stubbs, S., Daniell, H., Mingo-Castel, A.M. and Veramendi, J. (2004) High-yield expression of a viral peptide animal vaccine in transgenic tobacco chloroplasts. *Plant Biotechnology Journal* **2**, 141–153.

Monde, R.A., Schuster, G. and Stern, D.B. (2000) Processing and degradation of chloroplast mRNA. *Biochimie* **82**, 573–582.

Nemchinov, L.G., Liang, T.J., Rifaat, M.M., Mazyad, H.M., Hadidi, A. and Keith, J.M. (2000) Development of a plant-derived subunit vaccine candidate against hepatitis C virus. *Archives of Virology* **145**, 2557–2573.

Nicholson, L., Gonzalez-Melendi, P., van Dolleweerd, C., Tuck, H., Perrin, Y., Ma, JK-C., Fischer, R., Christou, P. and Stoger, E. (2005) A recombinant multimeric immunoglobulin expressed in rice shows assembly-dependent subcellular localization in endosperm cells. *Plant Biotechnology Journal* **3**, 115–127.

Padidam, M. (2003) Chemically regulated gene expression in plants. *Current Opinion in Plant Biology* **6**, 169–177.

Perez-Filgueira, D.M., Zamorano, P.I., Dominguez, M.G., Taboga, O., Del Medico, M.P., Puntel, M., Romera, S.A., Morris, T.J., Borca, M.V. and Sadir, A.M. (2003) Bovine herpes virus gD protein produced in plants using a recombinant tobacco mosaic virus (TMV) vector possesses authentic antigenicity. *Vaccine* **21**, 4201–4209.

Perrin, Y., Vaquero, C., Gerrard, I., Sack, M., Drossard, J., Stoger, E., Christou, P. and Fischer, R. (2000) Transgenic pea seeds as bioreactors for the production of a single-chain Fv fragment (scFv) antibody used in cancer diagnosis and therapy. *Molecular Breeding* **6**, 345–352.

Pizza, M., Giuliani, M.M., Fontana, M.R., Monaci, E., Douce, G., Dougan, G., Mills, K.H.G., Rappuoli, R. and Del Giudice, G. (2001) Mucosal vaccines: non toxic derivatives of LT and CT as mucosal adjuvants. *Vaccine* **19**, 2534–2541.

Rennermalm, A., Li, Y.H., Bohaufs, L., Jarstrand, C., Brauner, A., Brennan, F.R. and Flock, J.I. (2001) Antibodies against a truncated *Staphylococcus aureus* fibronectin-binding protein protect against dissemination of infection in the rat. *Vaccine* **19**, 3376–3383.

Ritcher, L.J., Thanavala, Y., Arntzen, C.J. and Mason, H.S. (2000) Production of hepatitis B surface antigen in transgenic plants for oral immunization. *Nature Biotechnology* **18**, 1167–1171.

Russell, D.A. (1999) Feasibility of antibody production in plants for human therapeutic use. *Current Topics in Microbiology and Immunology* **236**, 119–137.

Smith, M.L., Ritcher, L., Arntzen, C.J., Shuler, M.L. and Mason, H.S. (2003) Structural characterization of plant-derived hepatitis B surface antigen employed in oral immunization studies. *Vaccine* **21**, 4011–4021.

Staczek, J., Bendahmane, M., Gilleland, L.B., Beachy, R.N. and Gilleland, H.E., Jr. (2000) Immunization with a chimeric tobacco mosaic virus containing an epitope of outer membrane protein F of *Pseudomonas aeruginosa* provides protection against challenge with *P. aeruginosa*. *Vaccine* **18**, 2266–2274.

Stoger, E., Sack, M., Fischer, R. and Christou, P. (2002) Plantibodies: applications, advantages and bottlenecks. *Current Opinion in Biotechnology* **13**, 161–166.

Stoger, E., Vaquero, C., Torres, E., Sack, M., Nicholson, L., Drossard, J., Williams, S., Keen, D., Perrin, Y., Christou, P. and Fischer, R. (2000) Cereal crops as viable production and storage systems for pharmaceutical scFv antibodies. *Plant Molecular Biology* **42**, 583–590.

Streatfield, S.J., Jilka, J.M., Hood, E.E., Turner, D.D., Bailey, M.R., Mayor, J.M., Woodard, S.L., Beifuss, K.K., Horn, M.E., Delaney, D.E., Tizard, I.R. and Howard, J.A. (2001) Plant-based vaccines: unique advantages. *Vaccine* **19**, 2742–2748.

Streatfield, S.J., Lane, J.R., Brooks, C.A., Barker, D.K., Poage, M.L., Mayor, J.M., Lamphear, B.J., Drees, C.F., Jilka, J.M., Hood, E.E. and Howard, J.A. (2003) Corn as a production system for human and animal vaccines. *Vaccine* **21**, 812–815.

Svab, Z., Hajdukiewicz, P. and Maliga, P. (1990) Stable transformation of plastids in higher plants. *Proceedings of the National Academy of Sciences USA* **87**, 8526–8530.

Tackaberry, E.S., Dudani, A.K., Prior, F., Tocchi, M., Sardana, R., Altosaar, I. and Ganz, P.R. (1999) Development of biopharmaceuticals in plant expression systems: cloning, expression, and immunological reactivity of human cytomegalovirus glycoprotein B (UL55) in seeds of transgenic tobacco. *Vaccine* **17**, 3020–3029.

Tacket, C.O., Mason, H.S., Losonsky, G., Clements, J.D., Levine, M.M. and Arntzen, C.J. (1998) Immunogenecity in humans of a recombinant bacterial-antigen delivered in transgenic potato. *Nature Medicine* **4**, 607–609.

Tacket, C.O., Mason, H.S., Losonsky, G., Estes, M.K., Levine, M.M., Arntzen and C.J. (2000) Human immune responses to a novel Norwalk virus vaccine delivered in transgenic potatoes. *Journal of Infectious Diseases* **182**, 302–305.

Tacket, C.O., Pasetti, M.F., Edelman, R., Howard, J.A. and Streatfield, S. (2004) Immunogenicity of recombinant LT-B delivered orally to humans in transgenic corn. *Vaccine* **22**, 4385–4389.

Thanavala, Y., Mahoney, M., Pal, S., Scott, A., Ritcher, L., Natarajan, N., Goodwin, P., Arntzen, C.J. and Mason, H.S. (2005) Immunogenicity in humans of an edible vaccine for hepatitis B. *Proceedings of the National Academy of Sciences USA* **102**, 3378–3382.

Thanavala, Y., Yang, Y-F., Lyons, P., Mason, H.S. and Arntzen, C.J. (1995) Immunogenecity of transgenic plant-derived hepatitis B surface antigen. *Proceedings of the National Academy of Sciences USA* **92**, 3358–3361.

Torres, E., Vaquero, C., Nicholson, L., Sack, M., Stoger, E., Drossard, J., Christou, P. and Fischer, R. (1999) Rice cell culture as an alternative production system for functional diagnostic and therapeutic antibodies. *Transgenic Research* **8**, 441–449.

Tregoning, J.S., Nixon, P., Kuroda, H., Svab, Z., Clare, S., Bowe, F., Fairweather, N., Ytterberg, J., van Wijk, K.J., Dougan, G. and Maliga, P. (2003) Expression of tetanus toxin fragment C in tobacco chloroplasts. *Nucleic Acids Research* **31**, 1174–1179.

Tuboly, T., Yu, W., Bailey, A., Degrandis, S., Du, S., Erickson, L. and Nagy, E. (2000) Immunogenecity of procine transmissible gastroenteritis virus spike protein expressed in plants. *Vaccine* **18**, 2023–2028.

Vaquero, C., Sack, M., Chandler, J., Drossard, J., Schuster, F., Monecke, M., Schillberg, S. and Fischer, R. (1999) Transient expression of a tumor-specific single-chain fragment and a chimeric antibody in tobacco leaves. *Proceedings of the National Academy of Sciences USA* **96**, 11128–11133.

Vaquero, C., Sack, M., Schuster, F., Finnern, R., Drossard, J., Schumann, D., Reimann, A. and Fischer, R. (2002) A carcinoembryonic antigen-specific diabody produced in tobacco. *FASEB Journal* **16**, 408–410.

Verch, T., Yusibov, V. and Koprowski, H. (1998) Expression and assembly of a full-length monoclonal antibody in plants using a plant virus vector. *Journal of Immunological Methods* **220**, 69–75.

Voinnet, O., Rivas, S., Mestre, P. and Baulcombe, D. (2003) An enhanced transient expression system in plants based on suppression of gene silencing by the p19 protein of tomato bushy stunt virus. *Plant Journal* **33**, 949–956.

Watson, J., Koya, V., Leppla, S.H. and Daniell, H. (2004) Expression of *Bacillus anthracis* protective antigen in transgenic chloroplasts of tobacco, a non-food/feed crop. *Vaccine* **22**, 4374–4384.

Webster, D.E., Thomas, M.C., Huang, Z. and Wesselingh, S.L. (2005) The development of a plant-based vaccine for measles. *Vaccine* **23**, 1859–1865.

Wigdorovitz, A., Carrillo, C., Dus Santos, M.J., Trono, K., Peralta, A., Gomez, M.C., Rios, R.D., Franzone, P.M., Sadir, A.M., Escribano, J.M. and Borca, M.V. (1999a) Induction of a protective antibody response to foot and mouth disease in mice following oral or parenteral immunization virus with alfalfa transgenic plants expressing the viral structural protein VP1. *Virology* **255**, 347–353.

Wigdorovitz, A., Perez Filgueira, D.M., Robertson, N., Carrillo, C., Sadir, A.M., Morris, T.J. and Borca, M.V. (1999b) Protection of mice against challenge with foot and mouth disease virus (FMDV) by immunization with foliar extracts from plants infected with recombinant tobacco mosaic virus expressing the FMDV structural protein VP1. *Virology* **264**, 85–91.

Wu, L., Jiang, L., Zhou, Z., Fan, J., Zhang, Q., Zhu, H., Han, Q. and Xu, Z. (2003) Expression of foot-and-mouth disease virus epitopes in tobacco by a tobacco mosaic virus-based vector. *Vaccine* **21**, 4390–4398.

Yu, J. and Langridge, W.H. (2001) A plant-based multicomponent vaccine protects mice from enteric diseases. *Nature Biotechnology* **19**, 548–552.

Yusibov, V., Mett, V., Mett, V., Davidson, C. Musiychuk, K., Gilliam, S., Farese, A., MacVittie, T. and Mann, D. (2005) Peptide-based candidate vaccine against respiratory syncytial virus. *Vaccine* **23**, 2261–2265.

Yusibov, V., Modelska, A., Steplewski, K., Agadjanyan, M., Weiner, D., Hooper, D.C. and Koprowski, H. (1997) Antigens produced in plants by infection with chimeric plant viruses immunize against rabies virus and HIV-1. *Proceedings of the National Academy of Sciences USA* **94**, 5784–5788.

Zeitlin, L., Olmsted, S.S., Moench, T.R., Co, M.S., Martinell, B.J., Paradkhar, V.M., Russell, D.R., Queen, C., Cone, R.A. and Whaley, K.J. (1998) A humanized monoclonal antibody produced in transgenic plants for immunoprotection of the vagina against genital herpes. *Nature Biotechnology* **16**, 1361–1364.

Zerges, W. (2000) Translation in chloroplasts. *Biochimie* **82**, 583–601.

2.7

Prospects for Using Genetic Modification to Engineer Drought Tolerance in Crops

*Mundree S.G., Iyer R., Baker B., Conrad N., Davis E.J., Govender K.,
Maredza A.T. and Thomson J.A.*

*Department of Molecular and Cell Biology, University of Cape Town,
Private Bag, Rondebosch, 7701 South Africa*

Introduction

Abiotic stress is a leading factor that prevents the use of certain environments for cultivation and restricts agricultural productivity worldwide. These abiotic stresses include high/low temperature, drought and salinity, which may act independently or simultaneously (Cherry, Locy and Rychter, 1999). The world's rising population and decline in available arable land have led to the exploration of methods that could lead to the utilization of traditionally non-arable land. Ultimately, all the above mentioned stresses result in osmotic and oxidative stress at the cellular level by inhibiting the absorption and transport of water (Shi *et al.*, 2002). There are plants that contain desiccation-tolerant vegetative tissue. These include certain algae, bryophytes, lichens, ferns and a unique group of angiosperms known as 'resurrection' plants (Oliver *et al.*, 1998). These plants have the ability to survive extremes of dehydration and subsequent rehydration from the air-dried state (Gaff, 1971). *Xerophyta viscosa* (family Velloziaceae), endemic to southern Africa, Madagascar and southern America, is such a plant, and is the model plant used in our laboratory. *X. viscosa* can survive extremes of dehydration, down to 5 % relative water content (RWC), and upon rehydration reach full

Plant Biotechnology. Edited by Nigel Halford.
© 2006 John Wiley & Sons, Ltd.

turgor and regain all physiological activities within 80 h of re-watering (Sherwin and Farrant, 1996).

Traditional Crop Improvement

The worlds' demand for food and fiber has grown dramatically over the last few centuries (Huffman, 2004). Initially mankind met this demand by increasing the amount of land under cultivation and later on by improving crops to enhance their yield (Huffman, 2004). Traditional plant breeding is the act of plant improvement using genetic improvement (Lamkey, 2002). In plant improvement methods, genes are exchanged mainly between sexually compatible species using a method called backcrossing (Lamkey, 2002). In spite of current technology and the recent advances in molecular biology, almost nothing is known about what changes occur at the DNA level when backcrossing takes place (Lamkey, 2002). Plant breeders select for a desired phenotype and have no control over the resultant genotype (Lamkey, 2002). Crop improvement by plant breeding has seldom caused controversy. Regulations for plant breeding were not seen as being necessary, despite the fact that in several cases traditional methods have resulted in cultivars that have negative effects on the ecology (National Research Council, 2002).

Recombinant DNA technology, or genetic engineering, is the human-mediated construction and insertion of recombinant DNA molecules into an organism (Smith *et al.*, 1997). The exchange of genes between sexually incompatible species has caused debate and resulted in the implementation of regulations on transgenic plants (Lamkey, 2002). Genetic engineering offers the most ground-breaking advances to impact agricultural research in recent years (Grover *et al.*, 2001). Advantages of genetic engineering include:

(i) a wider gene pool at one's disposal. Characteristics from sexually incompatible plant species, as well as other organisms can be utilized. This is particularly important with respect to engineering pest and disease resistance. Many resistance genes transferred via traditional breeding become ineffective as these pests and disease-causing organisms undergo selection pressure in order to survive;

(ii) single, specific genes can be transferred. Only the specific minimal DNA necessary for the desired trait is transferred. This prevents linkage drag which is associated with traditional breeding in which genetically linked, undesired genes are also transferred;

(iii) new genes can be directly transferred into existing plant lines. This eliminates time-consuming plant breeding cycles, in which many generations are required in order to recover specific lines; and,

(iv) novel gene constructions can be purposely designed. For example, promoter regions that control spatial, temporal and quantitative gene expression can be introduced (Conner and Jacobs, 1999).

Utilizing an Extremophile as a Model Organism

Variations in gene expression of particular genes from stress-sensitive and stress-tolerant species show that stress tolerance is genetically encoded (Bohnert and Jensen, 1996). It is therefore important to study tolerant species, as understanding their genetic makeup could lead to understanding tolerance mechanisms.

X. viscosa is an extremophile as it can survive in environments that experience severe temperature, salinity, high light and drought. Plants that have been 'engineered' by nature to survive and flourish under such conditions are an important source of knowledge as they may utilize novel proteins or signal transduction cascades.

Most crop plants have not been bred to survive extreme environments, and therefore their ability to survive under these conditions is restricted (Grover *et al.*, 2001). Abiotic stresses adversely affect the survival, biomass production and yield of most crop plants and can result in reduced seed germination, seedling establishment and pollen viability. It is therefore important to develop plants that can produce sufficient crop and survive extreme environments, using traditional breeding and/or genetic engineering.

Approaches used to Isolate Genes of Interest

A number of approaches are currently used to isolate and identify stress-responsive genes and gene products. These include differential screening of cDNA libraries, the screening of expression libraries, subtractive hybridization techniques and insertion mutations (Mundree and Farrant, 2000). Our research group uses a number of different strategies to identify genes and gene products of interest.

Complementation by functional sufficiency (Singh, Mundree and Locy, 2000), which uses eukaryotic expression libraries to complement appropriate *Escherichia coli (E. coli)* mutants to identify genes conferring a specific phenotype on the bacteria, has been widely utilized (Delauney and Verma, 1990; Hoff *et al.*, 1995; Hu, Delauney and Verma, 1992; Ilag, Kumar and Soll, 1994; Ravanel *et al.*, 1996). It allows the identification of cDNA clones that independently confer tolerance to the host organism in an osmotically stressed environment. This technique identified nine genes from *X. viscosa* that could independently confer tolerance to osmotically stressed *E. coli* (srl::Tn10) cells (Mundree and Farrant, 2000).

For the differential screening of a cDNA library Ndima *et al.* (2001) constructed a λ ZAP cDNA library from poly(A)$^+$ RNA isolated from *X. viscosa* leaves. The plants were desiccated to 85 %, 37 % and 5 % RWC and the isolated RNA was separately pooled. This could identify genes that were expressed in the leaves during dehydration. The library was probed separately with cDNA probes synthesized from total RNA isolated from leaves at 100 % (hydrated) and 37 % (dehydrated) RWC. The result was that 30 cDNAs had higher levels of transcripts present when the plant was dehydrated, and 20 cDNAs appeared to be downregulated.

For microarray analysis, RNA extracted from dehydrating leaves (35 % RWC) of *X. viscosa* was used to construct a full-length cDNA library. Individual clones (9696) were amplified and spotted onto microarray slides. These were then screened with mRNA isolated from hydrated and dehydrated *X. viscosa* leaf tissue. Forty-five independent cDNA clones were observed to be upregulated more than twofold in response to dehydration. These cDNAs could be categorized broadly into chlorophyll synthesis, translation machinery and other known stress-responsive genes such as those encoding enzymes involved in antioxidant production, late embryogenesis abundant proteins (LEAs), dehydrins and other desiccation-related proteins.

Screening for changes in protein levels at varying stages of RWC is currently in progress. Total protein was isolated from *X. viscosa* plants at full turgor, 65 % and 35 % RWC and 18 h after rehydration. The samples were electrophoresed on two-dimensional gels, and proteins of interest are being identified by mass spectrometry. The total protein profile at full turgor is being compared to that at 65 % and 35 % RWC in order to identify candidate proteins that are upregulated in response to dehydration. Although most research involving resurrection plants indicates that cellular protection occurs during dehydration, cellular damage is present in rehydrating cells in these tissues, and so cellular repair must take place, presumably during rehydration. In order to determine whether or not there are rehydration-induced changes in expression, the total protein profile at 35 % RWC is being compared to the profile obtained at 18 h after rehydration to determine whether transcripts are stored during dehydration in preparation for recovery. It has been reported that rehydration-specific transcripts were observed 15 h after re-watering (Oliver *et al.*, 1998). Therefore, the protein profile 18 h after rehydration should include the products of such transcripts.

Candidate Genes Isolated from *X. viscosa* for Crop Transformation

A number of interesting genes have been identified as upregulated in response to various stresses. A few of these include *XvPer1, XvPrx2, XvGols, XvIno1, XvALDR4, XvERD15, XvSAP1, XvVHA-c"1 and XvCaM.*

Antioxidants

When desiccation is experienced, *X. viscosa* undergoes numerous changes to reduce oxidative stress that results in the production of reactive oxygen species (ROS). The loss of water affects the electron transport chain of chloroplasts and mitochondria. Some plants use structural changes to protect the leaves from absorbing excess light, thereby reducing photo-oxidative stress (Sherwin and Farrant, 1998). *X. viscosa* is poikilochlorophyllous and this means that it breaks down its chlorophyll and thylakoid membranes during the drying process (Gaff, 1977). It also produces several antioxidant enzymes of which two peroxiredoxins (Prxs), XvPer1 and XvPrx2, are being characterized as possible candidates for maize transformation. Prxs are a new type of antioxidant, unrelated to any other peroxidase families (Finkemeier *et al.*, 2005). They function in antioxidant defense in photosynthesis, respiration, stress response and redox signaling. Prxs are shown to detoxify ROS and reactive nitrogen species (RNS).

XvPer1 is a stress-inducible antioxidant enzyme (Mowla *et al.*, 2002). It has a characteristic conserved cysteine residue and forms part of the 1-Cys Prx grouping. Other 1-Cys Prxs have been reported in plants, but only in seeds and immature embryos (Mundree *et al.*, 2002). XvPer1 has been found in vegetative tissues exposed to stresses such as dehydration, high light, high/low temperature and abscisic acid, thus alluding to the fact that it has a protective function under these stress conditions.

XvPrx2, which encodes a type II Prx, was isolated from a low-temperature stress library (unpublished data). XvPrx2 displays highest identity to an ortholog from rice, also possessing only a single catalytic cysteinyl residue. A previous study (Dietz, 2003) has shown that generally type II Prxs contain two cysteine residues of which only one is catalytic. XvPrx2 was tested in a mixed-function oxidation system containing Fe^{2+}, O_2 and dithiothreitol (DTT) to initiate oxidative damage to macromolecules. Supplementation of the assay with *XvPrx2* suppressed the occurrence of this damage. Enzyme assays using XvPrx2 have shown that it is able to detoxify a wide range of substrates using suitable electron donors for regeneration (unpublished data).

These antioxidants may be useful in protecting plants from harmful damage caused by ROS/RNS produced by plants as a result of photosynthesis, respiration and stress response. Analyzing the effects of overexpressing these proteins in plants is important in determining whether the plants will be able to cope better with ROS/RNS stress.

Osmoprotectants

Resurrection plants in particular accumulate an assortment of carbohydrates, amino acids and polyols that are thought to be involved in protecting cell integrity upon water loss (Ramanjulu and Bartels, 2002). A significant set of compatible solutes is the 'raffinose family oligosaccharides' (RFOs). The RFOs play an important role in the physiology of plants, which includes acting as osmoprotectants during various osmotic stresses. Two genes involved in the synthesis of osmoprotectants have been isolated from *X. viscosa*, namely, *XvGols* and *XvIno1*.

A full-length *XvGols* was isolated from a dehydrated cDNA library. *XvGols* encodes a galactinol synthase and may be an important component in compatible solute biosynthesis. It was found to be upregulated in the leaves of *X. viscosa* during drought stress (unpublished data). Galactinol synthase enzymes represent the first step in the synthesis of RFOs. These not only have been extensively characterized as principle agents in carbon translocation in plants but also have been observed to accumulate under low temperature, drought or salinity stress, implying a role for RFOs in stress adaptation.

XvIno1 is a myo-inositol-1-phosphate synthase gene isolated from a low-temperature stress cDNA library. This gene encodes the enzyme that catalyzes the committed step in the biosynthesis of RFOs. The enzyme catalyzes the conversion of glucose-6-phosphate to myo-inositol-1-phosphate, which is subsequently dephosphorylated to myo-inositol. Myo-inositol is a precursor for a number of important metabolites, which include membrane components, storage molecules, phytohormones and a variety of osmoprotectants. *XvIno1* has been shown to be upregulated during various stresses (unpublished data).

XvALDR4 (formerly referred to as *ALDR4XV*; Mundree *et al.*, 2000) is a cDNA isolated from the dehydration library and codes for an aldose reductase. This enzyme catalyzes the reduction of sugars to their analogous alcohol, for example glucose-6-phosphate to sorbitol (Bohren *et al.*, 1989; Rakowitz *et al.*, 2002). Oberschall *et al.* (2000) also demonstrated that plant aldose reductase can detoxify cytotoxic aldehydes, such as 4-hydroxynon-2-enal that is a product of ROS-induced lipid peroxidation. Transcript and protein levels of XvALDR4 have been shown to increase within leaves in response to water deficit (Mundree *et al.*, 2000). Transgenic plants transformed with *XvALDR4* are being studied at the molecular and physiological level to verify whether it will increase drought tolerance in these plants.

Early Response to Dehydration (ERD)

This category of genes has only recently been elucidated and little is known about the role they play during the early stages of dehydration. Twenty-six *X. viscosa* cDNA clones were detected by a differential screen and these were analyzed further to identify clones involved in the 'ERD' (Kiyosue, Yamaguchi-Shinozaki and Shinozaki, 1993).

Arabidopsis thaliana produces *ERD15* (Kiyosue, Yamaguchi-Shinozaki and Shinozaki, 1994), a gene encoding an ortholog of which we have found in *X. viscosa* from a low-temperature stress cDNA library (unpublished data). ERD15 is hydrophilic, acidic and lacks cysteine residues. The *X. viscosa* ortholog proved to be stress responsive at the early stages of dehydration stress (unpublished data). Overexpression of *XvERD15* in *A. thaliana* is being studied to ascertain its function.

Cell Membrane Integrity

Maintaining intact cell membranes during osmotic stress is critical to stress-tolerance mechanisms. If the membrane is damaged, ion homeostasis and signal transduction will be adversely affected, creating metabolic imbalances. The disruption of the cell membrane by osmotic stress induces the production of protection proteins such as embryogenesis abundant proteins (LEAs) and heat shock proteins (Hoekstra, Golvina and Buitink, 2001).

XvSAP1 was isolated from a cDNA library synthesized from dehydration stressed *X. viscosa* leaves. XvSAP1 is a highly hydrophobic protein and has two membrane lipoprotein lipid attachment sites. It also displays high sequence similarity with the K^+ transporter family (Garwe, Thomson and Mundree, 2003). Expression of XvSAP1 in *E. coli* (*srl*::Tn10) conferred osmotic stress tolerance when the cells were grown in 1 M sorbitol. Transgenic *A. thaliana* and *Nicotiana tabacum* plants constitutively expressing *XvSAP1* displayed increased tolerance to osmotic, salt, heat and dehydration stress (unpublished data). These features suggest that XvSAP1 could play an important role in repair of damage to the cell membrane.

Ion Homeostasis

Hyperosmotic stress reduces the chemical attributes of water causing a decline in cell turgor. This can be brought about by drought as well as salinity stress and the latter particularly could disrupt ionic equilibrium in the cell by a cytotoxic buildup of Na^+ and Cl^- (Serrano *et al.*, 1996; 1999). Upon NaCl stress, the excess sodium is removed from the cytoplasm by the vacuole. Vacuoles are multifunctional as they provide structural support to the cell and cell wall, sequester and store toxic compounds, and maintain ion homeostasis. Na^+/H^+ antiporters on the vacuolar membrane transport the Na^+ into the vacuole using a proton motive force, which is generated by the V-ATPase (Apse *et al.*, 1999; Taiz, 1992).

XvVHA-c″1 is a cDNA coding for the subunit c″ protein of the V-ATPase isolated from a dehydration library. Southern blot analysis showed that the gene was present possibly as a single copy. Transcripts increased in response to NaCl, dehydration and low-temperature shock. *Saccharomyces cereviseae Vma3* (defective Ca^{2+} homeostasis) transformed with *XvVHA-c″1* was able to withstand a 100 mM CaCl$_2$ stress, verifying its functional significance (Marais *et al.*, 2004). It is postulated that XvVHA-c''*1* plays a role in creating a proton translocating pore and assisting in adapting to osmotic pressure fluctuations as well as having a housekeeping role to maintain luminal acidification.

Calcium-Binding Proteins

Plants utilize a number of messengers to allow the cells to process the multitude of stimuli they experience. Signaling pathways use secondary messengers, such as Ca^{2+}, pH, lipids, inositol triphosphate, cyclic guanosine monophosphate and activated oxygen species (Bowler and Fluhr, 2000; Sanders, Brownlee and Harper, 1999), yet Ca^{2+} responds to more stimuli than any other messenger. Transient increases in cytosolic Ca^{2+} concentration occur in response to signals from stresses induced by low temperature, drought, high light, phytohormones (abscisic acid) and even mechanical stimuli such as gravity and touch. Calmodulin (CaM) is highly conserved and ubiquitous as a Ca^{2+} receptor in plants. In rice seedlings, a CaM-like protein was isolated, which is induced by abscisic acid, salt stress and dehydration.

XvCaM was isolated from a cDNA library synthesized from dehydration stressed *X. viscosa* leaves and codes for a classic CaM with four EF hands. Northern blot analyses indicate that transcript levels only fluctuate under dehydration stress. Western blot analyses show that the protein accumulates at low RWC and is present during the rehydration of the plant. No protein is present at full turgor after rehydration. Under a 150 mM NaCl stress, *XvCaM* is present only 24 h after application of NaCl. Purified recombinant protein was subjected to an overlay assay using Ca^{45+} (Maruyama, Mikawa and Ebashi, 1984) and showed that 0.5 µg of protein was able to bind calcium, thus establishing its functional importance (unpublished data).

Use of Model Systems to Test Genes of Interest

Garwe (2003) introduced *XvSAP1*, under the control of a cauliflower mosaic virus 35S promoter and *nos* terminator, into *A. thaliana* and *N. tabacum* by *Agrobacterium tumefaciens*-mediated transformation. She noted that expression of *XvSAP1* in both Arabidopsis and *Nicotiana* plants led to the constitutive accumulation of the corresponding protein in the leaves. Transgenic *A. thaliana* grown on plant nutrient agar in petri dishes and *Nicotiana* grown hydroponically were more tolerant to salt and osmotic stress. Non-transgenic plants had shorter roots, leaf expansion was inhibited and leaves were more chlorotic than those of the transgenic plants. In addition, transgenic *Nicotiana* plants attained a higher fresh and dry weight than the untransformed controls. Transgenic *Nicotiana* also showed a significantly greater tolerance to desiccation stress when grown in soil. Untransformed plants were smaller, wilted earlier and were more chlorotic in response to desiccation stress. The transgenic plants expressing *XvSAP1* had greater membrane permeability to electrolytes, measured by electrolyte leakage, which resulted in less negative water potential and thus the ability to continue normal growth. The expression of *XvSAP1* had no effect on the efficiency of photosystem II under normal growth conditions or under drought stress. These experiments confirmed that expression of *XvSAP1* confers tolerance to NaCl and osmotic stress.

Transgenic *A. thaliana* plants constitutively expressing the *X. viscosa* aldose reductase, *XvALDR4*, exhibited increased tolerance to salt and dehydration stresses compared to the wild type (Colombia ecotype). Transgenic and wild-type *A. thaliana* seedlings exposed to salt stress were cultured on vertical plant nutrient agar plates. Unsupplemented plant nutrient agar was used as a control. Wild-type and transgenic plants were grown alongside each other to allow for comparison of root lengths that were used as a measure of growth. While growth of seedlings on unsupplemented agar was similar, the transgenic plants were better able to cope with salt stress. On 75 mM and 100 mM NaCl, wild-type *A. thaliana* attained on average 50 % of the root length obtained by the transgenic plants. Dehydration of 1-month-old *A. thaliana* plants showed that transgenic plants survived for more than a week longer than the wild type (unpublished data).

Maize Transformation—Current Status Internationally with the Shift from Biolistics to *A. tumefaciens*-mediated Transformation

Gene delivery by microparticle bombardment has become a widely used technique with broad applications in plant transformation (Klein *et al.*, 1989, 1998a). The bombardment method has been widely adopted by many researchers for use on different maize tissue types and on other crops (Armstrong, 1999; Fromm *et al.*, 1990; Gordon-Kamm *et al.*, 1990). An example of the successful use of this technology is maize resistant to European corn borer (Koziel *et al.*, 1993).

Many studies (Gordon-Kamm *et al.*, 1990; Klein *et al.*, 1998a, 1989; Register *et al.*, 1994; Vasil *et al.*, 1992; Wan and Lemaux, 1994) however, have reported that transformation by particle bombardment often leads to integration at one locus of complex arrays of multiple copies of the introduced genes including the plasmid vector, often fragmented and rearranged. Since the multiple copies inserted during biolistic transformation are usually genetically linked, they cannot be segregated during subsequent breeding. Multiple copies of transgenes may lead to instability of their expression (Matzke and Matzke, 1995). In addition, multiple copies of transgenes may interact to inactivate each other and related host genes by epigenetic mechanisms, and homologous recombination may cause genetic instability of multiple gene copies. For these reasons, reduction of the copy number of inserted transgenes and simplification of their arrangement should prove beneficial for maintaining the fidelity and expression of introduced genes (Hansen and Chilton, 1996).

In the last decade the application of *A. tumefaciens*-mediated transformation to monocotyledonous species, such as rice and maize, has been reported (Chan, Lee and Chang, 1992; Hiei *et al.*, 1994; Ishida *et al.*, 1996). Although not natural hosts for *A. tumefaciens*, monocotyledonous species appear under certain conditions to be susceptible to infection. Studies on *Agrobacterium* infection of maize were first reported by Grimsley *et al.* (1987) and Gould *et al.* (1991). However, the first evidence of the possibility of the application of *A. tumefaciens*-mediated transformation to cereal species comes from the work of Chan *et al.* (1992) and Hiei *et al.* (1994), who first obtained transgenic rice plants by means of transformation of immature embryos. Thereafter, the technique was successfully applied to maize, and transgenic plants have been obtained at high frequency (Ishida *et al.*, 1996).

The integration pattern of foreign genes introduced via *A. tumefaciens*-mediated transformation is, in general, strikingly different from the pattern resulting from particle bombardment of plant cells. There are fewer intact and re-arranged transgenes in transgenic plants introduced by the *Agrobacterium* system. The main characteristics of the *Agrobacterium* system in maize, as for dicotyledonous species, are (i) a greater proportion of stable, low-copy number transgenic events; (ii) proper integration of the foreign gene into the host genome; and, (iii) the possibility of transferring larger DNA segments into recipient cells, promoting correct expression of the transgene (Frame *et al.*, 2002; Ingham *et al.*, 2001; Miller *et al.*, 2002; Negrotto *et al.*, 2000).

Continuous development of the *A. tumefaciens*-mediated transformation technology has been reported. Concurrent with gene delivery methods, selectable marker development has been integral in developing efficient maize transformation. Kanamycin (Fromm, Taylor and Walbot, 1986; Lyznik *et al.*, 1989; Rhodes *et al.*, 1988) and hygromycin (Walters *et al.*, 1992) were two of the earliest antibiotics used as selection agents in maize. More recently, Li *et al.* (2003) developed protoporphyrinogen oxidase as a selectable marker. Reproducible protocols for *A. tumefaciens*-mediated maize transformation have used super-binary vectors, in which the bacterium carries extra copies of *VirB*, *VirC* and *VirG* (Komari, 1990) to infect immature zygotic embryos of the inbred line A188 (Ishida *et al.*, 1996; Negrotto *et al.*, 2000) or the hybrid line HiII (Zhao *et al.*, 1998, 1999). More recently, Frame *et al.* (2002) reported the use of an *A. tumefaciens* standard binary vector system to routinely produce fertile, stable transgenic maize.

Concluding Remarks

During the first half of the 20th century, African farmers transformed maize from a minor imported food crop into the continent's principal staple food. In the second half of the century, newly independent governments launched support programs that greatly expanded smallholder production, leading to substantial production surges of 10–20 years in duration. Today, after widespread adoption by both commercial farmers and smallholders, farmers now dedicate 58 % of all the maize-growing area in East and southern Africa to new high-yielding varieties, which on average out-perform traditional varieties by 40–50 % even without fertilizer (Smale and Jayne, 2004).

The sustained domestic breeding programs that underpin this transformation represent impressive technical and political commitments. During 1960, Zimbabwe (then Southern Rhodesia) released its famous SR-52, the first commercially grown single-cross maize hybrid in the world (Smale and Jayne, 2004). In recent years genetic engineering has offered the most ground-breaking advances to impact agricultural research (Grover *et al.*, 2001). With the advances in this technology it is anticipated that this trend should continue. Our strategy focuses on the:

(i) identification of genetic material from *X. viscosa* upregulated during dehydration stress;
(ii) in-depth characterization of these genetic elements to identify those that display the potential to reduce the impact of dehydration stress;
(iii) incorporation of characterized genetic elements into maize plants; and,

(iv) identification of those genetic elements that increase the ability of maize plants to tolerate dehydration stress.

The satisfactory completion of this research goal would ideally result in a maize variety that is able to withstand the harsh natural conditions, especially drought, experienced in southern Africa.

References

Apse, M.P., Aharon, G.S., Snedden, W.A. and Blumwald, E. (1999) Salt tolerance conferred by over-expression of a vacuolar Na$^+$/H$^+$ antiport in *Arabidopsis thaliana. Science* **285**, 1256–1258.

Armstrong, C.L. (1999) The first decade of maize transformation: a review and future perspective. *Maydica* **44**, 101–109.

Bohnert, H.J. and Jensen, R.G. (1996) Strategies for engineering water-stress tolerance in plants. *Trends in Biotechnology* **14**, 89–97.

Bohren, K.M., Bullock, B., Wermuth, B. and Gabbay, K.H. (1989) The aldo-keto reductase superfamily. *Journal of Biological Chemistry* **264**, 9547–9551.

Bowler, C. and Fluhr, R. (2000) The role of calcium and activated oxygens as signals for controlling cross-tolerance. *Trends in Plant Science* **5**, 241–246.

Chan, M-T., Lee, T-M. and Chang, H-H. (1992) Transformation of indica rice (*Oryza sativa* L.) mediated by *Agrobacterium tumefaciens. Plant and Cell Physiology* **33**, 577–583.

Cherry, J.H., Locy, R.D. and Rychter, A. (1999) Preface, in *Plant Tolerance to Abiotic Stresses in Agriculture: Role of Genetic Engineering* (eds J.H. Cherry, R.D. Locy and A. Rychter), Kluwer Academic Publishers, The Netherlands, p. ix.

Conner, A.J. and Jacobs, J.M.E. (1999) Genetic engineering of crops as potential source of genetic hazard in the human diet. *Mutation Research* **443**, 223–234.

Delauney, A.J. and Verma, D.P. (1990) A soybean gene encoding delta1-pyrroline-5-carboxylate reductase was isolated by functional complementation in *Escherichia coli* and is found to be osmoregulated. *Molecular and General Genetics* **221**, 299–305.

Dietz, K-J. (2003) Plant peroxiredoxins. *Annual Reviews of Plant Biology* **54**, 93–107.

Finkemeier, I., Goodman, M., Lamkemeyer, P., Kandlbinder, A., Sweetlove, L.J. and Dietz, K.J. (2005) The mitochondrial type II peroxiredoxin F is essential for redox homeostasis and root growth of *Arabidopsis thaliana* under stress. *Journal of Biological Chemistry*, in press.

Frame, B.R., Shou, H., Chikwamba, R.K., Zhang, Z., Xiang, C., Fonger, T.M., Pegg, S.E.K., Li, B., Nettleton, D.S., Pei, D. and Wang, K. (2002) *Agrobacterium tumefaciens*-mediated transformation of maize embryos using a standard binary vector system. *Plant Physiology* **129**, 13–22.

Frandsen, G., Muller-Uri, F., Nielsen, M., Mundy, J. and Skriver, K. (1996) Novel plant Ca^{2+}-binding protein expressed in response to abscisic acid and osmotic stress. *Journal of Biological Chemistry* **271**, 343–348.

Fromm, M.E., Morrish, F., Armstrong, C., Williams, R., Thomas, J. and Klein, T.M. (1990) Inheritance and expression of chimeric genes in the progeny of transgenic maize plants. *Biotechnology* **8**, 833–839.

Fromm, M.E., Taylor, L.P. and Walbot, V. (1986) Stable transformation of maize after gene transfer by electroporation. *Nature* **319**, 791–793.

Gaff, D.F. (1971) Desiccation tolerant flowering plants in Southern Africa. *Science* **174**, 1033–1034.

Gaff, D.F. (1997) Dessication-tolerant vegetative plants of Southern Africa. *Oceanlogia* **31**, 95–109.

Garwe, D. (2003) The characterisation of *XvSAP1*, a gene isolated from the resurrection plant *Xerophyta viscosa* Baker, and its expression in transgenic plants. Ph.D. Thesis, Department of Molecular and Cell Biology, University of Cape Town.

Garwe, D., Thomson, J.A. and Mundree, S.G. (2003) Molecular characterization of *XvSAP1*, a stress-responsive gene from the resurrection plant *Xerophyta viscosa* Baker. *Journal of Experimental Botany* **54**, 191–201.

Gordon-Kamm, W.J., Spencer, T.M., Mangano, M., Adams, T.R., Daines, R.J., Start, W.G., O'Brien, J.V., Chambers, S.A., Adams, W.R., Jr., Willetts, N.G., Rice, T.B., Mackey, C.J., Krueger, R.W., Kausch, A.P. and Lemaux, P.G. (1990) Transformation of maize cells and regeneration of fertile transgenic plants. *Plant Cell* **2**, 603–618.

Gould, J., Devey, M., Hasegawa, O., Ulian, E., Peterson, G. and Smith, R.H. (1991) Transformation of *Zea mays* L. using *Agrobacterium tumefaciens* and the shoot apex. *Plant Physiology* **95**, 426–434.

Grimsley, N.H., Hohn, T., Davies, J.W. and Hohn, B. (1987) *Agrobacterium*-mediated delivery of infectious maize streak virus into maize plants. *Nature* **325**, 177–179.

Grover, A., Kapoor, A., Lakshmi, O.S., Agarwal, S., Sahi, C., Katiyar-Agarwal, S., Agarwal, M. and Dubey, H. (2001) Understanding molecular alphabets of the plant abiotic stress responses. *Current Science* **80**, 206–216.

Hansen, G. and Chilton, M.D. (1996) "Agrolistic" transformation of plant cells: integration of T-strands generated in planta. *Proceedings of the National Academy of Sciences USA* **93**, 14978–14983.

Hiei, Y., Ohta, S., Komari, T. and Kumashiro, T. (1994) Efficient transformation of rice (*Oryza sativa* L.) mediated by *Agrobacterium* and sequence analysis of the boundaries of the T-DNA. *The Plant Journal* **6**, 271–282.

Hoekstra, F.A., Golvina, E.A. and Buitink, J. (2001) Mechanisms of plant desiccation tolerance. *Trends in Plant Science* **6**, 431–438.

Hoff, T., Schnorr, K.M., Meyer, C. and Caboche, M. (1995) Isolation of two *Arabidopsis* cDNAs involved in early steps of molybdenum cofactor biosynthesis by functional complementation of *E. coli* mutants. *Journal of Biological Chemistry* **271**, 6100–6107.

Hu, C.A., Delauney, A.J. and Verma, D.P. (1992) A bifunctional enzyme (delta 1-pyrroline-5-carboxylate synthetase) catalyses the first two steps in proline biosynthesis in plants. *Proceedings of the National Academy of Sciences USA* **89**, 9354–9358.

Huffman, W.E. (2004) Production, identity preservation, and labelling in a marketplace with genetically modified and non-genetically modified foods. *Plant Physiology* **134**, 3–10.

Ilag, L.L., Kumar, A.M. and Soll, D. (1994) Light regulation of chlorophyll biosynthesis at the level of 5-aminolevulinate chain reaction. *Plant Cell* **6**, 265–275.

Ingham, D.J., Beer, S., Money, S. and Hansen, G. (2001) Quantitative real-time PCR assay for determining transgene copy number in transformed plants. *Biotechniques* **31**, 132–140.

Ishida, Y., Saito, H., Ohta, S., Hiei, Y., Komari, T. and Kumashiro, T. (1996) High efficiency transformation of maize (*Zea mays* L.) mediated by *Agrobacterium tumefaciens*. *Nature Biotechnology* **14**, 745–750.

Kiyosue, T., Yamaguchi-Shinozaki, K. and Shinozaki, K. (1993) Characterization of two cDNAs (ERD11 and ERD13) for dehydration-inducible genes that encode putative glutathione S-transferases in *Arabidopsis thaliana* L. *FEBS Letters* **335**, 189–192.

Kiyosue, T., Yamaguchi-Shinozaki, K. and Shinozaki, K. (1994) Cloning of cDNAs for genes that are early-responsive to dehydration stress (ERDs) in *Arabidopsis thaliana* L.: identification of three ERDs as HSP cognate genes. *Plant Molecular Biology* **25**, 791–798.

Koziel, M.G., Beland, G.L., Bowman, C., Carozzi, N.B., Crenshaw, R., Crossland, L., Dawson, J., Desai, N., Hill, M., Kadwell, S., Launis, K., Lewis, K., Maddox, D., McPherson, K., Meghji, M.R., Merlin, E., Rhodes, R., Warren, G.W., Wright, M. and Evola, S.V. (1993) Field performance of elite transgenic maize plants expressing an insecticidal protein derived from *Bacillus thuringiensis*. *Nature Biotechnology* **4**, 194–200.

Klein, TM., Fromm, M., Weissinger, A., Tomes, D., Schaaf, S., Sletten, M. and Sanford, J.C. (1988a) Transfer of foreign genes into intact maize cells with high-velocity microprojectiles. *Proceedings of the National Academy of Sciences USA* **85**, 4305–4309.

Klein, T.M., Gradziel, T., Fromm, M.E. and Sanford, J.C. (1988b) Factors influencing gene delivery into *Zea mays* cells by high-velocity microprojectiles. *Nature Biotechnology* **6**, 559–563.

Klein, T.M., Kornstein, L., Sanford, J.C. and Fromm, M.E. (1989) Genetic transformation of maize cells by particle bombardment. *Plant Physiology* **91**, 440–444.

Komari, T. (1990) Transformation of callus cells of *Chenopodium quinoa* by binary vectors that carry a fragment of DNA from the virulence region of pTi Bo542. *Plant Cell Reports* **9**, 303–306.

Lamkey, K. (2002) GMO's Gene Flow: A Plant Breeding Perspective. Presented at Biotechnology, Gene Flow, and Intellectual Property Rights: An Agricultural Summit, Purdue University.

Li, X., Volrath, S.L., Nicholl, D.B.G., Chilcott, C.E., Johnson, M.A., Ward, E.R. and Law, M.D. (2003) Development of protporphyrinogen oxidase as an efficient selection marker for *Agrobacterium tumefaciens*-mediated transformation of maize. *Plant Physiology* **133**, 736–747.

Lyznik, L.A., Ryan, R.D., Ritchie, S.W. and Hodges, T.K. (1989) Stable co-transformation of maize protoplasts with gusA and neo genes. *Plant Molecular Biology* **13**, 151–161.

Marais, S., Thomson, J.A., Farrant, J.M. and Mundree, S.G. (2004) XvVHA-c″1—a novel stress-responsive V-ATPase subunit c'homologue isolated from the resurrection plant *Xerophyta viscosa*. *Physiologia Plantarum* **122**, 54–61.

Maruyama, K., Mikawa, T. and Ebashi, S. (1984) Detection of calcium binding proteins by ^{45}Ca autoradiography on nitrocellulose membrane after sodium dodecyl sulfate gel electrophoresis. *Journal of Biochemistry* **95**, 511–519.

Matzke, M.A. and Matzke, A.J.M. (1995) How and why do plants inactivate homologous transgenes? *Plant Physiology* **107**, 679–685.

Miller, M., Tagliani, L., Wang, N., Berka, B., Bidney, D. and Zhao, Z.Y. (2002). High efficiency transgene segregation in co-transformed maize plants using an *Agrobacterium tumefaciens* T-DNA binary system. *Transgenic Research* **11**, 381–396.

Mowla, S.B., Thomson, J.A., Farrant, J.M. and Mundree, S.G. (2002) A novel stress-inducible antioxidant enzyme identified from the resurrection plant *Xerophyta viscosa* Baker. *Planta* **215**, 716–726.

Mundree, S.G., Baker, B., Mowla, S., Peters, S., Marais, S., Vander Willigen, C., Govender, K., Maredza, A., Muyanga, S., Farrant, J.M. and Thomson, J.A. (2002) Physiogical and molecular insights into drought Tolerance. *African Journal of Biotechnology* **1**, 28–38.

Mundree, S.G. and Farrant, J.M. (2000) Some physiological and molecular insights into the mechanisms of desiccation tolerance in the resurrection plant *Xerophyta viscosa* Baker, in *Plant Tolerance to Abiotic Stresses in Agriculture: Role of Genetic Engineering* (eds J.H. Cherry, R.D. Locy and A. Rychter), Kluwer Academic Publishers, The Netherlands, pp. 210–222.

Mundree, S.G., Whittaker, A., Thomson, J.A. and Farrant, J.M. (2000) An aldose reductase homolog from the resurrection plant *Xerophyta viscosa* Baker. *Planta* **211**, 693–700.

Ndima, T., Farrant, J., Thomson, J. and Mundree, S. (2001) Molecular characterisation of *XvT8*, a stress-responsive gene from the resurrection plant *Xerophyta viscosa* Baker. *Plant Growth Regulation* **35**, 137–145.

Negrotto, D., Jolley, M., Beer, S., Wench, A.R. and Hansen, G. (2000) The use of phospho-mannose-isomerase as a selectable marker to recover transgenic maize plants (*Zea mays* L.) via *Agrobacterium* transformation. *Plant Cell Reports* **19**, 798–803.

National Research Council (2002) *Environmental Effects of Transgenic Plants: The Scope and Adequacy of Regulation*, National Academy Press, Washington DC.

Oberschall, A., Deak, M., Torok, K., Sass, L., Vass, I., Kovacs, I., Feher, A., Dudits, D. and Horvath, G.V. (2000) A novel aldose/aldehyde reductase protects transgenic plants against lipid peroxidation under chemical and drought stresses. *The Plant Journal* **24**, 437–446.

Oliver, M.J., Woods, A.J. and O'Mahoney, P. (1998) "To dryness and beyond"—preparation for the dried state and rehydration in vegetative desiccation-tolerant plants. *Plant Growth Regulation* **24**, 193–201.

Rakowitz, D., Matuszczak, B., Gritsch, S., Hofbauer, P., Krassnigg, A. and Costantino, L. (2002) On the prodrug potential of aldose reductase inhibitors with diphenylmethyleneamino-oxycarboxylic acid structure. *European Journal of Pharmaceutical Sciences* **15**, 11–20.

Ramanjulu, S. and Bartels, D. (2002) Drought and dessiction-induced modulation of gene expression in plants. *Plant Cell and Environment* **25**, 141–151.

Ravanel, S., Ruffel, M.L. and Douce, R. (1996) Cloning of an *Arabidopsis thaliana* cDNA encoding cystathione beta-lyase by functional complementation of *E. coli. Plant Molecular Biology* **29**, 875–882.

Register, J.C., Peterson, D.J., Bell, P.J., Bullock, W.P., Evans, I.J., Frame, B., Greenland, A.J., Higgs, N.S., Jepson, I. and Jiao, S. (1994) Structure and function of selectable and non-selectable transgenes in maize after introduction by particle bombardment. *Plant Molecular Biology* **25**, 951–961.

Rhodes, C.A., Pierce, D.A., Mettler, I.J., Mascarenhas, D. and Detmer, J.J. (1988) Genetically transformed maize plants from protoplasts. *Science* **240**, 204–207.

Sanders, D., Brownlee, C. and Harper, J.F. (1999) Communicating with calcium. *Plant Cell* **11**, 691–706.

Serrano, R. (1996) Salt tolerance in plants and microorganisms: toxicity targets and defense responses. *International Review of Cytology* **165**, 1–52.

Serrano, R., Mulet, J., Rios, G., Marquez, J., de Larrinoa, I., Leube, M., Mendizabal, I., Pascual-Ahuir, A., Prott, M., Ros, R. and Montesinos, C. (1999) A glimpse of the mechanisms of ion homeostasis during salt stress. *Journal of Experimental Botany* **50**, 1023–1036.

Sherwin, H.W. and Farrant, J.M. (1996) Differences in rehydration of three desiccation-tolerant angiosperm species. *Annals of Botany* **78**, 703–710.

Sherwin, H.W. and Farrant, J.M. (1998) Protection mechanisms against excess light in resurrection plants *Craterostgma wilmsii* and *Xerophyta viscosa*. *Plant Growth Regulation* **24**, 203–210.

Shi, H., Xiong, L., Stevenson, B., Lu, T. and Zhu, J-K. (2002) The *Arabidopsis* salt overly sensitive 4 mutants uncover a critical role for vitamin b6 in plant salt tolerance. *Plant Cell* **14**, 575–588.

Singh, N.K., Mundree, S.G. and Locy, R.D. (2000) Novel determinants of salinity tolerance, in *Plant Tolerance to Abiotic Stresses in Agriculture: Role of Genetic Engineering* (eds J.H. Cherry, R.D. Locy and A. Rychter), Kluwer Academic Publishers, The Netherlands, pp. 131–138.

Smale, M. and Jayne, T.S. (2004) Building on successes in African agriculture: Maize breeding in east and southern Africa, 1900–2000. International Food Policy Research Unit. Focus 12, Brief 4.

Smith, A.D., Datta, S.P., Howard Smith, G., Campbell, P.N., Bentley, R. and McKenzie, H.A. (eds) (1997) *Oxford Dictionary of Biochemistry and Molecular Biology*, Oxford University Press, Oxford, p. 259.

Taiz, L. (1992) The plant vacuole. *Journal of Experimental Biology* **172**, 113–122.

Vasil, V., Castillo, A.M., Fromm, M.E. and Vasil, I.K. (1992) Herbicide resistant fertile transgenic wheat plants obtained by microprojectile bombardment of regenerable embryogenic callus. *Nature Biotechnology* **10**, 667–674.

Walters, D.A., Vetsch, C.S., Potts, D.E. and Lundquist, R.C. (1992) Transformation and inheritance of a hygromycin phosphotransferase gene in maize plants. *Plant Molecular Biology* **18**, 189–200.

Wan, Y. and Lemaux, P.G. (1994) Generation of large numbers of independently transformed fertile barley plants. *Plant Physiology* **104**, 37–38.

Zhao, Z.Y., Gu, W., Cai, T. and Pierce, D.A. (1999) Methods for *Agrobacterium*-mediated transformation. United States Patent No. 5,981,840.

Zhao, Z.Y., Gu, W., Cai, T., Tagliani, L.A., Hondred, D.A., Bond, D., Krell, S., Rudert, M.L., Bruce, W.B. and Pierce, D.A. (1998) Molecular analysis of T0 plants transformed by *Agrobacterium* and comparison of *Agrobacterium*-mediated transformation with bombardment transformation in maize. *Maize Genetics Cooperation Newsletter* **72**, 34–37.

2.8

Salt Tolerance

Eduardo Blumwald

Department of Plant Sciences, University of California,
One Shields Avenue, Davis, CA 95616 USA

Anil Grover

Department of Plant Molecular Biology, University of Delhi South
Campus, Benito Juarez Road, New Delhi 110021, India

Introduction

Agricultural productivity is severely affected by soil salinity, and the damaging effects of salt accumulation in agricultural soils have influenced ancient and modern civilizations. Environmental stress due to salinity is one of the most serious factors limiting the productivity of agricultural crops, which are predominantly sensitive to the presence of high concentrations of salts in the soil. Levels of salt inimical to plant growth affect large terrestrial areas of the world. It is estimated that more than 20% of all the irrigated land in the world is presently affected by salinity. This is exclusive of the regions classified as arid and desert lands (which comprise 25 % of the total land of our planet). The loss of farmable land due to salinization is directly in conflict with the needs of the world population, projected to increase by 1.5 billion in the next 20 years, and the challenge of maintaining the world food supplies. Although famine in the world nowadays is originated by complex problems and not only by an insufficient production of food, there is no doubt that the gains in food production provided by the Green Revolution have reached their ceiling while the world population continues to rise. Therefore, increasing the yield of crop plants in normal soils and in less productive lands, including salinized lands, is an absolute requirement for feeding the world.

Plant Biotechnology. Edited by Nigel Halford.
© 2006 John Wiley & Sons, Ltd.

The degradation of agricultural land and water supplies is a result of the intensive agricultural practices employed in developed and developing countries. Ideally these practices should be changed to a more rational use of land and water resources, but this change will not occur in the foreseeable future. For example, mixed cropping with perennials and trees would alleviate the accumulation of sodium and other salts in the upper soil layers. Nonetheless, this kind of change in farming systems and the development of new products is likely a long and difficult process, since it will require the use of new land and will not address the problem of growing crops in land that is already compromised. The development and use of crops that can tolerate high levels of salinity in the soils is a practical solution, at least for the time-being.

The need to produce stress-tolerant crops was evident even in ancient times (Jacobsen and Adams, 1958). However, efforts to improve crop performance under environmental stresses have not been particularly fruitful because the fundamental mechanisms of stress tolerance in plants remain to be completely understood. Epstein *et al.* (1980) described technical and biological constraints to the problem of salinity. While there appears to have been some success with the technical solutions to the problem, the biological solutions have been more difficult to develop. A biological approach for the development of salt-tolerant crop varieties requires as a pre-requisite the identification of key genetic determinants of stress tolerance and the use of salt stress tolerance-related genes or QTLs (Quantitative Trait Loci). The existence of salt-tolerant plants (halophytes) and differences in salt tolerance between genotypes within salt-sensitive plant species (glyco-phytes) clearly indicate that there is a genetic basis to salt response. While differences in salt tolerance between varieties have been known for a long time (Epstein, 1977, 1983) and intra-specific selection for salt tolerance reported in rice (Akbar and Yabuno, 1977) and barley (Epstein *et al.*, 1980), there exists still a large gap in our understanding. Flowers and Yeo (1995) reviewed the evidence for the paucity of salt-tolerant cultivars and concluded that the number was likely to be less than 30. Flowers (2004) pointed out that since 1993 there have been just three registrations of salt-resistant cultivars in *Crop Science* (Al-Doss and Smith, 1998; Dierig *et al.*, 2001; Owen *et al.*, 1994).

Two basic genetic approaches are currently being utilized to improve stress tolerance: (1) exploitation of natural genetic variations, either through direct selection in stressful environments or through the mapping of QTLs and subsequent marker-assisted selection and (2) generation of transgenic plants to introduce novel genes or alter expression levels of the existing genes to affect the degree of salt stress tolerance. We discuss these approaches in some detail, focusing on the recent experimentation with transgenic plants that has led to increased salinity tolerance, with emphasis on the areas of ion home-ostasis, osmotic regulation and antioxidant protection. There is an emerging body of work in the area of signaling and transcriptional control that has been recently reviewed (Hasegawa *et al.*, 2000; Zhang, Creelman and Zhu, 2004; Zhu, 2001, 2002) and thus will not be dealt with here.

Genetics of Salt Tolerance

The direct selection of superior salt-tolerant genotypes under field conditions is hindered by the significant influence that environmental factors have on the response of plants to salinity (Richards, 1996). There is also evidence suggesting that salt tolerance in plants is developmentally regulated and that the tolerance of plants at one stage of development is not always correlated with tolerance at other stages (Foolad 2004; Greenway and Munns, 1980; Tal and Shannon, 1983). For example, in tomato, barley, corn, rice and wheat, salt tolerance of the plants tends to increase with age (Foolad, 2004 and references therein). QTLs associated with salt tolerance at the germination stage in barley (Mano and Takeda, 1997), tomato (Foolad, Lin and Chen, 1999) and Arabidopsis (Quesada *et al.*, 2002) were different from those associated with salt tolerance at the early stage of growth, and the plants selected by their ability to germinate at high salinity did not display similar salt tolerance during vegetative growth (Flowers, 2004; Foolad *et al.*, 1999; Lindsay *et al.*, 2004; Mano and Takeda, 1997; Quesada *et al.*, 2002).

The development of molecular biology techniques has allowed for the development of DNA markers that can be used to identify QTLs. A QTL is a region of the genome that is associated with the variation of a quantitative trait of interest. The use of QTLs has improved the efficiency of selection, in particular for those traits that are controlled by several genes and are highly influenced by environmental factors (Flowers, 2004). QTLs and marker-assisted selection provide several advantages over direct phenotypic screening, in particular because the PCR-based methodologies used to detect the markers reduce the time needed to screen individuals and reduce the environmental effects on the trait under study. There is considerable evidence to support the view that salt tolerance and its subtraits are determined by multiple QTLs and that both additive and dominance effects are important in the inheritance of many of the associated traits (Flowers, 2004; Foolad, 2004; Gregorio *et al.*, 2002). The development of high-density DNA maps that incorporate microsatellite markers, restriction fragment length polymorphisms (RFLPs) and amplified fragment length polymorphisms (AFLPs), and advances in marker-assisted selection techniques will facilitate the pyramiding of traits of interest to attain substantial improvement in crop salt tolerance.

Salt Tolerance Using Transgenic Approaches

Physiologically, salinity (a) imposes water deficit that results from the relatively high solute concentrations in the soil, (b) causes ion-specific stresses resulting from altered K^+/Na^+ ratios and (c) leads to buildup in Na^+ and Cl^- concentrations that are detrimental to plants. Plants respond to salinity using two different types of responses. Salt-sensitive plants restrict the uptake of salt and adjust their osmotic pressure by the synthesis of compatible solutes (proline, glycinebetaine, sugars etc.) (Greenway and Munns, 1980). Salt-tolerant plants sequester and accumulate salt into the cell vacuoles, controlling the salt concentrations in the cytosol and maintaining a high cytosolic K^+/Na^+ ratio in their cells (Glenn, Brown and Blumwald, 1999). Clearly, ion exclusion mechanisms could provide a degree of tolerance to relatively low NaCl concentrations but would not work at high salt concentrations, resulting in the inhibition of key metabolic processes with concomitant growth inhibition. Here, we discuss three key processes that contribute to salt tolerance at the cellular level: (a) the establishment of cellular ion homeostasis, (b) the synthesis of compatible solutes for osmotic adjustment and (c) the increased ability of the cells to neutralize reactive oxygen species (ROS) generated during the stress response.

Ion Homeostasis

Although Na^+ is required in some plants, particularly halophytes (Glenn *et al.*, 1999), a high NaCl concentration is a toxic factor for plant growth. The alteration of ion ratios in plants is due to the influx of Na^+ through pathways that function in the acquisition of K^+ (Blumwald, Aharon and Apse, 2000). The sensitivity to salt of cytosolic enzymes is similar in both glycophytes and halophytes, indicating that the maintenance of a high cytosolic K^+/Na^+ concentration ratio is a key requirement for plant growth in high salt (Glenn *et al.*, 1999). Strategies that plants could use in order to maintain a high K^+/Na^+ ratio in the cytosol include: (i) extrusion of Na^+ ions out of the cell and (ii) vacuolar compartmentation of Na^+ ions. Under typical physiological conditions, plants maintain a high cytosolic K^+/Na^+ ratio. Given the negative membrane potential difference at the plasma membrane ($-140\,mV$) (Higinbotham, 1973), a rise in extracellular Na^+ concentration will establish a large electrochemical gradient favoring the passive transport of Na^+ into the cells.

Three classes of low-affinity K^+ channels have been identified. Inward rectifying channels (KIRC), such as AKT1 (Sentenac *et al.*, 1992), activate K^+ influx upon plasma-membrane hyperpolarization and display a high K^+/Na^+ selectivity ratio. A knockout mutant of *AKT1* in Arabidopsis (akt1-1) displayed similar sensitivity to salt as the wild type, suggesting that this channel does not play a role in Na^+ uptake (Spalding *et al.*, 1999). K^+ outward rectifying channels (KORCs) could play a role in mediating the influx of Na^+ into plant cells. KORC channels showed a high selectivity for K^+ over Na^+ in barley roots (Wegner and Raschke, 1994) and a somewhat lower K^+/Na^+ selectivity ratio in Arabidopsis root cells (Maathuis and Sanders, 1995). These channels, which open during the depolarization of the plasma membrane (i.e. upon a shift in the electrical potential difference to more positive values), could mediate the efflux of K^+ and the influx of Na^+ ions (Maathuis and Sanders, 1995). Voltage-independent cation channels (VIC) in plant plasma membranes have been reported (de Boer and Wegner, 1997; Roberts and Tester, 1997). These channels have a relatively high Na^+/K^+ selectivity, are not gated by voltage and provide a pathway for the entry of Na^+ into plant cells (Maathuis and Amtmann, 1999).

Sodium ions can enter the cell through a number of low- and high-affinity K^+ carriers; among these is AtHKT1 from Arabidopsis, which was shown to function as a selective Na^+ transporter and, to a lesser extent, to mediate K^+ transport (Uozumi *et al.*, 2000). AtHKT1 was identified as a regulator of Na^+ influx in plant roots. This conclusion was based on the capacity of *hkt1* mutants to suppress Na^+ accumulation and sodium hypersensitivity in a *sos3* (salt-overly-sensitive) mutant background (Rus *et al.*, 2001), suggesting that AtHKT1 is a salt tolerance determinant that controls the entry of Na^+ into the roots.

Na^+ extrusion from plant cells is powered by the operation of the plasma membrane H^+-ATPase generating an electrochemical H^+ gradient that allows plasma membrane Na^+/H^+ antiporters to couple the passive movement of H^+ inside the cells, along its electrochemical potential, to the active extrusion of Na^+ (Blumwald *et al.*, 2000).

Table 2.8.1   Salt tolerance in transgenic plants expressing genes involved in ion transporters.

Gene	Gene product	Source	Cellular role(s)	Target plant	Parameter studied	Reference
AtNHX1	Vacuolar Na$^+$/H$^+$ antiporter	A. thaliana	Na$^+$ vacuolar sequestration	Arabidopsis	Biomass	Apse et al., 1999
AtNHX1	Vacuolar Na$^+$/H$^+$ antiporter	A. thaliana	Na$^+$ vacuolar sequestration	Tomato	Biomass and fruit production	Zhang and Blumwald, 2001
AtNHX1	Vacuolar Na$^+$/H$^+$ antiporter	A. thaliana	Na$^+$ vacuolar sequestration	B. napus	Biomass and oil production	Zhang et al., 2001
AtNHX1	Vacuolar Na$^+$/H$^+$ antiporter	A. thaliana	Na$^+$ vacuolar sequestration	Maize	Biomass	Yin et al., 2004
GhNHX1	Vacuolar Na$^+$/H$^+$ antiporter	Gossypium hirsutum	Na$^+$ vacuolar sequestration	Tobacco	Biomass	Wu et al., 2004
AgNHX1	Vacuolar Na$^+$/H$^+$ antiporter	Atriplex gmelini	Na$^+$ vacuolar sequestration	Rice	Biomass	Ohta et al., 2002
AtSOS1	Plasma membrane Na$^+$/H$^+$ antiporter	A. thaliana	Na$^+$ extrusion	Arabidopsis	Biomass	Shi et al., 2003
AVP1	Vacuolar H$^+$-pyrophosphatase	A. thaliana	Vacuolar acidification	Arabidopsis	Biomass	Gaxiola et al., 2001
HAL1	K$^+$/Na$^+$ transport regulation	S. cerevisiae	K$^+$/Na$^+$ homeostasis	Tomato, melon, Arabidopsis	Ion content, plant growth	Bordas et al., 1997; Gisbert et al., 2000; Yang et al., 2001

Recently, *AtSOS1* from Arabidopsis has been shown to encode a plasma membrane Na$^+$/H$^+$ antiporter with significant sequence similarity to plasma membrane Na$^+$/H$^+$ antiporters from bacteria and fungi (Shi *et al.*, 2000). The overexpression of *SOS1* improved the salt tolerance of Arabidopsis, demonstrating that improved salt tolerance can be attained by limiting Na$^+$ accumulation in plant cells (Shi *et al.*, 2003) (Table 2.8.1). The compartmentation of Na$^+$ ions into vacuoles also provides an efficient mechanism to avert the toxic effects of Na$^+$ in the cytosol. The transport of Na$^+$ into the vacuoles is mediated by a Na$^+$/H$^+$ antiporter that is driven by the electrochemical gradient of protons generated by the vacuolar H$^+$-translocating enzymes, the H$^+$-ATPase and the H$^+$-PPiase (Blumwald, 1987). The overexpression in Arabidopsis of *AtNHX1*, an Arabidopsis gene encoding a vacuolar Na$^+$/H$^+$ antiporter, resulted in transgenic plants that were able to grow in high salt concentrations (Apse *et al.*, 1999).

The paramount role of Na$^+$ compartmentation in plant salt tolerance has been further demonstrated in transgenic tomato plants overexpressing AtNHX1 (Zhang and

Blumwald, 2001). The transgenic tomato plants grown in the presence of 200 mM NaCl were able to grow, flower, and set fruit. Although the leaves accumulated high sodium concentrations, the tomato fruits displayed very low amounts of sodium (Zhang and Blumwald, 2001). Similar results were obtained with transgenic *Brassica napus* (Canola) overexpressing *AtNHX1* (Zhang *et al.*, 2001). Leaves of transgenic plants grown in the presence of 200 mM NaCl accumulated sodium to up to 6% of their dry weight, but the seed yields and oil quality were not affected, demonstrating the potential use of this technology for agricultural use in saline soils. Similar results have been reported in other species. The introduction of a vacuolar Na^+/H^+ antiporter from the halophyte *Atriplex gmelini* conferred salt tolerance in rice (Ohta *et al.*, 2002). Most recently, the over-expression of *GhNHX1* from cotton in tobacco plants (Wu *et al.*, 2004) and the overexpression of *AtNHX1* in maize (Yin *et al.*, 2004) resulted in enhanced salt tolerance.

Additional evidence supporting the role of vacuolar transport in salt tolerance has been provided by Arabidopsis plants overexpressing a vacuolar H^+-PPiase (Gaxiola *et al.*, 2001). Transgenic plants overexpressing *AVP1*, coding for the vacuolar H^+-pyropho-sphatase, displayed enhanced salt tolerance that was correlated with an increased ion content of the plants. These results suggest that the enhanced vacuolar H^+-pumping in the transgenic plants provided additional driving force for vacuolar sodium accumulation via the vacuolar Na^+/H^+ antiporter.

Synthesis of Compatible Solutes

The cellular response of salt-tolerant organisms to both long- and short-term salinity stress includes the synthesis and accumulation of a class of osmoprotective compounds known as compatible solutes. These relatively small, organic osmolytes include amino acids and derivatives, polyols and sugars, methylamines etc. These osmolytes stabilize proteins and cellular structures and can increase the osmotic pressure of the cell (Yancey *et al.*, 1982). This response is homeostatic for cell water status and protein integrity, which is perturbed in the face of soil solutions containing higher amounts of NaCl and the consequent loss of water from the cell. The accumulation of osmotically active compounds in the cytosol increases the osmotic potential to provide a balance between the apoplastic solution, which itself becomes more concentrated with Na^+ and Cl^- ions, and the vacuolar lumen, which in halophytes can accumulate up to 1 M Na^+ (and Cl^-). For a short-term stress, this may provide the cells with the ability to prevent water loss. However, for continued growth under salinity stress, an osmotic gradient (toward the cytosol) must be kept in order to maintain turgor, water uptake and facilitate cell expansion.

The enhancement of proline and glycinebetaine syntheses in target plants has received much attention (Rontein, Basset and Hanson, 2002). Two themes have emerged from the results of these efforts: (i) there are metabolic limitations on the absolute levels of the target osmolyte that can be accumulated and (ii) the degree to which the transformed plants are able to tolerate salinity stress is not necessarily correlative with the amounts of the osmoprotectants attained.

The metabolic limitations on increasing the concentration of a given osmoprotectant are well illustrated with both proline and glycinebetaine. Initial strategies aimed at engineering higher concentrations of proline began with the overexpression of genes encoding the biosynthetic enzymes pyrroline-5-carboxylate (P5C) synthase (P5CS) and P5C reductase (P5CR) that catalyze the two steps between the substrate, glutamic acid, and the product, proline. P5CS overexpression in transgenic tobacco dramatically elevated free proline (Kishor *et al.*, 1995) (Table 2.8.2). However, the regulation of free proline does not appear to be straightforward. Proline catabolism, via proline dehydrogenase (ProDH), is upregulated by free proline and there is strong evidence that free proline inhibits P5CS (Roosens *et al.*, 1999). Further, a two-fold increase in free proline was achieved in tobacco plants transformed with a *P5CS* modified by site-directed mutagenesis (Hong *et al.*, 2000). This modification alleviated the feedback inhibition by proline on the P5CS activity and resulted in improved germination and growth of seedlings under salt stress. Free cellular proline levels are also transcriptionally and translationally controlled. P5CR promoter analysis revealed that P5CR transcripts have reduced translational initiation. A 92-bp segment of the 5' UTR of P5CR was sufficient to provide increased mRNA stability and translational inhibition under salt stress to the GUS reporter gene that was ligated at the 3' end of this small region (Hua *et al.*, 2001). These results highlighted the complex regulation of P5CR during stress and emphasized the importance of stability and translation of P5CR mRNA during salt stress.

An alternative approach to attain significant free proline levels, where antisense cDNA transformation was used to decrease ProDH expression, has been utilized (Nanjo *et al.*, 1999). Levels of proline in the transgenic Arabidopsis were twice (100 µg/g fresh weight) that of control plants when grown in the absence of stress, and three times higher (600 µg/g fresh weight) than in control plants when grown under stress. The high levels of proline were correlated with an improvement in tolerance to salinity, albeit for a short duration exposure to 600 mM NaCl.

There has been considerably more experimentation directed at the engineering of glycinebetaine synthesis than for any other compatible solute. Unlike proline, glycine-betaine degradation is not significant in plants (Nuccio *et al.*, 2000), but the problems of metabolic fluxes, compounded with the compartmentation of the substrate and product pools, has made the engineering of appreciable levels of glycinebetaine problematic. In plants that are naturally glycinebetaine accumulators (e.g., spinach and sugar beet), synthesis of this compound occurs in the chloroplast, with two oxidation reactions from choline to glycinebetaine. The first oxidation to betaine aldehyde is catalyzed by choline mono-oxygenase (CMO), an iron–sulfur enzyme. Betaine aldehyde oxidation to glyci-nebetaine is catalyzed by betaine aldehyde dehydrogenase (BADH), a non-specific soluble aldehyde dehydrogenase (Rathinasabapathi, 2000). In *Escherichia coli*, these reactions are cytosolic; the first reaction is catalyzed by choline dehydrogenase (CDH), an NAD^+-dependent enzyme that is encoded by the *betA* locus; the second reaction is catalyzed by BADH, which is encoded by the *betB* locus. In *Arthrobacter globiformis*, the two oxidation steps are catalyzed by one enzyme, choline oxidase (COD), which is encoded by the *codA* locus (Sakamoto and Murata, 2000).

The *CodA* gene of *A. globiformis* offers an attractive alternative to the engineering of glycinebetaine synthesis as it necessitates only a single gene transformation event. This strategy was employed for engineering glycinebetaine synthesis in Arabidopsis (Hayashi *et al.*, 1997). The 35S promoter-driven construct for transformation included the transit peptide for the small subunit of Rubisco so that the COD protein would be targeted to the chloroplast. Improved salinity tolerance was obtained in transgenic

Table 2.8.2 Salt tolerance in transgenic plants expressing genes involved in osmolyte biosynthesis.

Gene	Gene product	Source	Cellular role(s)	Target plant	Parameter studied	Reference
betA	Choline dehydrogenase	E. coli	Glycinebetaine	Tobacco	Dry weight	Lilius, Holmberg and Bulow, 1996; Holmstrom et al. (2000)
BADH	Betaine dehydrogenase	Atriplex hortensis	Glycinebetaine	Tomato	Root growth	Jia et al., 2002
EctA, EctB, EctC	L-2,4-diaminobutyric acid acetyltransferase, L-2,4-diaminobutyric acid trans- aminase, L-ectoine synthase	Halomonas elongata	Ectoyne	Tobacco	Salinity tolerance	Nakayama et al., 2000
OtsA OtsB	Trehalose-6-P synthase Trehalose-6-P phosphatase	E. coli	Trehalose	Tobacco Rice	Increased biomass, morphogenesis, growth	Pilon-Smits et al., 1998; Garg et al., 2002
TPS1	Trehalose-6-P synthase	S. cerevisiae	Trehalose	Tobacco	Improved drought tolerance	Romero et al., 1997
P5CS	Δ^1-Pyrroline-5-carboxylate synthase	V. aconitifolia	Proline	Tobacco	Increased proline; plant growth	Kishor et al., 1995
ProDH	Proline dehydrogenase	Arabidopsis thaliana	Proline	Arabidopsis	Inflorescence lodging in response to NaCl stress	Nanjo et al., 1999
IMT1	Myo-inositol-O-methyl transferase	M. chrystallinum	D-Ononitol	Tobacco	Seed germination	Vernon et al., 1993; Sheveleva et al., 1997
COD1; COX	Choline oxidase	A. globiformis; A. panescens	Glycinebetaine	Arabidopsis, rice, Brassica	Seed germination; plant growth	Hayashi et al., 1997; Alia et al., 1999; Sakamoto et al., 1998; Huang et al., 2000; Prasad et al., 2000
HAL3	FMN-binding protein	S. cerevisiae	K^+/Na^+ homeostasis	Arabidopsis	Seedlings	Espinosa-Ruiz et al., 1999

Arabidopsis that accumulated, as a result of the transformation, 1 μmol/g fresh weight glycinebetaine.

The same construct was used for transformation of *B. juncea* (Prasad *et al.*, 2000) and tolerance to salinity during germination and seedling establishment was improved markedly in the transgenic lines. *COX* from *A. panescens*, which is homologous to the *A. globiformis COD*, was used to transform Arabidopsis, *B. napus* and tobacco (Huang *et al.*, 2000). This set of experiments differs from those above in that the COX protein was directed to the cytoplasm and not to the chloroplast. Improvements in tolerance to salinity, drought and freezing were observed in some transgenics from all three species, but the tolerance was variable. The levels of glycinebetaine in the transgenic plants were not significantly higher than those of wild-type plants, but increased significantly with the exogenous supply of choline to plants, suggesting that the supply of choline is a significant constraint on the synthesis of glycinebetaine (Huang *et al.*, 2000).

Two important issues emerge from the results of the above discussion. The first is that the concentrations of glycinebetaine in the transgenic plants were much lower than the concentrations noted in natural accumulators. Despite the fact that these levels are not high enough to be osmotically significant, a moderate (but significant) increase in tolerance to salinity and other stresses was conferred. This raises the possibility that the protection offered by glycinebetaine is not only osmotic, which is a point raised by several of the above groups and by Bohnert and Shen (1999). Compatible solutes, including mannitol, may also function as scavengers of oxygen radicals; this may be supported by the results of Alia *et al.* (1999), where the protection to photosystem II in plants expressing *CodA* was observed. An alternative possibility, not necessarily exclusive of the first, is that the increased level of peroxide generated by the COD/COX oxidation of choline causes an upregulation of ascorbate peroxidase (APX) and catalase (CAT) (Holmstrom *et al.*, 2000), which may also improve tolerance to salinity stress (Rontein *et al.*, 2002).

The second issue is that the level of glycinebetaine production in the transgenics is limited by choline. Betaine synthesis takes place in the chloroplast and the level of the free choline pool may not reflect the availability of choline to the chloroplast, which may be limited by the activity and/or abundance of choline transporters. However, a dramatic increase in glycinebetaine levels (to 580 μmol/g dry weight in Arabidopsis) was shown in the transgenic plants when they were supplemented with choline in the growth medium (Huang *et al.*, 2000). This limitation was not explored in the transgenic tobacco expressing *E. coli* enzymes CDH and BADH in the cytoplasm (Holmstrom *et al.*, 2000). Although these transgenic plants demonstrated an improved tolerance to salinity, glycinebetaine levels were on the order of those mentioned above. Sakamoto and Murata (2000) also asserted that, despite the similarities in tolerance exhibited by transgenic plants engineered to synthesize betaine in either the chloroplast or cytoplasm, the site of synthesis of betaine may play a role in the degree of tolerance shown. Indeed, if the betaine present in these plants is localized primarily in the chloroplast, it may be present at significant concentrations (50 mM) (Hayashi *et al.*, 1997). However, Sakamoto and Murata (2000) downplayed the limitation of the metabolic pool of choline on the levels of glycinebetaine obtained in the engineered plants by suggesting that the choline oxidizing activity may be the limiting factor. This argument seems to be supported by Huang *et al.* (2000) who found that the levels of glycinebetaine correlated with the levels of COX activity measured in each plant. The increase in glycinebetaine with exogenous choline argues against this notion.

Stronger evidence for the limitations of choline metabolism has been presented by McNeil *et al.* (2001). By overexpressing spinach phosphoethanolamine *N*-methyltransferase (PEAMT), which catalyzes the three methylation reactions required for the conversion of phosphoethanolamine to phosphocholine, up to a 50-fold increase in free choline was obtained. This led to an increase in glycinebetaine levels (+60 %) in plants that were expressing spinach CMO and BADH in the chloroplast. Further, the addition of ethanolamine to the plant growth medium caused an increase in choline and glycinebetaine levels, showing that the metabolic flux through this pathway is also limited by the supply of ethanolamine. As PEAMT is itself inhibited by phosphocholine, further engineering efforts need to include (a) the modification of PEAMT to remove this inhibition (McNeil *et al.*, 2001), (b) increasing the supply of ethanolamine by over-expression of serine decarboxylase and (c) resolving the compartmentation problem of choline supply and choline oxidation, either by use of choline oxidation in the cytoplasm or by finding the appropriate transporters to improve choline supply to the chloroplast (Rontein *et al.*, 2002).

Finally, as the compatible solutes are non-toxic, the interchangeability of these compounds between species has held much interest (Table 2.8.1). The recent examples include the engineering of (a) ectoine synthesis with enzymes from the halophylic bacterium *Halomonas elongata* (Nakayama *et al.*, 2000; Ono *et al.*, 1999) and (b) trehalose synthesis into potato (Yeo *et al.*, 2000) and rice (Garg *et al.*, 2002) (trehalose occurs in bacteria and yeast; while trehalose metabolism may have an important role in metabolic signaling in plants, trehalose itself only accumulates in extremely desiccation-tolerant plants (Goddijn and van Dun, 1999)).

An intriguing report on the improved tolerance to salinity in tobacco expressing yeast invertase in the apoplast highlights the potential of manipulating sucrose metabolism (Fukushima *et al.*, 2001). The authors reported improved salt tolerance of transgenic tobacco plants expressing a yeast invertase in their apoplastic space, and concluded that the changes in sucrose metabolism in the transgenic plants protected the photosynthetic apparatus under salt stress. The overexpression of polyols, such as mannitol (Tarczynski, Jensen and Bohnert, 1993) and D-ononitol (Sheveleva *et al.*, 1997) have been shown to contribute to enhanced drought and salt tolerance in transgenic tobacco plants.

Antioxidant Protection

An important aspect of salinity stress in plants is the stress-induced production of ROS including superoxide radicals (O_2^-), hydrogen peroxide (H_2O_2) and hydroxyl radicals (OH). ROS are a product of altered chloroplastic and mitochondrial metabolism during stress. These ROS cause oxidative damage to different cellular components including membrane lipids, protein and nucleic acids (Halliwell and Gutteridge, 1986). The alleviation of this oxidative damage could provide enhanced plant resistance to salt stress. Plants use low-molecular-mass antioxidants such as ascorbic acid and reduced glutathione and employ a diverse array of enzymes such as superoxide dismutases (SOD), CAT, APX, glutathione-*S*-transferases (GST) and glutathione peroxidases (GPX)

Table 2.8.3   *Salt tolerance in transgenic plants expressing genes involved in redox reactions.*

Gene	Gene product	Source	Cellular role(s)	Target plant	Parameter studied	Reference
MnSOD	Superoxide dismutase	S. cerevisiae	Reduction of O_2 content	Rice	Photosynthetic electron transport	Tanaka et al., 1999
GlyI	Glyoxylase	B. juncea	S-D-Lactoylglu-tathi-one	Tobacco	Chlorophyll content of detached leaves	Veena, Reddy and Sopory, 1999
TPX2	Peroxidase	N. tabaccum	Change cell wall properties	Tobacco	Germination; water retention in seed walls	Amaya et al., 1999
GST GPX	Glutathione S-transferase Glutathione peroxidase	N. tabaccum N. tabaccum	ROS scavenging	Tobacco	Germination and growth	Roxas et al., 2000

to scavenge ROS. Transgenic rice overexpressing yeast mitochondrial Mn-dependent SOD displayed enhanced salt tolerance (Tanaka *et al.*, 1999) (Table 2.8.3). The over-expression of a cell wall peroxidase in tobacco plants improved germination under osmotic stress (Amaya *et al.*, 1999). Transgenic tobacco plants overexpressing both GST and GPX displayed improved seed germination and seedling growth under stress (Roxas *et al.*, 1997). Subsequent studies (Roxas *et al.*, 2000) demonstrated that, in addition to increased GST/GPX activities, the transgenic seedlings contained higher levels of glutathione and ascorbate than wild-type seedlings and showed higher levels of mono-dehydroascorbate reductase activity, while the glutathione pools were more oxidized. These results would indicate that the increased glutathione-dependent peroxidase scavenging activity and the associated changes in glutathione and ascorbate metabolism led to reduced oxidative damage in the transgenic plants and contributed to their increased salt tolerance.

During salt stress, plants display an increase in the generation of H_2O_2 and other ROS (Gueta-Dahan *et al.*, 1997; Roxas *et al.*, 2000). The major substrate for the reductive detoxification of H_2O_2 is ascorbate, which must be continuously regenerated from its oxidized form. A major function of glutathione in protection against oxidative stress is the reduction of ascorbate via the ascorbate-glutathione cycle, where GSH acts a recycled intermediate in the reduction of H_2O_2 (Foyer and Halliwell, 1976). Ruiz and Blumwald (2002) investigated the enzymatic pathways leading to glutathione synthesis during the response to salt stress of wild-type and salt-tolerant *B. napus* L. (Canola) plants overexpressing a vacuolar Na^+/H^+ antiporter (Zhang *et al.*, 2001). Wild-type plants showed a marked increase in the activity of enzymes associated with cysteine synthesis (the crucial step for assimilation of reduced sulfur into organic compounds such as glutathione) resulting in a significant increase in GSH content. On the other hand, these activities were unchanged in the transgenic salt-tolerant plants and their GSH content did not change with salt stress. These results clearly showed that salt stress induced an increase in the assimilation of sulfur and the biosynthesis of cysteine and GSH aimed to mitigate the salt-induced oxidative stress. The small changes seen in the transgenic plants overexpressing the vacuolar Na^+/H^+ antiporter (Zhang and Blumwald, 2001) suggested that the accumulation of excess Na^+ in the vacuoles (and the maintenance of a high cytosolic K^+/Na^+ ratio) greatly diminished the salt-induced oxidative stress, highlighting the important role of Na^+ homeostasis during salt stress.

Assessment of Salt Tolerance in Transgenic Plants

The assessment of salt tolerance in transgenic experiments as described above has mostly been carried out using a limited number of seedlings or mature plants in laboratory experiments. In most of the cases, the experiments were carried out in greenhouse conditions where the plants were not exposed to those conditions that prevail in high-salinity soils (alkaline soil pH, high diurnal temperatures, presence of other sodic salts, elevated concentrations of selenium and/or boron etc.). There is a clear need for the evaluation of salt tolerance of plants in the field, and more importantly salt tolerance should be evaluated as a function of yield. The evaluation of field performance under salt stress is difficult because of the variability of salt levels in field conditions (Daniells *et al.*, 2001; Richards, 1983) and the potential for interactions with other environmental factors, including soil fertility, temperature, light intensity and water loss due to transpiration. Evaluating tolerance is made more complex by variation in sensitivity to salt during the life cycle. For example, in rice grain yield is much more affected by salinity than vegetative growth (Khatun and Flowers, 1995). In tomato, the ability of the plants to germinate under high salinity is not correlated with the plant's ability to grow under salt stress because both are controlled by different mechanisms (Foolad and Lin, 1997), although some genotypes may display similar tolerance at germination and during vegetative growth (Foolad and Chen, 1999).

It thus needs to be recognized that the assessment of stress tolerance in the laboratory often has little correlation to tolerance in the field. Although there have been many successes in developing stress-tolerant transgenics in model plants such as tobacco, Arabidopsis or rice (Grover *et al.*, 2003), there is an urgent need to test these successes in crops. Rice has the advantage that it is both the model monocot and an important crop. However, the same is not the case when transgenes are tested with tobacco or Arabidopsis (reviewed in Flowers, 2004; Grover *et al.*, 2003). This brings a number of technical and financial challenges associated with transforming many of the crop plants, particularly the monocots. First, transformation of some monocots is still not routine and to develop a series of independent homozygous T2 lines is costly, both in money and in time. Second, the stress tolerance screens will need to include a field component since many of the stress tolerance assays used by basic researchers involve using rich nutrient media, which include sucrose. This type of screen is unlikely to have much relationship to field performance. Finally, since saline soils are often complex and may include NaCl, $CaCl_2$, $CaSO_4$ and Na_2SO_4, high boron concentrations, alkaline pH etc., plants that show particular promise will eventually have to be tested in all of these environments.

Conclusions and Perspectives

Conventional breeding programs for raising salt-tolerant genotypes have met with limited success. This lack of success is in part because breeders prefer to evaluate their genetic material in ideal conditions. This issue is getting more urgent to deal with because of the growing interest of commercial seed companies in making salt-tolerant crops. From a business perspective, in order for plant breeding companies to invest in the development of new varieties with enhanced stress tolerance, there will always be the question raised as to whether the investment in the development of these cultivars is worth the effort. There is no benefit in developing salinity-tolerant plants unless there are the economic drivers that will allow the plant to be competitively productive with non-saline-tolerant plants growing on uncompromised soil. The viewpoints of basic researchers might differ; for the researchers the actual though small increase in salt tolerance is worth the efforts.

In evaluating the possibility of improving stress tolerance in plants, there are a number of considerations that we propose that the research community should consider.

While it has been recognized by many researchers that there are dramatic changes in gene expression associated with all types of stresses, the promoters that are most commonly used for transgene introductions are primarily constitutively expressed, including the CaMV35S promoter, ubiquitin and actin promoters (Grover *et al.*, 2003). Recent studies have noted that the overexpression of specific stress-induced genes under the control of stress-induced or tissue-specific promoters often results in a better phenotype than overexpressing the same genes under a constitutive promoter (Kasuga *et al.*, 1999; Zhu *et al.*, 1998).

Second, while there have been a number of successes in the production of abiotic stress-tolerant plants using tobacco or Arabidopsis, there is a clear need to begin to introduce these tolerance genes into crop plants. Moreover, even though researchers tend to focus on a few basic plant systems, with Arabidopsis, tobacco and rice being the major species of choice, there has been no attempt to choose specific genetic backgrounds.

Third, it is likely that the effectiveness of a specific transgene will be based on the specific genetic background into which it is transformed. One component of this is the well-known phenomenon of 'position effect'. However, in addition, the ability of a transgene to work may well be determined by the overall genetic background, independent of the chromosomal location of the transgene, referred to as 'Transgene Combining Ability' (TCA).

Finally, we also need to establish better comparative systems. At the same time, we need to look at rational concepts for combining genes, just as the disease resistance researchers are now doing with gene stacking (Chapter 2.9). For example, the overexpression of *AtSOS1* in meristems (non-vacuolated cells) and *AtNHX1* (for vacuolar Na^+ accumulation), together with the overproduction of compatible solutes, would provide not only the ability of using NaCl as an osmoticum during vegetative growth but also the seedlings with the ability to reduce Na^+ toxicity during early growth and seedling establishment. Wherever applicable, genes for protection against oxidative stress must be combined, particularly in actively photosynthesizing cells that are prone to more chloroplast damage due to ROS.

While progress in improving stress tolerance has been slow, there are a number of opportunities and reasons for optimism. Over the last 10 years there has been the development of a number of the functional tools that can allow us to dissect many of the fundamental questions associated with stress tolerance. These include: (a) the development of molecular markers for gene mapping and the construction of associated maps, (b) the development of EST libraries, (c) the complete sequencing of plant genomes including Arabidopsis, rice and maize, (d) the production of T-DNA or transposon-tagged mutagenic populations of Arabidopsis and (e) the development of a number of forward genetics tools that can be used in gene function analysis such as Tilling (Colbert *et al.*, 2001). Thus, we need to focus on looking at the comparative effects and interactions of specific transgenes within a defined genetic background and determine the efficacy of these approaches in the field.

Over the last 50 years, many researchers have argued for the development of salt-tolerant crops from true halophytes. Although halophytes are present in a wide diversity of plant forms, to date very few halophytic crops can compete effectively with glycophytic crops (Glenn *et al.*, 1999). Moreover, research on the physiology of tolerance suggests that the overall trait is determined by a number of subtraits, any of which might, in turn, be determined by any number of genes. We believe that by comparing different genes and genetic combinations, researchers will be able to advance the field more quickly and develop stress-tolerant germplasms.

References

Akbar, M. and Yabuno, T. (1977) Breeding saline-resistant varieties of rice. IV. Inheritance of delayed type panicle sterility induced by salinity. *Japanese Journal of Breeding* **27**, 237–240.

Al-Doss, A.A. and Smith, S.E. (1998) Registration of AZ-97MEC and AZ-97MEC-ST very non-dormant alfalfa germplasm pools with increased shoot weight and differential response to saline irrigation. *Crop Science* **38**, 568–568.

Alia, Kondo, Y., Sakamoto, A., Nonaka, H., Hayashi, H., Saradhi, P.P., Chen, T.H.H. and Murata, N. (1999) Enhanced tolerance to light stress of transgenic *Arabidopsis* plants that express the *codA* gene for a bacterial choline oxidase. *Plant Molecular Biology* **40**, 279–288.

Amaya, I., Botella, M.A., De La Calle, M., Medina, M.I., Heredia, A., Bressan, R.A., Hasegawa, P.M., Quesada, M.A. and Valpuesta, V. (1999) Improved germination under osmotic stress of tobacco plants overexpressing a cell wall peroxidase. *FEBS Letters* **457**, 80–84.

Apse, M.P., Aharon, G.S., Snedden, W.A. and Blumwald, E. (1999) Salt tolerance conferred by overexpression of a vacuolar Na^+/H^+ antiport in *Arabidopsis*. *Science* **285**, 1256–1258.

Blumwald, E. (1987) Tonoplast vesicles for the study of ion transport in plant vacuoles. *Physiologia Plantarum* **69**, 731–734.

Blumwald, E., Aharon, G.S. and Apse, M.P. (2000) Sodium transport in plant cells. *Biochimica et Biophysica Acta* **1465**, 140–151.

Bohnert, H.J. and Shen, B. (1999) Transformation and compatible solutes. *Scientia Horticulturae* **78**, 237–260

Bordas, M., Montesinos, C., Dabauza, M., Salvador, A., Roig, L.A., Serrano, R. and Moreno, V. (1997) Transfer of the yeast salt tolerance gene *HAL* 1 to *Cucumis melo* L. cultivars and *in vitro* evaluation of salt tolerance. *Transformation Research* **6**, 41–50.

Colbert, T., Till, B.J., Tompa, R., Reynolds, S., Steine, M.N., Yeung, A.T., McCallum, C.M., Comai, L. and Henikoff, S. (2001) High-throughput screening for induced point mutations *Plant. Physiology* **126**, 480–484.

Daniells, I.G., Holland, J.F., Young, R.R., Alston, C.L. and Bernardi, A.L. (2001) Relationship between yield of grain sorghum (*Sorghum bicolor*) and soil salinity under field conditions. *Australian Journal of Experimental Agriculture* **41**, 211–217.

de Boer, A.H. and Wegner, L.H. (1997) Regulatory mechanisms of ion channels in xylem parenchyma cells. *Journal of Experimental Botany* **48**, 441–449.

Dierig, D.A., Shannon, M.C. and Grieve, C.M. (2001) Registration of WCL-SL1 salt-tolerant *Lesquerella fendleri* germplasm. *Crop Science* **41**, 604–605.

Epstein, E. (1977) Genetic potentials for solving problems of soil mineral stress: adaptation of crops to salinity, in *Plant Adaptation to Mineral Stress in Problem Soils* (ed M.J. Wright), Cornell University Agricultural Experiment Station, Ithaca, NY, pp. 73–123.

Epstein, E., Norlyn, J.D., Rush, D.W., Kingsbury, R., Kelley, D.B. and Wrana, A.F. (1980) Saline culture of crops: a genetic approach. *Science* **210**, 399–404.

Epstein, E. (1983) Crops tolerant to salinity and other mineral stresses, in *Better Crops for Food, Ciba Foundation Symposium* (eds. J. Nugent and M. O'Connor), Pitman, London, pp. 61–82.

Espinosa-Ruiz, A., Belles, J.M., Serrano, R. and Culiañez-Macia, F.A. (1999) *Arabidopsis thaliana* AtHAL3: a flavoprotein related to salt and osmotic tolerance and plant growth. *The Plant Journal* **20**, 529–539.

Flowers, T.J. (2004) Improving crop salt tolerance. *Journal of Experimental Botany* **55**, 307–319.

Flowers, T.J. and Yeo, A.R. (1995) Breeding for salinity resistance in crop plants—where next? *Australian Journal of Plant Physiology* **22**, 875–884.

Foolad, M.R. (2004) Recent advances in genetics of salt tolerance in tomato. *Plant Cell Tissue and Organ Culture* **76**, 101–119.

Foolad, M.R. and Chen, F.Q. (1999) RFLP mapping of QTLs conferring salt tolerance during the vegetative stage in tomato. *Theoretical and Applied Genetics* **99**, 235–243.

Foolad, M.R. and Lin, G.Y. (1997) Absence of a genetic relationship between salt tolerance during seed germination and vegetative growth in tomato. *Plant Breeding* **116**, 363–367.

Foolad, M.R., Lin, G.Y. and Chen, F.Q. (1999) Comparison of QTLs for seed germination under non-stress, cold stress and salt stress in tomato. *Plant Breeding* **118**, 167–173.

Foyer, C. and Halliwell, B. (1976) The presence of glutathione and glutathione reductase in chloroplasts: a proposed role in ascorbic acid metabolism. *Planta* **133**, 21–25.

Fukushima, E., Arata, Y., Endo, T., Sonnewald, U. and Sato, F. (2001) Improved salt tolerance of transgenic tobacco expressing apoplastic yeast-derived invertase. *Plant and Cell Physiology* **42**, 245–249.

Garg, A.K., Kim, J.K., Owens, T.G., Ranwala, A.P., Do Choi, Y., Kochian, L.V. and Wu, R.J. (2002) Trehalose accumulation in rice plants confers high tolerance leveles to different abiotic stresses. *Proceedings of the National Academy of Sciences USA* **99**, 15898–15903.

Gaxiola, R.A., Li, J., Unurraga, S., Dang, L.M., Allen, G.J., Alper, S.L. and Fink, G.R. (2001). Drought- and salt-tolerant plants result form overexpression of the AVP1 H^+-pump. *Proceedings of the National Academy of Sciences USA* **98**, 11444–11449.

Glenn, E.P., Brown, J.J. and Blumwald, E. (1999) Salt tolerance and crop potential of halophytes. *Critical Reviews in Plant Science* **18**, 227–255.

Gisbert., C., Rus, A.M., Bolarin, M.C., Lopez-Coronado, J.M., Arrillaga, I., Montesinos, C., Caro, M., Serrano, R. and Moreno, V. (2000) The yeast *HAL1* gene improves salt tolerance of transgenic tomato. *Plant Physiology* **123**, 393–402.

Goddijn, O.J.M. and Van Dun, K. (1999) Trehalose metabolism in plants. *Trends in Plant Sciences* **4**, 315–319.

Gueta-Dahan, Y., Yaniv, Z., Zilinskas, A. and Ben-Hayyinm, G. (1997) Salt and oxidative stress: similar and specific responses and their relation to salt tolerance in *Citrus. Planta* **203**, 460–469.

Greenway, H. and Munns, R. (1980) Mechanisms of salt tolerance in non-halophytes. *Annual Review of Plant Physiology* **31**, 149–190.

Gregorio, G.B., Senadhira, D., Mendoza, R.D., Manigvas, N.L., Roxas, J.P. and Guerta, C.Q. (2002) Progress in breeding for salinity tolerance and associated abiotic stresses in rice. *Field Crop Research* **76**, 91–101.

Grover, A., Aggarwal, P.K., Kapoor, A., Katiyar-Agarwal, S., Agarwal, M. and Chandramouli, A. (2003) Addressing abiotic stresses in agriculture through transgenic technology. *Current Science* **84**, 355–367.

Halliwell, B. and Gutteridge, J.M.C. (1986) Oxygen free radicals and iron in relation to biology and medicine: some problems and concepts. *Archives of Biochemistry and Biophysics* **246**, 501–514.

Hasegawa, P.M., Bressan, R.A., Zhu, J.K. and Bohnert, H.J. (2000) Plant cellular and molecular responses to high salinity. *Annual Reviews of Plant Physiology and Plant Molecular Biology* **51**, 463–499.

Hayashi, H., Alia Mustardy, L., Deshnium, P.G., Ida, M. and Murata, N. (1997) Transformation of *Arabidopsis thaliana* with the *codA* gene for choline oxidase; accumulation of glycinebetaine and enhanced tolerance to salt and cold stress. *Plant Journal* **12**, 133–142.

Higinbotham, N. (1973) Electropotentials of plant cells. *Annual Reviews of Plant Physiology* **24**, 25–46.

Holmstrom, K.O., Somersalo, S., Manda, A., Palva, T.E. and Welin, B. (2000) Improved tolerance to salinity and low temperature in transgenic tobacco producing glycine betaine. *Journal of Experimental Botany* **51**, 177–185.

Hong, Z.L., Lakkineni, K., Zhang, Z.M. and Verma, D.P.S. (2000) Removal of feedback inhibition of delta(1)-pyrroline-5-carboxylate synthetase results in increased proline accumulation and protection of plants from osmotic stress. *Plant Physiology* **122**, 1129–1136.

Hua, X.J., Van De Cotte, B., Van Montagu, M. and Verbruggen, N. (2001) The $5'$ untranslated region of the *At-P5R* gene is involved in both transcriptional and post-transcriptional regulation. *Plant Journal* **26**, 157–169.

Huang, J., Hirji, R., Adam, L., Rozwadowski, K.L., Hammerlindl, J.K., Keller, W.A. and Selvaraj, G. (2000) Genetic engineering of glycinebetaine production toward enhancing stress tolerance in plants: metabolic limitations. *Plant Physiology* **122**, 747–756.

Jacobsen, T. and Adams, R.M. (1958) Salt and silt in ancient Mesopotamian agriculture. *Science* **128**, 1251–1258.

Jia, G.X., Zhu, Z.Q., Chang, F.Q. and Li, Y.X. (2002) Transformation of tomato with the *BADH* gene from Atriplex improves salt tolerance. *Plant Cell Reports* **21**, 141–146.

Kasuga, M., Liu, Q., Miura, S., Yamaguchi-Shinozaki, K. and Shinozaki, K. (1999) Improving plant drought, salt, and freezing tolerance by gene transfer of a single stress-inducible transcription factor. *Nature Biotechnology* **17**, 287–291.

Khatun, S. and Flowers, T.J. (1995) Effects of salinity on seed set in rice. *Plant Cell and Environment* **18**, 61–67.

Kishor, P.B.K., Hong, Z., Miao, G.H., Hu, C.A.A. and Verma, D.P.S. (1995) Overexpression of Δ-1-pyrroline-5-carboxylate synthetase increases proline production and confers osmotolerance in transgenic plants. *Plant Physiology* **108**, 1387–1394.

Lilius, G., Holmberg, N. and Bulow, L. (1996) Enhanced NaCl stress tolerance in transgenic tobacco expressing bacterial choline dehydrogenase. *Biotechnology* **14**, 177–180.

Lindsay, M.P., Lagudah, E.S., Hare, R.A. and Munns, R. (2004) A locus for sodium exclusion (Nax1), a trait for salt tolerance, mapped in durum wheat. *Functional Plant Biology* **31**, 1105–1114.

Mano, Y. and Takeda, K. (1997) Mapping quantitative trait loci for salt tolerance at germination and the seedling stage in barley (*Hordeum vulgare* L.). *Euphytica* **94**, 263–272.

Maathuis, F.J.M. and Sanders, D. (1995) Contrasting roles in ion transport of two K^+-channel types in root cells of *Arabidopsis thaliana*. *Planta* **197**, 456–464.

Maathuis, F.J.M. and Amtmann, A. (1999) K^+ nutrition and Na^+ toxicity: the basis of cellular K^+/Na^+ ratios. *Annals of Botany* **84**, 123–133.

McNeil, S.D., Nuccio, M.L., Ziemak, M.J. and Hanson, A.D. (2001) Enhanced synthesis of choline and glycine betaine in transgenic tobacco plants that overexpress phosphoethanolamine *N*-mehtyltransferase. *Proceedings of the National Academy of Sciences USA* **98**, 10001–10005.

Nakayama, H., Yoshida, K., Ono, H., Murooka, Y. and Shinmyo, A. (2000) Ectoine, the compatible solute of *Halomonas elongata*, confers hyperosmotic tolerance in cultured tobacco cells. *Plant Physiology* **122**, 1239–1247.

Nanjo, T., Kobayashi, M., Yoshiba, Y., Kakubari, Y., Yamaguchi-Shinozaki, K. and Shinozaki, K. (1999) Antisense suppression of proline degradation improves tolerance to freezing and salinity in *Arabidopsis thaliana*. *FEBS Letters* **461**, 205–210.

Nuccio, M.L., McNeil, S.D., Ziemak, M.J., Hanson, A.D., Jain, R.K. and Selvaraj, G. (2000) Choline import into chloroplasts limits glycine betaine synthesis in tobacco: analysis of plants engineered with a chloroplastic or a cytosolic pathway. *Metabolic Engineering* **2**, 300–311.

Ohta, M., Hayashi, Y., Nakashima, A., Hamada, A., Tanaka, A., Nakamura, T. and Hayakawa, T. (2002) Introduction of a Na^+/H^+ antiporter gene from *Atriplex gmelini* confers salt tolerance in rice. *FEBS Letters* **532**, 279–282.

Ono, H., Sawads, K., Khunajakr, N., Tao, T., Yamamoto, M., Hiramoto, M., Shinmyo, A., Takano, M. and Murooka, Y. (1999) Characterization of biosynthetic enzymes for ectoine as a compatible solute in a moderately halophilic eubacterium, *Halomonas elongata*. *Journal of Bacteriology* **181**, 91–99.

Owen, P.A., Nickell, C.D., Noel, G.R., Thomas, D.J. and Frey, K. (1994) Registration of 'saline' soybean. *Crop Science* **34**, 1689.

Pilon-Smits, E.A.H., Terry, N., Sears, T., Kim, H., Zayed, A., Hwang, S.B., Van Dun, K., Voogd, E., Verwoerd, T.C., Krutwagen, R.W.H. and Goddijn, O.J.M. (1998) Trehalose-producing transgenic tobacco plants show improved growth performance under drought stress. *Journal of Plant Physiology* **152**, 525–532.

Prasad, K.V.S.K., Sharmila, P., Kumar, P.A. and Saradhi, P.P. (2000) Transformation of *Brassica juncea* (L.) Czern with bacterial *codA* gene enhances its tolerance to salt stress. *Molecular Breeding* **6**, 489–499.

Quesada, V., Garcia-Martinez, S., Piqueras, P., Ponce, M.R. and Micol, J.L. (2002) Genetic architecture of NaCl tolerance in *Arabidopsis*. *Plant Physiology* **130**, 951–963.

Rathinasabapathi, B. (2000) Metabolic engineering for stress tolerance: installing osmoprotectant synthesis pathways. *Annals of Botany* **86**, 709–716.

Richards, R.A. (1983) Should selection for yield in saline regions be made on saline or non-saline soils. *Euphytica* **32**, 431–438.

Richards, R.A. (1996) Defining selection criteria to improve yield under drought. *Plant Growth Regulation* **20**, 157–166.

Roberts, S.K. and Tester, M. (1997) Permeation of Ca^{2+} and monovalent cations through an outwardly rectifying channel in maize root stelar cells. *Journal of Experimental Botany* **48**, 839–846.

Romero, C., Belles, J.M., Vaya, J.L., Serrano, R. and Culianez-Macia, F.A. (1997) Expression of the yeast trehalose-6-phosphate synthase gene in transgenic tobacco plants: pleiotropic phenotypes include drought tolerance. *Planta* **201**, 293–297.

Rontein, D, Basset, G. and Hanson, A.D. (2002) Metabolic engineering of osmoprotectant accumulation in plants. *Metabolic Engineering* **4**, 49–56.

Roosens, N.H., Willem, R., Li, Y., Verbruggen, I.I., Biesemans, M. and Jacobs, M. (1999) Proline metabolism in the wild-type and in a salt-tolerant mutant of *Nicotiana plumbaginifolia* studied by (13)C-nuclear magnetic resonance imaging. *Plant Physiology* **121**, 1281–1290.

Roxas, V.P., Smith, R.K., Allen, E.R. and Allen, R.D. (1997) Overexpression of glutathione S-transferase/glutathione peroxidase enhances the growth of transgenic tobacco seedlings during stress. *Nature Biotechnology* **15**, 988–991.

Roxas, V.P., Lodhi, S.A., Garrett, D.K., Mahan, J.R. and Allen, R.D. (2000) Stress tolerance in transgenic tobacco seedlings that overexpress glutathione S-transferase/glutathione peroxidase. *Plant Cell Reports* **41**, 1229–1234.

Ruiz, J.M. and Blumwald, E. (2002) Salinity-induced glutathione synthesis in *Brassica napus*. *Planta* **214**, 965–969.

Rus, A., Yokoi, S., Sharkhuu, A., Reddy, M., Lee, B., Matsumoto, T.K., Koiwa, H., Zhu, J.K., Bressan R.A. and Hasegawa, P.M. (2001) AtHKT1 is a salt tolerance determinant that controls Na^+ entry into plant roots. *Proceedings of the National Academy of Sciences USA* **98**, 14150–14155.

Sakamoto, A., Alia, Murata, N. and Murata, A. (1998) Metabolic engineering of rice leading to biosynthesis of glycinebetaine and tolerance to salt and cold. *Plant Molecular Biology* **38**, 1011–1019.

Sakamoto, A. and Murata, N. (2000) Genetic engineering of glycinebetaine synthesis in plants: current status and implications for enhancement of stress tolerance. *Journal of Experimental Botany* **51**, 81–88.

Sentenac, H., Bonneaud, N., Minet, M., Lacroute, F., Salmon, J.M., Gaymard, F. and Grignon, C. (1992) Cloning and expression in yeast of a plant potassium ion transport system. *Science* **256**, 663–665.

Sheveleva, E., Chmara, W., Bohnert, H. and Jensen, R. (1997) Increased salt and drought tolerance by D-oninitol production in transgenic *Nicotiana tabaccum* L. *Plant Physiology* **115**, 1211–1219.

Shi, H., Ishitani, M., Kim, C. and Zhu, J.K. (2000) The *Arabidopsis thaliana* salt tolerance gene *SOS1* encodes a putative Na^+/H^+ antiporter. *Proceedings of the National Academy of Sciences USA* **97**, 6896–6901.

Shi, H., Lee, B.H., Wu, S.J. and Zhu, J.K. (2003) Overexpression of a plasma membrane Na^+/H^+ antiporter gene improves salt tolerance in *Arabidopsis thaliana*. *Nature Biotechnology* **21**, 81–85.

Spalding, E.P., Hirsch, R.E., Lewis, D.R., Zhi, Q., Sussman, M.R. and Lewis, B.D. (1999) Potassium uptake supporting plant growth in the absence of AKT1 channel activity. Inhibition by ammonium and stimulation by sodium. *Journal of General Physiology* **113**, 909–918.

Tal, M. and Shannon, M.C. (1983) Salt tolerance in two wild relatives of the cultivated tomato: responses of *Lycopersican esculentum, L. cheesmani, L. peruvianum, Solanum pennelli*, and F_1 hybrids of high salinity. *Australian Journal of Plant Physiology* **10**, 109–117.

Tanaka, Y., Hibino, T., Hayashi, Y., Tanaka, A., Kishitani, S., Takabe, T., Yokota, S. and Takabe, T. (1999) Salt tolerance of transgenic rice overexpressing yeast mitochondrial Mn-SOD in chloroplasts. *Plant Science* **148**, 131–138.

Tarczynski, M., Jensen, R. and Bohnert, H. (1993) Stress protection of transgenic tobacco by production of the osmolyte mannitol. *Science* **259**, 508–510.

Uozumi, N., Kim, E.J., Rubio, F., Yamaguchi, T., Muto, S., Tsuboi, A., Bakker, E.P., Nakamura, T. and Scroeder, J.I. (2000) The *Arabidopsis HKT1* gene homolog mediates inward Na^+ currents in *Xenopus laevis* oocytes and Na^+ uptake in *Saccaharomyces cerevisiae*. *Plant Physiology* **122**, 1249–1259.

Veena, J., Reddy, V.S. and Sopory, S.K. (1999) Glyoxalase I from *Brassica juncea*: molecular cloning, regulation and its over-expression confer tolerance in transgenic tobacco under stress. *The Plant Journal* **17**, 385–395.

Vernon, D.M., Tarczynski, M.C., Jensen, R.G. and Bohnert, H.J. (1993) Cyclitol production in transgenic tobacco. *The Plant Journal* **4**, 199–205.

Wegner, L.H. and Raschke, K. (1994) Ion channels in the xylem parenchyma of barley roots. *Plant Physiology* **105**, 799–813.

Wu, C.A., Yang, G.D., Meng, Q.W. and Zheng, C.C. (2004) The cotton *GhNHX1* gene encoding a novel putative tonoplast Na^+/K^+ antiporter plays an important role in salt stress. *Plant and Cell Physiology* **45**, 600–607.

Yancey, P.H., Clark, M.E., Hand, S.C., Bowlus, R.D. and Somero, G.N. (1982) Living with water stress: evolution of osmolyte systems. *Science* **217**, 1214–1222.

Yang, S.X., Zhao, Y.X., Zhang, Q., He, Y.K., Zhang, H. and Luo, H. (2001) HAL 1 mediate salt adaptation in *Arabidopsis thaliana*. *Cell Research* **11**, 142–148.

Yeo, E.T., Kwon, H.B., Han, S.E., Lee, J.T., Ryu, J.C. and Byu, M.O. (2000) Genetic engineering of drought resistant potato plants by introduction of the trehalose-6-phosphate synthase (TPS1) gene from *Saccharomyces cerevisiae*. *Molecules and Cells* **10**, 263–268.

Yin, X.Y., Yang, A.F., Zhang, K.W. and Zhang, J.R. (2004) Production and analysis of transgenic maize with improved salt tolerance by the introduction of *AtNHX1* gene. *Acta Botanica Sinica* **46**, 854–861.

Zhang, H.X. and Blumwald, E. (2001) Transgenic salt-tolerant tomato plants accumulate salt in foliage but not in fruit. *Nature Biotechnology* **19**, 765–768.

Zhang, H.X., Hodson, J.N., Williams, J.P. and Blumwald, E. (2001) Engineering salt-tolerant *Brassica* plants: characterization of yield and seed oil quality in transgenic plants with increased vacuolar sodium accumulation. *Proceedings of the National Academy of Sciences USA* **98**, 6896–6901.

Zhang, J.Z., Creelman, R.A. and Zhu, J.K. (2004) From laboratory to field. Using information from *Arabidopsis* to engineer salt, cold and drought tolerance in crops. *Plant Physiology* **135**, 615–621.

Zhu, B., Su, J., Chang, M. Verma, D.P.S., Fan, Y.-L. and Wu, R. (1998) Overexpression of a delta1-pyrroline-5-carboxylate synthetase gene and analysis of tolerance to water- and salt-stress in transgenic rice. *Plant Science* **139**, 41–48.

Zhu, J.K. (2001) Cell signaling under salt, water and cold stress. *Current Opinion in Plant Biology* **4**, 401–406.

Zhu, J.K. (2002) Salt and drought stress signal transduction in plants. *Annual Reviews of Plant Biology* **53**, 247–273.

2.9

Engineering Fungal Resistance in Crops

Maarten Stuiver

BASF Plant Science GmbH, Agricultural Center,
Carl Bosch Strasse 64, 67117 Limburgerhof, Germany

Introduction

With an established 6B$ worldwide market for fungicides, and losses due to fungal infection of at least that number, it comes as no surprise that engineering fungal resistance in plants through modern biotechnology has been among the earliest of research programs in plant biotechnology. Since 1988, field trials have been conducted in the US with genetically modified (GM) crops engineered to withstand fungal infection. However, at this moment, no GM fungal resistant crop is on the market. No simple, Bt-like, solution has been identified which makes plants resistant to fungal infection.

One of the major reasons for this is that plant pathogenic fungi are widely divergent in the way that they infect plants, and usually more than one fungus is of importance for each crop. No simple solution exists for multiple fungi. In addition, any engineered resistance needs to provide resistance at a very high level, that is, almost no infection should be visible on the produce. A number of fungal species are known for their ability to acquire resistance to fungicides and to the monogenic resistance that is introduced by breeders in new cultivars. With the current development timelines for engineered crops (from 10 years onwards) and the significant costs for developing and deregulating events, it is key that the engineered resistance is durable.

Plant Biotechnology. Edited by Nigel Halford.
© 2006 John Wiley & Sons, Ltd.

One other aspect that will become apparent is that plants have a complex pathogen defense system: there are tight balances between induction of specific defense pathways, and a balance between induction and prevention of cell death has to be maintained as well as induced specifically.

In this review I will discuss the approaches that were and are being followed, their limitations and what is most likely to get to products in the market.

Crops

What crops are the most suitable targets for engineering fungal resistance? Fungal diseases are widespread in almost any crop or ornamental plant species, and therefore there are seemingly endless possibilities. However, the deregulation costs of GM crops are significant, and therefore it will come as no surprise that the major field crops, like corn, wheat, potato and soybean, are among the crops which have been most extensively studied based on field trial applications of GM crops in the USA (see www.isb.vt.edu/cfdocs/fieldtests1.cfm for an overview of all applications). Overall, more than 20 plant species, including even unlikely candidates like *Pelargonium*, rhododendron, persimmon and *Dendrobium*, have been in such field trials, many of them only on small trials, though.

What is also clearly seen from this field trial overview is that in many cases the targets for engineering disease resistance are different from those normally treated with fungicides. Corn, for example, is not a typical fungicide market, but is the most significantly studied plant. Ear and stalk rot in corn cannot be treated easily with fungicides and genetic engineering is used to address this need. Likewise in wheat, most trials done in the US are directed toward preventing infection by *Fusarium graminearum*, the causal agent of head blight, which cannot be adequately prevented by fungicides currently available to the farmer.

Approaches to Engineering Resistance

It is often said that plants are able to withstand the infection of most fungi that land on them, and are trying to infect. The enormous diversity in infection modes that fungi use makes this a truly amazing feat.

Plant defense is adapted to stop infection, on the one hand, by necrotrophs (pathogens that bring toxic enzymes and metabolites and kill the tissue directly upon invasion; Walton, 1996) and, on the other hand, by hemibiotrophs or biotrophs (pathogens that feed on the plants parasitically in the initial stages of their life cycle, where the plant cells are kept alive, sometimes followed by a more necrotrophic infection style in later stages of the fungal life cycle). Plants are able to stop most fungi from successfully completing their life cycle, with usually only a handful of different fungi that really cause economic damage.

Most knowledge we have on the plant–fungal interaction originates from work on plant defense. It is known that plants bring an impressive arsenal of defense-related proteins. A crude classification of Arabidopsis (*Arabidopsis thaliana*) genes into functional categories indicates that 5–10 % of the genes may have a role in defense. Some of these defense-related genes are expressed constitutively and others can be induced upon pathogen infection. Interestingly, most plants can reach a high level of resistance even against their most damaging pathogens. This systemic acquired resistance (SAR) can be induced by treatment of the plants with high doses of an incompatible pathogen or elicitors of the hypersensitive response, for example. This finding has prompted a lot of work on the anti-fungal components of this acquired resistance, and the signaling networks behind it.

Anti-Fungal Protein Overexpression

One of the earliest strategies to engineer pathogen resistance was based on the findings of the SAR (Stintzi *et al.*, 1993). Overexpression of a small number of the most abundant anti-fungal proteins induced during SAR was thought to mimic the state of SAR in a continuous manner.

Although this approach is still followed frequently, the technology has not been proven extremely successful (e.g. Anand *et al.*, 2003). What has been shown clearly is that the effect of introduced anti-fungal proteins does depend strongly on the plant in which they are introduced. Punja and Raharjo did show, in 1996, the difficulty of this approach by transferring a chitinase gene to two different crops, which results in a resistance-elevating effect in carrot, but not in cucumber, even when the exact same pathogen was used to challenge the plants. This indicates that there is interplay between endogenous and transgene defense compounds. Because the exact endogenous defense complement is rarely known, and when known, poorly understood in function, the approach does not lend itself well to optimization.

Further, the SAR is not an absolute resistance, but an increase in partial resistance or increased tolerance to infection. Therefore, the approach to overexpress components of this SAR is unlikely to provide complete control. As another hurdle, the observed resistance is usually specific for a few pathogens, and generally does not provide a broad-spectrum control (Alexander *et al.*, 1993; Kim *et al.*, 2003).

The last concern with anti-fungal proteins is the durability of the engineered resistance. When challenged with isolated anti-fungal proteins *in vitro*, many fungi have been able to adapt, or become resistant to the anti-fungal protein. It is unclear whether the same might happen in transgenic crops. The transgenically produced anti-fungal protein may act in concert with endogenous anti-fungal compounds, and the pathogen may have more difficulties adapting to this mixture. It is currently impossible to understand the interplay of endogenous and transgenic anti-fungal proteins. There is hardly any knowledge of the complement of anti-fungal proteins present in crops. Very few crops have been sequenced on a genomic level. However, with the rice and Arabidopsis sequences available, and the genome of some other crops being elucidated in the next decade, it might not be impossible to get a better grip on the exact defense complement, and to employ this strategy more successfully.

Anti-Fungal Small Molecule Overexpression

It is not just proteins that are induced during a defense response and SAR. Also anti-fungal small chemicals are synthesized constitutively and induced during pathogen infection. These phytoalexins have been shown to play a key role in plant defense against fungal attack (e.g. in the case of resveratrol in grapes). Attempts to engineer resistance by overexpression of the biosynthetic enzymes in heterologous plants have led to sporadic successes (Hain *et al.*, 1993; He and Dixon, 2000; Hipskind and Paiva, 2000).

In general, since the specific anti-fungal activity of phytoalexins is relatively low, the amounts needed for proper anti-fungal activity are extremely high. These high levels are not always reachable in heterologous crop plants (Hain *et al*, 1993). At the concentrations reached in the heterologous plant, most phytoalexins lack the broad-spectrum and high-level anti-fungal activity. It is questionable whether this strategy will be widely used.

Using General Defense Pathways or Compounds to Engineer Resistance

Instead of using isolated anti-fungal proteins or small molecules, which are induced during SAR, it is also possible to exploit entire defense pathways. There are a number of distinct signaling pathways that induce a co-ordinated expression of sets of anti-fungal proteins and/or compounds (Delaney *et al.*, 1994; Thomma *et al.*, 2001; Pieterse and van Loon, 1999). Interestingly, induction of each of these pathways appears to induce resistance to some, but not to all pathogens (Thomma *et al.*, 2001). These pathways are commonly denominated by the small signal molecules that play a central role in the induction of the defense proteins. The best known compound that induces such a pathway is salicylic acid (SA). Treatment of plants with SA and analogs thereof does induce expression of a subset of plant defense responses (Uknes *et al.*, 1992). Also here, increased resistance is found only to a limited set of pathogens (Gorlach *et al.*, 1996; Thomma *et al.*, 1998). Other pathways appear to use the small plant signaling molecules, jasmonic acid (JA), ethylene or certain reactive oxygen species (ROS), such as superoxide and hydrogen peroxide, for their activation (Feys and Parker, 2000; Grant *et al.*, 2000). Several other pathways inducing defense exist in addition (e.g., Nakashita *et al.*, 2002; Pieterse *et al.,* 1998; Ton and Mauch-Mani, 2004).

Induction of each of these signaling pathways leads to the induction of various defense components, including sets of anti-fungal proteins, phytoalexins and enzymes involved in plant cell wall reinforcement. Some defense components are only induced when multiple pathways are stimulated simultaneously (Penninckx *et al.*, 1998), but it should be noted that there is significant antagonism observed between most pathways as well (Doares *et al.*, 1995; Leslie and Romani, 1988).

The finding that induction needs to be pathogen specific, and that there is antagonism between the pathways makes all engineering with these pathways hazardous, at least.

Our understanding of the specific induction of targeted pathways in different crops is limited. Nor do we understand for most crops which pathway is causing resistance to which fungus. Upon adjustment of the balances between the pathways, one is likely to end up with plants that show increased resistance to some pathogens, but are clearly more vulnerable to others (Hoffman *et al.*, 1999).

Increased susceptibility to minor pathogens may not always be a drawback. It might even be advantageous in modern agriculture to adjust the balance between susceptibility to different pathogens. Farmers have many tools at hand to combat diseases. Some can be well treated with existing fungicides. At this moment, however, we understand too little from the pathways and their interplay in agriculturally important crops. Most of our knowledge comes only from Arabidopsis. Knowledge on these pathways in crop plants is only just developing.

Another aspect that warrants attention is that constitutive overexpression of the pathway through, for example synthesis of SA analogs (Yalpani *et al.*, 2001) or key signaling compounds in the pathway (Bowling *et al.*, 1994; Cao *et al.*, 1998) usually causes lack of vigor or impaired development in the plants (Maleck *et al.*, 2002). It appears that plants operate on a tight energy balance and too frequent or too high induction of defense pathways impacts plant growth or development readily. Because of these limitations, engineering resistance through this approach will most likely be infrequent.

Even application of a chemical inducer of the SA pathway (benzothiodiazole, also known as BTH/Bion/Actiguard, Gorlach *et al.*, 1996) that can be employed more flexibly, dependent on environmental conditions and imminent threats of fungal infection, has not proven to be a great success in the marketplace.

The Hypersensitive Response, Cell Death and Apoptosis

The hypersensitive response, which occurs in plants after recognition of a viral, bacterial, insect- or nematode-derived or fungal elicitor, causes a cascade of defense responses. Locally, at the site of infection by the pathogen, cell death occurs, which in itself is sufficient to stop biotrophic pathogens from completing their life cycle. In addition, high local accumulation of phenolics and cell wall reinforcements are observed in cells surrounding the area of cell death, preventing further spread of the pathogen. More distal to the site of infection, plants induce SAR, the state of induced expression of anti-fungal proteins and compounds, which has been mentioned above (Costet *et al.*, 1999; Dorey *et al.*, 1997; Melchers and Stuiver, 2000; Morel and Dangl, 1997). Most likely, this co-ordinated response is sufficient to stop all microbial pathogens. It certainly is extremely well suited to stop biotrophic pathogens. For necrotrophic fungal pathogens there is limited information available, suggesting that also these are effectively halted in their life cycle by a hypersensitive response (Bonnet *et al.*, 1996).

Recognition of the elicitor was determined genetically to occur through a resistance (R) gene product made by the plants. It should be mentioned, however, that it is still unclear how the R gene product recognizes the elicitor, and exactly what it is that R gene products recognize (Dixon *et al.*, 2000). Whatever the mechanism, the recognition is highly specific and depends on the presence of the exact R gene and its cognate elicitor. R genes are commonly found in modern agricultural crops, where they have been bred in (through conventional breeding) from wild relatives of the crop. Invariably, crops with such introgressed R genes go through boom-bust cycles. That is, the crop performs very well for several years in warding off the pathogen (boom), but after a few years of widespread cultivation, new varieties of the pathogen appear to overcome the resistance, and the introgressed resistance becomes useless (bust). Only very few exceptions to that boom-bust cycle are found. A very limited number of R genes have survived a long time in agricultural use (Pink and Puddephat, 1999).

The reason for the breakdown of plant resistance is that pathogens can readily mutate or shed the corresponding elicitor, avoiding recognition by the R gene product (Gassmann *et al.*, 2000; Joosten *et al.*, 1994). When they have mutated or lost the gene, they can regain their original pathogenicity. In most cases studied, no effect on pathogenicity of the fungus or bacterium can be measured.

Breeders have usually responded to breakdown of resistance by breeding in novel R genes. Interestingly, older R genes, which were once overcome by the pathogen, can sometimes regain usefulness. This suggests that there is some negative effect on the pathogen to mutate or lose these elicitors.

The finding that 'old' R genes might still be employed has triggered the idea of stacking or pyramiding R genes (Datta *et al.*, 2002; Rommens and Kishore, 2000). With breeding, stacking is only possible to some extent. Some R genes are allelic, and thus in diploid plants maximally two R genes can be represented from a sometimes huge allelic pool of possible R genes. Especially in crops where R genes for key diseases are found in allelic locations, therefore, very limited possibilities exist for non-GM breeding of durable resistance.

Stacks or 'Smarties'

Genetic engineering offers a unique possibility to stack/pyramid R genes, including allelic variants, and one approach currently being followed is the transformation of multiple effective R genes on one construct.

The high level of resistance, which is one of the key indicators of success, will be guaranteed by using R genes. One of the big questions is certainly the durability of the resistance. Are pathogens able to overcome the loss or mutation of several elicitors at the same time, and at what rate is this possible? How many R genes need to be stacked to get a truly durable resistance, and is there any qualitative difference in the sets of R genes selected? Answers to these questions will hopefully come once such products are released onto the market.

However, stacking is still a challenge due to the size of R genes and the limitations that exist in transferring large DNA fragments. Other hurdles remain as well, such as proving that all R genes introduced are truly working simultaneously. This will require the availability of elicitors and development of R gene assay systems. Last, but not least, protein production of R gene products, as a necessary step in deregulation of a GM product, might be needed. Expression of R gene proteins in heterologous protein production systems has been attempted by many, but has failed consistently, and is obviously far from trivial.

Having said that, this approach is certainly one of the most promising for engineering high-level and durable resistance to fungi (and/or other pathogens) in plants. It will be used primarily in crops where only one single (hemi)-biotrophic pathogen is causing significant economic loss, and is worth the substantial investment in development of the product.

Another interesting solution to the problem of creating durable resistant plants with R genes was identified by Jonathan Jones (Dangl and Jones, 2001). His 'smarties' model also entails the transfer of multiple R genes to a crop, but instead of stacked R genes in one plant, a multitude of single R gene transformants are made. Then the various different R gene-containing plants are to be interplanted on the field. Epidemics should then be significantly delayed on such fields.

One of the major drawbacks of this approach is the necessary deregulation of multiple lines simultaneously, which may make the approach unattractive to all but the most economically important pathogens.

Non-host R Genes

Plants are called non-hosts when no accession of that plant species shows susceptibility to a given pathogen, based on the inability of that pathogen to complete its life cycle. The cause of non-host resistance cannot be pinpointed to one particular defense reaction, and ranges from complex aspects like leaf architecture to active recognition of pathogen elicitors, leading to a hypersensitive response (Heath, 2000; Holub and Cooper, 2004; Kamoun *et al.*, 1998). Interestingly, many of the components needed for non-host resistance are also needed for the normal (host–pathogen) hypersensitive response (Aarts *et al.*, 1998; Parker *et al.*, 1996; Vleeshouwers *et al.*, 2000).

The finding that at least some forms of non-host resistance appear to be based on R genes is intriguing. Obviously, in the case of non-host plant-mediated recognition, this particular pathogen is unable to shed or mutate its elicitor genes, and to become infectious on this plant species. Whether the R genes are unusually stable/durable due to recognition of an essential pathogenicity factor, or whether a specifically durable combination of R genes is found in such non-host plants, is not known at present (Kamoun, 2001). It is clear that some elicitors causing part of the non-host resistance are also recognized by R genes in host plants (Laugé *et al.*, 2000). This information supports the hypothesis that a specific, durable combination of R genes is present in the non-host plants.

One of the best studied non-host–pathogen interactions provides additional evidence for this theory. The *Phytophthora infestans* INF1 elicitor gene is recognized by several members of the tobacco family (*Nicotiana* ssp.), which are non-hosts. Interestingly, targeted deletion of the INF1 gene does not lead to a reduced pathogenicity on the host plant, potato. On most tobacco family members, still no infection can occur, suggesting that other factors are recognized in addition (Kamoun *et al.*, 1998).

Interestingly, one subspecies of this tobacco family, *N. benthamiana*, allows some infection and completion of the fungal life cycle when infected with the INF1 mutant. Because the INF1 deletion was also detected at a very low frequency in natural populations of *P. infestans*, it is surprising to see *N. benthamiana* classified as a non-host. Possibly it is a potential host, but since this plant species is not grown widely, and the INF1 mutation is rare, no infection has been noticed to this date. Presumed non-host plants could therefore be hosts in reality. To find proof that all accessions of the plant species are resistant to all pathogen accessions is not possible. This point may prove to be one of the potential pitfalls of this approach. It will require targeted experimentation, like described above, to develop knowledge on the molecular components of non-host resistance. With these components it may be possible to discriminate true non-host from host plants.

Since non-host plants are characterized by the fact that no single accession of that genus can be infected by the pathogen, genetic characterization of the resistance in these plants is not straightforward. Wide hybridization, sometimes using somatic cell hybrids between non-host and host plants (Zimnoch-Guzowska *et al*., 2003) is one way to identify and clone these interesting sets of genes. Recently, some useful tools for high-throughput screening of cloned R genes (Brigneti *et al*., 2004; Kamoun *et al*., 2003) have also been developed, and this will certainly speed up this research area considerably.

With the development of these tools, the identification of non-host R genes will advance rapidly in the coming years. No doubt this is one of the approaches that will be used in the near future to engineer fungal resistance successfully. Since the recognition of the pathogen by R genes is highly specific, it is very likely that only one pathogen can be tackled at any one time. The approach is best suited to crops with one or a very limited number of economically important pathogens.

Non-Pathogen-Specific HR: Avirulence Gene Strategy and Hairtrigger Responses

It is extremely tempting to make use of the power of the hypersensitive response pathway and try to trigger this sort of defense after infection by many different pathogens. Pierre de Wit (1992) has proposed to use plants transformed with a construct containing a pathogen-inducible promoter driving expression of an elicitor gene. In this case, when plants sense pathogen infection, they respond by synthesizing the elicitor and then by the accumulation of the elicitor through the triggering of the hypersensitive response. The hypersensitive response, as mentioned before, is a powerful and multi-layered defense response which is likely to stop most, if not all pathogens. Ricci and co-workers (Keller *et al*., 1999) have successfully shown this 'avirulence gene strategy' to work in tobacco.

The key issue in this strategy is the tight regulation of the pathogen-inducible promoter. The hypersensitive response includes local cell death and a large array of defense responses. Any leakiness of the promoter will cause detrimental effects to plant growth and yield. Keller *et al*. (1999) have created transgenic tobacco plants with no indications of inadvertent triggering of the hypersensitive response, at least under greenhouse conditions. Conditions during growth on the field might be more variable, making the tight regulation of promoter activity even more difficult. However, with the possibilities to do genome-wide transcriptional profiling, it should be possible to find promoters satisfying these criteria.

Alternatively, a mix-and-match approach has been shown to improve the specificity and response of pathogen-inducible promoters (Rushton *et al*., 2002); this should also help solve the hurdle of specifically responding promoters.

Interestingly, the results of Keller *et al*. (1999) show that the approach works not only against biotrophic fungi (which are known to be stopped efficiently by HR) but also against the true necrotroph, *Botrytis cinerea*, the cause of gray mould on various economically important crops. This is surprising, since Govrin and Levine (2000) had shown that in Arabidopsis an induced hypersensitive response facilitates rather than inhibits infection by this pathogen. Whether the findings of Keller *et al*. or those of Govrin and Levine are the exception to the rule still remains to be determined. When, indeed, a hypersensitive response would contribute to infection by necrotrophic pathogens, it will be crucial to identify pathogen-inducible promoters that do not respond to this class of pathogens.

This approach should be one of the most likely candidates to get products to the market, which have a broad-spectrum disease resistance.

Most modern barley cultivars contain a specific allele of the *Mlo* gene, making them stably and durably resistant to powdery mildew, one of the major pathogens of barley. This specific allele of *Mlo* was bred in conventionally. Interestingly, several alleles of *Mlo* cause a race-unspecific type of resistance, protecting against all known isolates of powdery mildew. The best functioning *Mlo* alleles were found to encode a mutant or truncated form of a novel protein, which causes a 'hair-trigger' HR response upon powdery mildew infection. Interestingly, the resistance is specific for powdery mildew. Barley lines having the mutant *Mlo* gene are as susceptible to other pathogens as barley lines having the fully functional *Mlo* gene. In fact, the mutation may make barley slightly more susceptible to another pathogen, *Magneporthe grisea* (Jarosch *et al.*, 1999), which does not cause any significant economic damage in barley. This is a clear example where shifting the balance in resistance between pathogens is acceptable.

The fact that such a hair-trigger HR may cause durable and high-level resistance to all accessions of an important pathogen makes an exciting finding. Unfortunately, very little is known about the molecular mechanisms, and especially whether it is possible to transfer or recreate such a resistance in other plant species. Powdery mildew is one of the major pathogens of many economically important crops, and therefore this research has a very significant potential. Because of the poor knowledge on this effect, however, it is unlikely to provide approaches to engineer fungal resistance in the short term.

Inhibition of Cell Death

Necrotrophic pathogens cause cell death in their hosts immediately upon infection. However, in many cases it is not completely understood how cell death is induced in the hosts and a multitude of toxic molecules is used (Walton, 1996). It is increasingly becoming clear that fungal toxins induce an apoptotic program in their hosts (Jones *et al.*, 2001; Wang *et al.*, 1996). Dickman *et al.* (2001) and Lincoln *et al.* (2002) have overexpressed anti-apoptotic proteins successfully to make plants resistant against these toxins. They observed not only toxin resistance, but also increased resistance to the necrotrophic pathogens.

Unfortunately, this approach suffers from serious side effects. It has become apparent that an efficient apoptotic program is essential for the functioning of plant R genes. Interference with apoptosis leads to an incomplete hypersensitive response, thereby allowing biotrophic pathogens, also the ones to which a plant has the R genes, to infect (Matsumura *et al.*, 2003). Furthermore, plant development processes also rely on apoptosis. Structure formation is highly dependent on well-functioning apoptotic machinery, and interference can lead to serious defects in development (Lam, 2004).

For the success of this approach, necrotrophic pathogen-specific induction of expression is required. Whether that induction can ever be fast enough is a serious question. It will not be easy to use this anti-apoptotic processes in engineering resistance.

Interference with Pathogenesis

Most fungal (and bacterial) pathogens bring a diverse set of enzymes, proteins and metabolites for the process of infecting plants. Because it is such a complex mixture, it has been found that many single components are dispensable for infection (e.g. Schafer, 1993). However, it is safe to assume these compounds contribute to pathogenicity in some way.

Interestingly, some compounds appear to be essential pathogenicity factors. Loss of such compounds makes pathogens unable to infect their hosts or complete their life cycle (Daub and Ehrenshaft, 2000; Johal and Briggs, 1992; Proctor *et al.*, 1995; Tanaka *et al.*, 1999). When these factors are secreted from the fungus (to modulate host responses), a good opportunity may exist to design strategies to neutralize or inactivate these compounds. For example, the necrotrophic fungus, *Sclerotinia sclerotiorum,* synthesizes oxalic acid upon plant infection. This compound is a clear pathogenicity factor for *Sclerotinia.* Mutants that have lost the ability to synthesize oxalic acid are invariably non-pathogenic, and revertants that regain the ability to make this compound also recover their virulence (Cessna *et al.*, 2000). Oxalic acid may bind cell-wall-associated calcium needed to degrade plant cell wall structures, enabling the fungus to colonize the plant, but it was recently also found to inhibit the onset of plant defense (Cessna *et al.*, 2000). The dependence of the fungus on oxalic acid can be used to engineer efficient resistance (Hartman *et al.*, 1992; Hu *et al.*, 2003).

Barley and wheat oxalic acid oxidase breaks down oxalic acid into carbon dioxide and hydrogen peroxide, which disable the main fungal pathogenicity factor. The formed hydrogen peroxide is a plant signal molecule, inducing a plant defense pathway. Thus, oxalic acid oxidase serves a dual role: it disables fungal pathogenicity and, with the breakdown product, it boosts plant defense (Hu *et al.*, 2003).

Similarly, the pathogen *F. graminearum*, which causes ear or head scab in wheat, has such a crucial pathogenicity factor. The fungal mycotoxin, deoxynivalenol (DON), plays a key role in plant infection and also poses a very significant health concern to humans and animals. Engineering wheat to contain an enzyme that converts DON to a non-active form has helped create wheat plants with significant resistance to this fungus (Okubara *et al.*, 2002). Syngenta indicate on their website that launch of a *Fusarium*-resistant wheat might be as early as 2007. A similar approach in maize against *F. moniliforme,* targeting the important mycotoxin, fumonisin, has been worked on by Pioneer-DuPont (Duvick, 2001).

The identification of other such essential pathogenicity factors is prompting strategies to counteract or interfere with their functions (Laugé *et al.*, 1998). It must be added, however, that the currently available knowledge of the molecules involved in pathogenicity represents only the tip of the iceberg, and for many fungi, even those of significant economic importance, the molecular basis of pathogenicity is poorly understood. With ongoing genomic sequencing of microbes and fungi, and possibilities to do functional genomics in pathogens, this approach might well become the most preferred strategy to deal with necrotrophic fungi.

Conclusions

Engineering resistance to diseases has taken much more time than engineering resistance to insects. Whereas a significant proportion of crops grown today have been engineered to express insect resistance (through the use of Bt insect toxin genes), no commercial transgenic product with enhanced disease resistance is available currently. Several attempts to engineer resistance have succeeded and are being developed into products (Duvick, 2001; Hu *et al.*, 2003; Okubara *et al.*, 2002), but rather poor knowledge of the molecular components in plant–fungus interactions is still holding back efficient product development.

Also, the studies of plant defense pathways and of the regulation of cell death and apoptosis have clearly shown interesting aspects. Plants carefully balance different defense pathways and keep a strict control over cell death and cell survival, in order to maximize resistance to pathogens. Improving any aspect of this defense to some pathogens impairs resistance to others.

The most promising strategies, therefore, in my opinion, are those that do not interfere with these tight controls. For (hemi)-biotrophic fungi, the approaches to use stacks of carefully chosen R genes, either from the host itself or from non-host plant species, is likely to be a preferred approach. The presence of a few extra R genes, on the massive amounts that plant cells already have (Monosi *et al.*, 2004), will probably not interfere with resistance to non-target pathogens. The identification of the R genes and the preferred combination of them will still take time. Therefore, no product will be delivered to the market shortly, within the next 5 years.

Alternatively, plants containing avirulence gene constructs might be developed. This is clearly an approach with a higher risk, but with the potential of broad pathogen resistance. The tight control of the expression of the trait needed in the field is still an issue, though. Therefore, this approach may deliver products later.

For necrotrophic fungi, the most promising approach is based on inactivation of essential pathogenicity factors. We know very few essential pathogenicity factors, but where identified, approaches for inactivation or breakdown can be developed. As indicated above, this approach is underlying some of the most developed products in the pipeline.

With increased public concerns about mycotoxins in food crops, there is a clear need for alternative or additional control strategies, which may include biotechnology. Many of the approaches mentioned could provide a high level of protection, comparable to what might be achieved today with fungicides, and provide durable resistance as well.

Acknowledgment

I would like to thank Sietske Hoekstra for critical reading of the manuscript.

References

Aarts, N., Metz, M., Holub, E., Staskawicz, B.J., Daniels, M.J., and Parker, J.E. (1998) Different requirements for EDS1 and NDR1 by disease resistance genes define at least two R gene-mediated signaling pathways in *Arabidopsis. Proceedings of the National Academy of Sciences USA* **95**, 10306–10311.

Alexander, D., Goodman, R.M., Gut-Rella, M., Glascock, C., Weymann, K., Friedrich, L., Maddox, D., Ahl-Goy, P., Luntz, T., Ward, E. and Ryals, J. (1993) Increased tolerance to two oomycete pathogens in transgenic tobacco expressing pathogenesis-related protein 1a. *Proceedings of the National Academy of Sciences USA* **90**, 7327–7331.

Anand, A., Zhou, T., Trick, H.N., Gill, B.S., Bockus, W.W. and Muthukrishnan, S. (2003) Greenhouse and field testing of transgenic wheat plants stably expressing genes for thaumatin-like protein, chitinase and glucanase against *Fusarium graminearum. Journal of Experimental Botany* **54**, 1101–1111.

Bonnet, P., Bourdon, E., Ponchet, M., Blein, J.P. and Ricci, P. (1996) Acquired resistance triggered by elicitins in tobacco and other plants. *European Journal of Plant Pathology* **102**, 181–192.

Bowling, S.A., Guo, A., Cao, H., Gordon, A.S., Klessig, D.F. and Dong, X. (1994) A mutation in *Arabidopsis* that leads to constitutive expression of systemic acquired resistance. *Plant Cell* **6**, 1845–1857.

Brigneti, G., Martin-Hernandez, A.M., Jin, H., Chen, J., Baulcombe, D.C., Baker, B. and Jones, J.D. (2004) Virus-induced gene silencing in *Solanum* species. *The Plant Journal* **39**, 264–272.

Cao, H., Li, X. and Dong, X. (1998) Generation of broad-spectrum disease resistance by over-expression of an essential regulatory gene in systemic acquired resistance. *Proceedings of the National Academy of Sciences USA* **95**, 6531–6536.

Cessna, S.G., Sear, V.E., Dickman, M.B. and Low, P.S. (2000) Oxalic acid, a pathogenicity factor for *Sclerotinia sclerotiorum*, suppresses the oxidative burst of the host plant. *Plant Cell* **12**, 2191–2200.

Costet, L., Cordelier, S., Dorey, S., Baillieul, F., Fritig, B. and Kauffmann, S. (1999) Relationship between localized acquired resistance (LAR) and the hypersensitive response (HR): HR is necessary for LAR to occur and salicylic acid is not sufficient to trigger LAR. *Molecular Plant-Microbe Interactions* **12**, 655–662.

Dangl, J.L. and Jones, J.D.G. (2001) Plant pathogens and integrated defence responses to infection. *Nature* **411**, 826–833.

Datta, K., Baisakh, N., Thet, K.M., Tu, J. and Datta, S.K. (2002) Pyramiding transgenes for multiple resistance in rice against bacterial blight, yellow stem borer and sheath blight. *Theoretical and Applied Genetics* **106**, 1–8.

Daub, M.E. and Ehrenshaft, M. (2000) The photoactivated *Cercospora* toxin cercosporin: contributions to plant disease and fundamental biology. *Annual Review of Phytopathology* **38**, 461–490.

Delaney, T.P., Uknes, S., Vernooij, B., Friedrich, L. and Weymann, K. (1994) A central role of salicylic acid in plant disease resistance. *Science* **266**, 1247–1250.

Dickman, M.B., Park, Y.K., Oltersdorf, T., Li, W., Clemente, T. and French, R. (2001) Abrogation of disease development in plants expressing animal antiapoptotic genes. *Proceedings of the National Academy of Sciences USA* **98**, 6957–6962.

Dixon, M.S., Golstein, C., Thomas, C.M., van Der Biezen, E.A. and Jones, J.D. (2000) Genetic complexity of pathogen perception by plants: the example of Rcr3, a tomato gene required specifically by Cf-2. *Proceedings of the National Academy of Sciences USA* **97**, 8807–8814.

Doares, S.H., Narvaez-Vazquez, J., Conconi, A. and Ryan, C.A. (1995) Salicylic acid inhibits synthesis of proteinase inhibitors in tomato leaves induced by systemin and jasmonic acid. *Plant Physiology* **108**, 1741–1746.

Dorey, S., Baillieul, F., Pierrel, M., Saindrenan, P., Fritig, B. and Kauffmann, S. (1997) Spatial and temporal induction of cell death, defense genes, and accumulation of salicylic acid in tobacco leaves reacting hypersensitively to a fungal glycoprotein elicitor. *Molecular Plant-Microbe Interactactions* **10**, 646–655.

Duvick, J. (2001) Prospects for reducing fumonisin contamination of maize through genetic modification. *Environmental Health Perspectives* **109**, 337–342.

Feys, B.J. and Parker, J.E. (2000) Interplay of signalling pathways in plant disease resistance. *Trends in Genetics* **16**, 449–455.

Gassmann, W., Dahlbeck, D., Chesnokova, O., Minsavage, G.V., Jones, J.B. and Staskawicz, B.J. (2000) Molecular evolution of virulence in natural field strains of *Xanthomonas campestris* pv. *vesicatoria. Journal of Bacteriology* **182**, 7053–7059.

Gilchrist, D.G., Lincoln, J., Richael, C., Pan, Z., Overduin, B., Fan, X., Smith, K., Li, J., Haworth, R. and Bostock, R. (2000) Disease after death: concept and consequence, in *Biology of Plant–Microbe Interactions* (eds P.J.M. de Wit, T. Bisseling and W.J. Stiekema), International Society for Molecular Plant-Microbe Interactions St. Paul, pp. 406–410.

Gorlach, J., Volrath, S., Knauf-Beiter, G., Hengy, G., Beckhove, U., Kogel, K.H., Oostendorp, M., Staub, T., Ward, E., Kessmann, H., and Ryals, J. (1996) Benzothiadiazole, a novel class of inducers of systemic acquired resistance, activates gene expression and disease resistance in wheat. *Plant Cell* **8**, 629–643.

Govrin, E.M. and Levine, A. (2000) The hypersensitive response facilitates plant infection by the necrotrphic pathogen *Botrytis cinerea*. *Current Biology* **10**, 751–757.

Grant, J.J., Yun, B.W. and Loake, G.J. (2000) Oxidative burst and cognate redox signalling reported by luciferase imaging: identification of a signal network that functions independently of ethylene, SA and Me-JA but is dependent on MAPKK activity. *The Plant Journal* **24**, 569–582.

Hain, R., Reif, H.J., Krause, E., Langbartels, R., Kindl, H., Vornam, B., Wiese, W., Schmelzer, E., Schreier, P.H., Stocker, R.H. and Stenzel, K. (1993) Disease resistance results from foreign phytoalexin expression in a novel plant. *Nature* **361**, 153–156.

Hartman, C.L., Sarjit, J. and Schmitt, M.S. (1992) Newly characterised oxalate oxidase and uses therefore. Patent application WO 92/14824.

He, X.Z. and Dixon, R.A. (2000) Genetic manipulation of isoflavone 7-*O*-methyltransferase enhances biosynthesis of 4′-*O*-methylated isoflavonoid phytoalexins and disease resistance in alfalfa. *Plant Cell* **12**, 1689–1702.

Heath, M.C. (2000) Nonhost resistance and non-specific plant defenses. *Current Opinion in Plant Biology* **3**, 315–319.

Hipskind, J.D. and Paiva, N.L. (2000) Constitutive accumulation of a resveratrol-glucoside in transgenic alfalfa increases resistance to *Phoma medicaginis*. *Molecular Plant-Microbe Interactions* **13**, 551–562.

Hoffman, T., Schmidt, J.S., Zheng, X. and Bent, A.F. (1999) Isolation of ethylene-insensitive soybean mutants that are altered in pathogen susceptibility and gene-for-gene disease resistance. *Plant Physiology* **119**, 935–949.

Holub, E.B. and Cooper, A. (2004) Matrix, reinvention in plants: how genetics is unveiling secrets of non-host disease resistance. *Trends in Plant Science* **9**, 211–214.

Hu, X., Bidney, D.L., Yalpani, N., Duvick, J.P., Crasta, O., Folkerts, O. and Lu, G. (2003) Overexpression of a gene encoding hydrogen peroxide-generating oxalate oxidase evokes defense responses in sunflower. *Plant Physiology* **133**, 170–181.

Jarosch, B., Kogel, K.-H. and Schaffrath, U. (1999) The ambivalence of the barley Mlo locus: mutations conferring resistance against powdery mildew (*Blumeria graminis f. sp. hordei*) enhance susceptibility to the rice blast fungus *Magnaporthe grisea*. *Molecular Plant-Microbe Interactions* **12**, 508–514.

Johal, G.S. and Briggs, S.P. (1992) Reductase activity encoded by the Hm1 resistance gene in maize. *Science* **258**, 985–987.

Jones, C., Ciacci-Zanella, J.R., Zhang, Y., Henderson, G. and Dickman, M. (2001) Analysis of fumonisin B1-induced apoptosis. *Environmental Health Perspectives* **109** (Suppl 2), 315–320.

Joosten, M.H.A.J., Cozijnsen T.J. and De Wit P.J.G.M. (1994) Host resistance to fungal tomato pathogen lost by a single base pair change in an avirulence gene. *Nature* **367**, 384–386.

Kamoun, S. (2001) Nonhost resistance to *Phytophthora*: novel prospects for a classical problem. *Current Opinion in Plant Biology* **4**, 295–300.

Kamoun, S., van West, P., Vleeshouwers, V.G.A.A., de Groot, K.E. and Govers, F. (1998) Resistance of *Nicotiana benthamiana* to *Phytophthora infestans* is mediated by the recognition of the elicitor protein INF1. *Plant Cell* **10**, 1413–1425.

Kamoun, S., Hamada, W. and Huitema, E. (2003) Agrosuppression: a bioassay for the hypersensitive response suited to high-throughput screening. *Molecular Plant-Microbe Interactions* **16**, 7–13.

Keller, H., Pamboukjian, N., Ponchet, M., Poupet, A., Delon, R., Verrier, J.L., Roby, D. and Ricci, P. (1999) Pathogen induced elicitin production in transgenic tobacco generates a hypersensitive response and nonspecific disease resistance. *Plant Cell* **11**, 223–235.

Kim, J.K., Jang, I.C., Wu, R., Zuo, W.N., Boston, R.S., Lee, Y.H., Ahn, I.P. and Nahm, B.H. (2003) Co-expression of a modified maize ribosome-inactivating protein and a rice basic chitinase gene in transgenic rice plants confers enhanced resistance to sheath blight. *Transgenic Research* **4**, 475–484.

Lam, E. (2004) Controlled cell death, plant survival and development. *Nature Reviews Molecular Cell Biology* **5**, 305–315.

Laugé, R., Joosten, M.H., Haanstra, J.P., Goodwin, D.H., Lindhout, P. and de Wit, P.J. (1998) Successful search for a resistance gene in tomato targeted against a virulence factor of a fungal pathogen. *Proceedings of the National Academy of Sciences USA* **95**(15), 9014–9018.

Laugé, R., Goodwin, H. de Wit, P.J. and Joosten, M.H. (2000) Specific HR-associated recognition of secreted proteins from *Cladosporium fulvum* occurs in both host and non-host plants. *The Plant Journal* **23**, 735–745.

Leslie, C.A. and Romani, R.J. (1988) Inhibition of ethylene biosynthesis by salicylic acid. *Plant Physiology* **88**, 833–837.

Lincoln, J.E., Richael, C., Overduin, B., Smith, K., Bostock, R. and Gilchrist, D.G. (2002) Expression of the antiapoptotic baculovirus p35 gene in tomato blocks programmed cell death and provides broad-spectrum resistance to disease. *Proceedings of the National Academy of Sciences USA* **99**, 15217–15221.

Maleck, K., Neuenschwander, U., Cade, R.M., Dietrich, R.A., Dangl, J.L. and Ryals, J.A. (2002) Isolation and characterization of broad-spectrum disease-resistant *Arabidopsis* mutants. *Genetics* **160**, 1661–1671.

Matsumura, H., Nirasawa, S., Kiba, A., Urasaki, N., Saitoh, H., Ito, M., Kawai-Yamada, M., Uchimiya, H., Terauchi, R. (2003) Over-expression of Bax inhibitor suppresses the fungal elicitor-induced cell death in rice (*Oryza sativa* L) cells. *The Plant Journal* **33**, 425–434.

Melchers, L.S. and Stuiver, M.H. (2000) Novel genes for disease-resistance breeding. *Current Opinion in Plant Biology* **3**, 147–152.

Monosi, B., Wisser, R.J., Pennill, L. and Hulbert, S.H. (2004) Full-genome analysis of resistance gene homologues in rice. *Theoretical and Applied Genetics* **109**, 1434–1447.

Morel, J.-B. and Dangl, J.L. (1997) The hypersensitive response and the induction of cell death in plants. *Cell Death and Differentiation* **4**, 671–683.

Nakashita, H., Yasuda, M., Nishioka, M., Hasegawa, S., Arai, Y., Uramoto, M., Yoshida, S. and Yamaguchi, I. (2002) Chloroisonicotinamide derivative induces a broad range of disease resistance in rice and tobacco. *Plant Cell Physiology* **43**, 823–831.

Okubara, P.A., Blechl, A.E., McCormick, S.P., Alexander, N.J., Dill-Macky, R., Hohn, T.M.L. (2002) Engineering deoxynivalenol metabolism in wheat through the expression of a fungal trichothecene acetyltransferase gene. *Theoretical and Applied Genetics* **106**, 74–83.

Parker, J.E., Holub, E.B., Frost, L.N., Falk, A., Gunn, N.D. and Daniels, M.J. (1996) Characterization of eds1, a mutation in *Arabidopsis* suppressing resistance to *Peronospora parasitica* specified by several different *RPP* genes. *Plant Cell* **8**, 2033–2046.

Penninckx, I.A.M.A, Thomma, B.P.H.J., Buchala, A., Metraux, J.-P. and Broekaert, W.F. (1998) Concomitant activation of jasmonate and ethylene response pathways is required for induction of a plant defensin gene in *Arabidopsis*. *Plant Cell* **10**, 2103–2113.

Pieterse, C.M., van Wees, S.C., van Pelt, J.A., Knoester, M., Laan, R., Gerrits, H., Weisbeek, P.J., and van Loon, L.C. (1998) A novel signaling pathway controlling induced systemic resistance in *Arabidopsis*. *Plant Cell* **10**, 1571–1580.

Pieterse, C.M. and van Loon, L.C. (1999) Salicylic acid-independent plant defence pathways. *Trends in Plant Science* **4**, 52–58.

Pink, D. and Puddephat, I. (1999) Deployment of disease resistance genes by plant transformation— a 'mix and match' approach. *Trends in Plant Science* **4**, 71–75.

Proctor, R.H., Hohn, T.M. and McCormick, S.P. (1995) Reduced virulence of *Gibberella zeae* caused by disruption of a trichothecene toxin biosynthetic gene. *Molecular Plant-Microbe Interactions* **8**, 593–601.

Punja, Z.K. and Raharjo, S.H.T. (1996) Response of transgenic cucumber and carrot plants expressing different chitinase enzymes to inoculation with fungal pathogens. *Plant Disease* **80**, 999–1005.

Rommens, C.M. and Kishore, G.M. (2000) Exploiting the full potential of disease-resistance genes for agricultural use. *Current Opinion in Biotechnology* **11**, 120–125.

Rushton, P.J., Reinstadler, A., Lipka, V., Lippok, B. and Somssich, I.E. (2002) Synthetic plant promoters containing defined regulatory elements provide novel insights into pathogen- and wound-induced signaling. *Plant Cell* **14**, 749–762.

Schafer, W. (1993) The role of cutinase in fungal pathogenesis. *Trends in Microbiology* **1**, 69–71.

Stintzi, A., Heitz, T., Prasad, V., Wiedemann-Merdinoglu, S., Kauffmann, S., Geoffroy, P., Legrand, M. and Fritig, B. (1993) Plant 'pathogenesis-related' proteins and their role in defense against pathogens. *Biochimie* **75**, 687–706.

Tanaka, A., Shiotani, H., Yamamoto, M. and Tsuge, T. (1999) Insertional mutagenesis and cloning of the genes required for biosynthesis of the host-specific AK-toxin in the Japanese pear pathotype of *Alternaria alternata*. *Molecular Plant-Microbe Interactions* **12**, 691–702.

Thomma, B.P., Penninckx, I.A., Broekaert, W.F. and Cammue, B.P. (2001) The complexity of disease signaling in *Arabidopsis*. *Current Opinion in Immunology* **13**, 63–68.

Thomma, B.P., Eggermont, K., Penninckx, I.A., Mauch-Mani, B., Vogelsang, R., Cammue, B.P. and Broekaert, W.F. (1998) Separate jasmonate-dependent and salicylate-dependent pathways in *Arabidopsis* are essential for resistance to distinct microbial pathogens. *Proceedings of the National Academy of Sciences USA* **95**, 15107–15111.

Ton, J. and Mauch-Mani, B. (2004) Beta-amino-butyric acid-induced resistance against necrotrophic pathogens is based on ABA-dependent priming for callose. *The Plant Journal* **38**, 119–130.

Uknes, S., Mauch-Mani, B. Moyer, M. Potter, S. and Williams, S. (1992) Acquired resistance in *Arabidopsis*. *Plant Cell* **4**, 645–656.

Vleeshouwers, V.G., van Dooijeweert, W., Govers, F., Kamoun, S. and Colon, L.T. (2000) The hypersensitive response is associated with host and non-host resistance to *Phytophthora infestans*. *Planta* **210**, 853–864.

Walton, J.D. (1996) Host-selective toxins: agents of compatibility. *Plant Cell* **8**, 1723–1733.

Wang, W., Jones, C., Ciacci-Zanella, J., Holt, T., Gilchrist, D.G. and Dickman, M.B. (1996) Fumonisins and *Alternaria alternata lycopersici* toxins: sphinganine analog mycotoxins induce apoptosis in monkey kidney cells. *Proceedings of the National Academy of Sciences USA* **93**, 3461–3465.

de Wit, P.J.G.M. (1992) Molecular characterization of gene-for-gene systems in plant-fungus interactions and the application of avirulence genes in control of plant pathogens. *Annual Review of Phytopathology* **30**, 391–418.

Yalpani, N., Altier, D.J., Barbour, E., Cigan, A.L. and Scelonge, C.J. (2001) Production of 6-methylsalicylic acid by expression of a fungal polyketide synthase activates disease resistance in tobacco. *Plant Cell* **13**, 1401–1409.

Zimnoch-Guzowska, E., Lebecka, R., Kryszczuk, A., Maciejewska, U., Szczerbakowa, A. and Wielgat, B. (2003) Resistance to *Phytophthora infestans* in somatic hybrids of *Solanum nigrum* L. and diploid potato. *Theoretical and Applied Genetics* **107**, 43–48.

PART III
SAFETY AND REGULATION

3.1

Plant Food Allergens

E.N. Clare Mills and John A. Jenkins

Institute of Food Research, Norwich, NR4 7UA, United Kingdom

Peter R. Shewry

Rothamsted Research, Harpenden, Herts AL5 2JQ, United Kingdom

Allergenicity—What is It?

Antibodies (also known as immunoglobulins; Ig) are produced by the body as part of the humoral immune responses that defend the body against infections. In general, antibodies are developed toward macromolecules, which must have a mass of M_r at least 4000–5000 in order to stimulate antibody production. The effectiveness with which a given molecule can do this is known as its immunogenicity. One particular kind of antibody, known as IgE, is generated as part of the normal immune reaction to parasitic infections, such as malaria. However, under certain circumstances predisposed individuals can develop IgE responses to macromolecules (usually proteins) present in environmental agents, such as pollen, dusts and foods. In such cases the macromolecules recognized by the IgE are known as allergens. IgE can bind to the surface of certain cells, such as basophils or mast cells, which contain granules of preformed inflammatory mediators, including histamine. When a multi-valent allergen encounters such cells, it can cross-link the cellular-bound IgE, an event which causes the cells to release their mediators. These mediators then go on to cause the physiological changes which manifest themselves as an allergic reaction. Allergic symptoms are quite diverse in nature, and can include respiratory (wheezing, shortness of breath), skin (rashes such as hives, eczema and atopic dermatitis) and gastrointestinal (vomiting, diarrhea) reactions, alone or in various combinations. In some

Plant Biotechnology. Edited by Nigel Halford.
© 2006 John Wiley & Sons, Ltd.

cases systemic reactions can be severe, even life-threatening, with the most severe being anaphylaxis.

IgE-mediated reactions generally occur quite rapidly following exposure to an allergen and have been classically defined as type I hypersensitivity reactions. They are quite distinct from food intolerances, which may take much longer (even days) to manifest themselves and do not necessarily have an immunological basis. Thus, lactose intolerance is a food intolerance, which actually results from the loss of a digestive enzyme meaning that individuals cannot consume cows' milk and some derived products. In contrast, the gluten intolerance syndrome known as coeliac disease is thought to involve a cellular-based immune reaction, but without the development of an IgE-mediated response, to wheat gluten proteins. Although the gluten proteins which trigger coeliac disease are not IgE-binding proteins, some workers refer to them as allergens, and they are treated as such in the FAO-WHO Codex Alimentarius Commission Food Labeling (FAO-WHO, 1999).

IgE-mediated food allergies fall into two distinct types. The first is foods which are thought to sensitize individuals via the gastrointestinal tract and can often result in severe symptoms, including anaphylaxis. Second, several types of allergy to fresh fruits and vegetables develop as a consequence of individuals first developing inhalant allergies to agents such as pollen and latex. These have been called cross-reactive allergy syndromes because of the high levels of homology between the sensitizing allergens found in pollen and latex and allergens in foods such as apples, peaches, pears and celery. This similarity means that IgE developed to pollen can bind homologous allergens in plant foods. Consequently, consumption of fresh fruits can often trigger allergic reactions in individuals sensitized to pollen. Severe reactions can occur but generally the symptoms triggered in cross-reactive allergy syndromes are much milder (often being confined to the oral cavity) than in foods believed to sensitize via the oral route. This has led to pollen-fruit/vegetable allergies being associated with oral allergy syndrome (OAS).

Of the foods classically described as causing the majority of food allergies, four originate from plants: peanuts, tree nuts, wheat and soy (Bush and Hefle, 1996). The present paper summarizes the relationships, structures and properties of the plant food allergens involved in these two types of allergy and discusses the approaches being used to identify potential allergens, an important part of identifying and managing allergenic risks posed by novel foods. Finally, we also discuss the potential applications of GM technology in removing allergenic proteins from plant food tissues.

Known Plant Food Allergens

Allergen Nomenclature

An official list of allergens is compiled by the World Health Organization and the International Union of Immunological Societies, designated by the Allergen Nomenclature subcommittee (Hoffmann *et al.*, 1994; Larson and Lowenstein, 1996). Allergens included in this listing must induce IgE-mediated allergy in humans with a prevalence of IgE reactivity above 5 %. However, many other IgE-binding proteins have also been identified in the literature which do not comply with this rigorous definition but are commonly known as allergens. An allergen is termed 'major' if it is recognized by IgE from at least 50 % of a group of allergic individuals but this is not related to allergenic potency. Other allergens are termed 'minor'. Allergens from the same species with greater than 67 % sequence identity are defined as iso-allergens. The allergen designation itself is based on the Latin name of the species from which it originates, being composed of the first three letters of the genus, followed by the first letter of the species finishing with an Arabic number. Thus, for example, Jug r 1 relates to an allergen from *Juglans regia* (walnut).

Classification of Plant Food Allergens

Most of the major allergens in the main allergenic foods have now been identified and sequenced, providing information which has allowed the structural and evolutionary relationships of plant food allergens to be considered (Breinteneder and Ebner, 2000; Breinteneder and Redauer, 2004; Mills *et al.*, 2003, 2004a). It is now becoming evident that the vast majority of plant food allergens that sensitize via the gastro-intestinal tract belong to only two major plant protein superfamilies, the prolamin and cupin super-families, with others such as the cysteine protease family being less prevalent. In contrast, inhalant allergens that can trigger food allergies are more diverse in nature and belong to a larger number of protein families, although they are dominated by one family, which is known as the Bet v 1 family based on the major inhalant allergen in birch pollen.

Many plant food allergens can also be defined as 'pathogenesis–related' (PR) proteins, being expressed as part of a defense response following infection by microbes and fungi or in response to abiotic stress (van Loon and van Strien, 1999). Currently 17 different classes of PR proteins have been defined, with known food allergens belonging to seven of these (Hoffmann-Sommergruber, 2002). The relationship between allergenic potential and PR family membership is intriguing, and while no molecular basis for this has been defined, many PR allergens are small proteins which are resistant to denaturation at low pH and proteolysis *in vitro* (Hoffman-Sommergruber, 2002).

A summary of the three main allergen families: the prolamin and cupin superfamilies that sensitize via the gastrointestinal tract and the Bet v 1 family, together with a broad overview of the many other plant food allergens identified till date, is given below.

The Prolamin Superfamily

Members of the prolamin superfamily, originally defined by Kreis *et al.* (1985), share a characteristic framework of cysteine residues, containing Cys Cys and Cys X Cys motifs, where X represents any other residue. Subfamilies within this superfamily include the alcohol-soluble (prolamin) storage proteins of cereals, the 2S storage albumins, non-specific lipid transfer proteins (nsLTP) and cereal inhibitors of α-amylase and/or trypsin, all but the nsLTPs being restricted to seeds. The conservation of structure extends to the protein fold, at least for the 2S albumins, nsLTP and α-amylase/trypsin inhibitors, which all share a related fold consisting of bundles of four α-helices stabilized by disulphide bonds. In the prolamin storage proteins this structural motif and the conserved cysteine skeleton has been disrupted by the insertion of a repetitive domain comprising short motifs rich in proline and glutamine.

Seed storage prolamins. While generally associated with triggering the food intolerance syndrome, coeliac disease, the seed storage prolamins of cereals have also been found to trigger allergies to cereals, both by ingestion and through inhalation (Bakers' asthma). Wheat can cause atopic dermatitis (Varjonen *et al.*, 1995, 2000) and a condition known as exercise-induced anaphylaxis (EIA) (Varjonen *et al.*, 1997). The latter is a severe allergic reaction that certain patients experience only when taking exercise following consumption of a problem food. Several allergens have been described as triggering such reactions, including a γ- and an α-gliadin (Palosuo *et al.*, 1999, 2001a) and an ω5-gliadin (Palosuo *et al.*, 2001b). Other prolamin storage proteins have been identified as major cereal allergens, including both the polymeric HMW and LMW subunits of glutenin, as well as the monomeric γ- and α-gliadins (Maruyama *et al.*, 1998; Simonato *et al.*, 2001b; Tanabe *et al.*, 1996; Watanabe *et al.*, 1995). Cooking appears to affect the type of symptoms observed, (Guenard-Bilbault, Kanny and Moneret-Vautrin, 1999) and recent studies have indicated that baking may be essential for allergenicity of cereal prolamins (Simonato *et al.*, 2001a).

2S storage albumins. These proteins are abundant in the seeds of many plant species and have been identified as major allergens in a number of plant foods including peanut (Ara h 2, 6, 7; Burks *et al.*, 1992b, Kleber *et al.*, 1999), oriental and yellow mustards (Bra j 1, Sin a 1; Menedez-Arias *et al.*, 1988; Monsalve *et al.*, 1993), oilseed rape (the napin, Bra n 1), castor bean (Ric c 1, 3), walnut (Jug r 1), sesame (Ses i 1; Pastorello *et al.*, 2001b and Ses i 2; Beyer *et al.*, 2002a), Brazil nut (Ber e 1; Pastorello *et al.*, 1998b), almond (Poltronieri *et al.*, 2002) and sunflower (Kelly, Hlywka and Hefle, 2000). It has also been reported that the 2S albumins of soy (Shibasaki *et al.*, 1980) and chickpea (Vioque *et al.*, 1999) are allergens.

nsLTPs. The nsLTPs are distributed even more widely in plants than 2S albumins, with over 100 sequences being available from a variety of plant organs including seeds, fruit and vegetative tissues. Major allergens have been identified in many Rosacea fruits including peach (Pru p 3, initially named Pru p 1; Pastorello *et al.*, 1999a), apple (Mal d 3; Pastorello, 1999c; Sanchez-Monge *et al.*, 1999), apricot (Pruarl; Pastorello *et al.*, 2000a) and cherry (Pru av 3; Scheurer *et al.*, 2001) as well as a variety of other fruits, nuts and seeds such as grape (Pastorello *et al.*, 2003), hazelnut (Pastorello *et al.*, 2002), maize (Zea m 14; Pastorello *et al.*, 2000b) and asparagus (Diaz-Perales *et al.*, 2002). This distribution has led to the family being termed as 'pan-allergens' (Sanchez-Monge *et al.*, 1999). Like the 2S albumins, nsLTPs can sensitize by both inhalation and/or the oral route, since they are present in pollen and dusts as well as in foods. An individual has been reported with an occupational allergy to spelt (a type of wheat) who experienced symptoms on consuming a variety of fruits and where the major allergen was a nsLTP, indicating some true IgE cross-reactivity (Pastorello *et al.*, 2001a). nsLTPs have also been identified as pollen allergens (Duro *et al.*, 1996; Tejera *et al.*, 1999) but as yet there is no clear evidence of a link between sensitization to pollen nsLTP and subsequent development of allergy to fruit.

Cereal α-amylase inhibitors. These proteins can also sensitize individuals via the lungs by inhalation of dusts and are responsible for occupational allergies such as Bakers' asthma (wheat, barley and rye) as well as via the gastrointestinal tract (wheat, barley and rice). A single M_r ~15 000 subunit of wheat has been identified as a food allergen (James *et al.*, 1997) while a second wheat inhibitor, termed CM3, has been identified as an allergen triggering atopic dermatitis (Kusaba-Nakayama, *et al.*, 2000). An M_r 16 000 beer allergen has also been described, which originates from barley and appears to belong to the α-amylase inhibitor family (Curioni *et al.*, 1999) together with an M_r 16 000 allergen in maize (Pastorello *et al.*, 2000b). Other well-characterized allergenic inhibitors occur in rice with M_r s of about 14 000–16 000. In particular an M_r ~16 000 subunit termed RA 17, having been identified (Nakase *et al.*, 1996).

The Cupin Superfamily

The name 'cupin' is derived from the Latin for a small barrel or cask (Dunwell, Purvis and Khuri, 2004), and refers to the common basic β-barrel structure of these proteins, which has been duplicated in certain proteins (bicupins), such as the 7S and 11S globulin seed storage proteins. Although the subunits of the two types of globulin have no obvious sequence similarity, they do have remarkably similar three-dimensional structures (Adachi *et al.*, 2001). Major allergens include the 7S and 11S globulins of legumes including soybean (Burks, Brooks and Sampson, 1988; Helm *et al.*, 2000), peanut (Ara h 1 and Ara h 3; Burks *et al.*, 1991, 1992a), one of the subunits of the proteolytically processed 7S globulin of lentil (Lopez-Torrejon *et al.*, 2003; Sanchez-Monge *et al.*, 2000) and the 7S globulin of pea (Wensing *et al.*, 2003). In addition, a number of allergenic globulins have been identified in tree nuts, including the 7S globulins of walnut (Jug r 2) (Teuber *et al.*, 1999), sesame (Beyer *et al.*, 2002a), cashew nut (Ana c 1) (Wang *et al.*, 2002) and hazelnut (Beyer *et al.*, 2002b). 11S globulins have been reported to be allergenic in almond (also known as almond major protein, AMP) (Roux *et al.*, 2001) and are probable allergens in coconut and walnut (Teuber and Petersen, 1999). There is also an indication that at least one member of the germin subfamily of cupins, namely from bell pepper, is an allergen (Leitner *et al.*, 1998).

Other Allergens

A number of other plant food allergens, many considered minor allergens, are scattered across a wide range of protein families. These include two members of the C1 protease superfamily, the major allergen from kiwi fruit, actinidin (Act c 1) (Fahlbusch *et al.*, 1998; Möller *et al.*, 1997; Pastorello *et al.*, 1998a) and a major allergen involved in soybean-induced atopic dermatitis (Ogawa *et al.*, 1993) known variously as Gly m Bd 30K, Gly m 1 or P34. The latter protein does not exhibit protease activity as the active site cysteine residue has been replaced with a glycine (Kalinski *et al.*, 1990). Another minor plant food allergen family is the Kunitz protease inhibitors, which have been identified as allergens in soybean (Moroz and Yang, 1980; Herian *et al.*, 1980; Burks *et al.*, 1994) and potato (Seppala *et al.*, 2001). The peanut lectin, aglutinnin, has also been identified as a minor allergen in peanut (Burks *et al.*, 1994).

Cross-Reactive Food Allergens

Bet v 1 family. The major birch pollen allergen, Bet v 1, is related to a class of putative plant defense proteins (PR10). Sensitization to Bet v 1 in birch pollen often results in cross-reactive allergies triggered by homologs in fruit (including Mal d 1 of apple and Pru av 1 of cherry), vegetables (including Api g 1 of celery and Dau c 1 of carrot) and seeds (such as Gly m 4, formerly known as SAM22, from soy). Although the biological role of PR10 proteins has not been established, they are presumed to be defensive (Hoffman-Sommergruber, 2002). The published 3-D structure of the birch pollen allergen Bet v 1 (Gajhede *et al.*, 1996) shows that it folds into a single domain formed by a single twisted seven-stranded anti-parallel β-sheet and three α-helices. A tunnel is present in the center of the molecule which can bind either two molecules of deoxycholine or plant steroids, such as brasinolides, which may be part of the physiological function of Bet v 1 in plants (Markovic-Housley *et al.*, 2003).

The IgE cross-reactivity observed across the Bet v 1 family members from various plant species and plant tissues is a direct consequence of their high level of sequence similarity (Hofmann-Sommergruber *et al.*, 1999a; Kleine-Tebbe *et al.*, 2002; Mittag *et al.*, 2004). Thus, apple Mal d 1b has 56 % of identical residues relative to Bet v 1, while soybean Gly m 4 has 47 %, celery Api g 1.0201 has 38 % and carrot Dau c 1 has 36 %. While the majority of birch pollen allergic patients suffer allergic reactions to apple, only 14 % react to celery (Jankiewicz *et al.*, 1997). This is because Api g 1 binds to only a fraction of the spectrum of the anti-Bet v 1 IgE antibodies (Bohle *et al.*, 2003).

Other cross-reactive allergens. A second major group of pollen allergens involved in cross-reactive allergies are the profilins, a group of soluble proteins which bind to the actin cytoskeleton of cells (Redauer and Hoffman-Sommergruber, 2003). Sensitization to this protein in birch pollen (known as Bet v 2) results in IgE responses, which show cross-reactivity with homologs in a wide range of fresh fruits and vegetables, including apples (Mal d 4, van Ree *et al.*, 1995) and celery (Redauer and Hoffman-Sommergruber, 2003; Api g 4, Scheurer *et al.*, 2000). However, profilins are often considered minor allergens and while they can bind IgE, there is some debate as to their ability to actually trigger allergic symptoms, as indicated by a recent study of pollen allergic subjects with 14 derived plant foods (Wensing *et al.*, 2002).

Thaumatin-like protein allergens have been identified in fruits such as cherry (Pru av 2, Inschlag *et al.*, 1998), apple (Mal d 2, Krebitz *et al.*, 2003), and kiwi fruit (Gavrovic-Jankulovic *et al.*, 2002). These are small, highly disulphide-bonded proteins (with eight intermolecular disulphide bonds) involved in the protection of plants against pathogens. Few pollen analogs have been identified and it is currently an open question as to whether they may be involved in oral sensitization to fruits and vegetables, or involved in cross-reactive allergy syndromes, possibly as a result of their being glycosylated (Breiteneder, 2004).

Inhalation of two other PR-type proteins from latex can also result in dietary allergies to foods such as avocado, chestnut, banana, fig and kiwi (Sanchez-Monge *et al.*, 2004). One important group of latex allergens involved in these allergy syndromes is the class I chitinases, the IgE cross-reactivity resulting from a common hevein-like domain (Diaz-Perales *et al.*, 2002; Posch *et al.*, 1999) which has a high level of homology across plant species. Another allergen involved is the PR-2 family member Hev b 2, a protein with β1,3-glucanase activity. However, the protein is glycosylated and it has been suggested that the observed IgE reactivity may result from cross-reactive carbohydrate determinants, which are not usually able to trigger full allergic reactions (Yagami *et al.*, 2002).

Identifying Potential Allergens in Foods

An assessment relating to potential allergenicity has to be made as part of the safety assessment of any novel food (including genetically modified organisms (GMOs)). However, this is not a straightforward process because we do not have a complete understanding of what makes some people become allergic and not others, and what makes some foods, such as peanuts, more allergenic than foods such as peas. It is also limited by the lack of adequate animal models for studying food allergenicity (Penniks and Knippels, 2001). Initially decision-tree approaches were proposed to facilitate this assessment, first by ILSI in conjunction with the Food Biotechnology Council in 1996 (Metcalf 1996; Figure 3.1.1a), which was followed by a revised version proposed by a joint FAO/WHO Expert Consultation on Foods Derived from Biotechnology (WHO-FAO, 2001; Figure 3.1.1b). However, in 2002, the third session of the FAO-WHO Codex Alimentarius Commission Ad Hoc Intergovernmental task force on Foods Derived from Biotechnology (ALINORM 03/34) decided not to elaborate the decision tree approach. Instead they recommended that the risk assessment process should adopt an integrated step-wise, case-by-case approach which, since no single criterion is sufficiently predictive of allergenicity, could then take account of information of several types. This includes characterizing the relationships between novel proteins and known allergens, and using bioinformatics tools and information regarding IgE cross-reactivity defined using patient allergic sera to provide information on the potential of novel proteins to elicit an allergic reaction in an individual who is already sensitized to another agent (e.g. birch pollen). These are complimented by *in vitro* measures of protein digestibility which relate to both elicitation and sensitization, together with information on *in vivo* sensitization using animal models, which for all their limitations are the best tools currently available.

Bioinformatic Approaches to Determining Similarity Between Novel Proteins and Known Allergens

In silico methods are very useful for giving a quantitative indication as to the similarity of a novel protein to a known allergen. The sequences of more than 100 different food allergens have now been published, corresponding to 70–80 % of the identified food allergens and, when isoallergens are included, the number of reported sequences rises to over 200. Bioinformatic approaches require the assembly of such allergen sequences into appropriate databases. Unfortunately the term 'allergen' is used loosely in databases such as ExPASy (www.expast.org) and is largely dependent on the information supplied by those submitting sequences. Thus, many proteins sequenced 20 years ago are now known to be allergens but do not include the term allergen in their description. In contrast allergen has been included in the description of other proteins simply because of their homology to known allergens, despite a lack of data on clinical reactivity. It is therefore difficult to extract a set of allergen sequences from the general sequence databases (Gendel, 2002). As a consequence, curated lists and databases of allergen sequences have

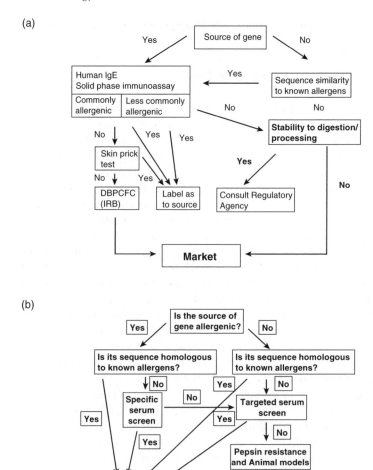

Figure 3.1.1 *Decision tree approaches for allergenic risk assessment proposed by ILSI in conjunction with the Food Biotechnology Council (a) and the joint FAO/WHO Expert Consultation on Foods Derived from Biotechnology (b) Based on ILSI, 1996 and FAO-WHO, 2001.*

been compiled with a clear definition as to their allergenicity, and include that developed by the Allergen Nomenclature Committee (http://www.allergen.org/), that maintained at the US FDA (http://www.iit.edu/~sgendel/foodallr.htm; Gendel, 1998), the Food Allergy Research and Resource Program (FARRP) at the University of Nebraska (http://www.allergenonline.com; Hileman *et al.*, 2002) and the SDAP site which contains a database of allergen sequences and information on IgE epitopes (http://fermi.utmb.edu/SDAP; Ivanciuc *et al.*, 2003).

Such databases form a starting point for analyzing similarities between novel food proteins and known allergens. The FAO/WHO Expert Consultation on Foods Derived from Biotechnology (FAO-WHO, 2001) considered that a novel protein would exhibit IgE cross-reactivity with a known allergen if it shared either $\geq 35\%$ sequence identity within a window of 80 amino acids (determined using a suitable gap penalty) or an identical stretch of six contiguous amino acids, such matches then being verified using allergic sera. This is based on our current knowledge of the size of B-cell and T-cell epitopes, features which are known to be important in developing humoral immune responses (Becker and Reese, 2001; Metcalfe *et al.*, 1996). However, using default values in alignment programs designed to detect evolutionary relationships may not identify the short identical sequences postulated to result in cross-reactivity, which may be addressed by only scoring identities and ignoring the statistical significance of the alignment (Gendel, 2002).

While pair-wise screening for sequence homology is very effective at identifying proteins which belong to families with highly conserved sequences, such as the Bet v 1 family of pollen allergens, it is less effective when levels of sequence identity drop below 40%. Large numbers of false positive allergens are identified using this approach, with 41 of 50 maize proteins being identified as potential allergens using a peptide length of six residues. Furthermore, even when the peptide length was increased to eight amino acids it falsely predicted seven of 50 maize proteins as allergens (Hileman *et al.*, 2002). More satisfactory results were obtained using the FASTA algorithm, which identified nine maize proteins with $>35\%$ identity over 80 amino acids and only six which were similar over the length of the proteins.

Other computational methods are being developed to 'predict' allergenicity. Recently a method based on a computerized learning system was developed. This extracts features from sequence alignments produced using FASTA3 and the k-Nearest-Neighbor (kNN) algorithm and can correctly classify about 81% of allergens and 98% of non-allergens (Soeria-Atmadja *et al.*, 2004; Zorzet, Gustafsson and Hammerling, 2002). Sequence motifs representing allergens have also been identified using an interactive motif discovery approach (Stadler and Stadler, 2003). While this process did not use a curated database of authenticated allergen sequences, it identified 52 allergen motifs and proved to be around 20 times more precise than the approach described in the FAO-WHO recommendations when used to scan query sequences. Finally, the SDAP online database and associated tools have been assembled in order to identify whether a query sequence matches any known IgE epitopes. Using, among others, the thaumatin-like PR5 protein allergens from cedar pollen (Jun a 3), Ivanciuc *et al.*, 2003 searched for potential cross-reactive IgE in proteins within their allergen sequence database and identified several candidates including homologs from cherry (Pru av 2), bell-pepper (Cap a 1) and apple (Mal d 2).

Stability of Potential Allergens to Digestion

One of the common properties identified for many allergenic proteins is their resistance to digestion by proteases such as pepsin, when compared with non-allergenic proteins. As peptides require a molecular weight of M_r greater than 3000 in order to stimulate an immune response, large stable fragments, as well as intact proteins, have the potential to act as sensitizers. Consequently, resistance to pepsin digestion has become enshrined in the approaches for assessing the allergenic potential of novel proteins.

In an early study of pepsin resistance, a number of allergenic proteins tested remained either undigested or gave stable fragments, which persisted for 8–60 min (depending upon the allergen), while the non-allergens were completely digested after less than 15 s (Astwood, Leach and Fuchs, 1996). However, subsequent studies using a larger number of plant proteins, including a number involved in metabolic processes with no history of food allergenicity, have thrown doubt on the usefulness of pepsin resistance for predicting allergenicity. IgE cross-reactive allergens present in protein extracts of latex, avocado, kiwi fruit and banana together with the main allergens associated with OAS in melon and peach have, in general, been shown to be broken down rapidly by pepsin, with the potato allergens being more stable (Yagami *et al.*, 2000). An extensive study of 22 allergens and 16 non-allergens showed that 12/22 compared with 8/16 non-allergens were degraded almost instantly, while 5/22 allergens and 3/16 non-allergens were broken down very rapidly (Fu, Abbott and Hatzos, 2002). Both groups of proteins were more stable to simulated intestinal fluid containing pancreatin, a mixture of pancreatic enzymes, with only 2/22 allergens compared with 3/16 non-allergens being broken down instantly, and 5/22 allergens and 3/16 non-allergens being degraded rapidly.

However, the apparent stability of a protein can be very dependent on the experimental conditions employed. The results are strongly dependent on the substrate:protease ratio, and the same protein can be shown to be either stable or labile to pepsin (Fu *et al.*, 2002). The pepsin digestion protocols that have been employed typically involve pepsin:substrate ratios in the range 1:5–1:10, which are probably far in excess of those found physiologically. Furthermore, they do not account for the pH titrations that occur *in vivo* nor do they contain phospholipids and other enzymes, such as gastric lipase found in the human stomach. Thus, the simulated intestinal fluid used by Fu *et al.* (2002) neither contained bile salts nor involved prior digestion of the proteins by pepsin, as is the case *in vivo*, where proteins first experience the low pH conditions of the stomach before entering the small intestine. More mobile regions of a protein are known to be more accessible to proteases such as pepsin, which has a preference for attacking peptide bonds in more mobile, flexible segments (Dunn and Hung, 2000). It therefore seems that rather than indicating the potential for a protein to be resistant to digestion *in vivo* in humans, the simulated digestion systems reported to date probably act more as bioassays of intrinsic protein stability.

Sensitization in pollen-fruit and latex-fruit cross-reactive allergy syndromes is thought to occur via inhalation, with reactions to fruits tending to be confined to the oral cavity. In such instances stability to digestion may not be a relevant property. This is also supported by the fact that nsLTP allergens from a wide range of plant species, which are resistant to pepsinolysis (Asero *et al.*, 2000) and able to cause systemic symptoms, are thought to sensitize via the GI tract.

IgE Reactivity and Sensitization Potential of Proteins

The methods for assessing cross-reactivity using human allergic sera are straightforward, although there are ethical issues regarding the development of serum banks on which such studies rely, together with the unequivocal diagnosis of food allergy in those

individuals contributing sera. Cross-reactivity studies using suitable antibody prepara-
tions from experimental animal sources can also be used to substantiate the potential for
cross-reactivity. The WHO/FAO decision tree recommends that combinations of allergic
sera from both food and inhalant (e.g. pollen) are used, which in the case of GMOs for
food use encompass allergies to both monocotyledonous and dicotyledonous plant
species.

Measures of sensitizing potential rely on animal models, the choice of animal species
and strain being important. Dogs have been found to be more amenable to oral
sensitization than small mammals (Ermel *et al.*, 1997), but require larger doses.
Furthermore, the extensive use of dogs in such toxicological assessments is not well
accepted in some societies. Small mammals such as mice (using the high-IgE responder
strain Balb/c), rats (especially the Brown Norway strain) and, to a more limited extent,
guinea pigs (Dearman *et al.*, 2001; Penninks and Knippels, 2001) are generally more
favored. However, these require the use of adjuvants (such as cholera toxin or
polysaccharides such as carageenan) in conjunction with oral dosing or a combination
of both oral and intraperitoneal exposure in order to generate an IgE response (Penninks
and Knippels, 2001). Another approach has been simply to determine the inherent
sensitizing potential of a protein independently of the effects of the GI tract using
parenteral administration (Dearman *et al.*, 2001). While there are indications that not all
proteins behave in a similar fashion in stimulating IgE and IgG responses in experimental
animals, the predictive value of such behavior in identifying potential allergens is
still debated.

Practical Application of Allergenic Risk Assessment

There have been several reported trials of the allergenic risk assessment process
including an ice structuring protein (ISP) from an arctic fish, ocean pout (Baderschneider
et al., 2002; Bindeslev-Jensen *et al.*, 2003), and an analysis of the Micronesian Nangai
nut which, because it has no previous history of consumption in the EU, would be
considered as a novel food (Sten *et al.*, 2002). Using the decision tree approach proposed
in the FAO-WHO 2001 consultation, an *in silico* analysis was first performed which
showed a complete absence of any sequence similarity between ISP type III isofom
HPLC12 and any known allergen, with 515 exact matches being found with unrelated
proteins when using hexamers for searching. Most of the exact matches found using
seven contiguous amino acids for searching were with other ISPs (Baderschneider *et al.*,
2003). The ISP was also broken down by pepsin with a half life of <28 min, the products
of digestion having M_r of 500–2000, with some evidence of minor components with M_r of
<4000 as determined by HPLC electrospray mass spectrometry. The main digestion
product was found to have an M_r about 1276. Further analysis of the ISP showed a lack of
IgE reactivity by immunoassay using allergic sera from individuals with codfish allergy,
proven by oral challenge, and an inability to trigger histamine release from basophils in
three of the cod allergic subjects (Bindeslev-Jensen, *et al.*, 2003). As a result, it was
concluded that the ISP was safe for those who already had fish allergy and showed that
the FAO-WHO 2001 approach was practicable.

In the case of the Nangai nut, IgE binding studies aimed at identifying potential cross-
reactive allergies using sera from individuals with three commonly occurring allergies to

pollen (mugwort, grass and birch) were undertaken (Sten *et al.*, 2002). Most (9/11) of the individuals showing positive IgE binding to Nangai nut were allergic to all three pollens, and in some instances Nangai nut extracts also stimulated histamine release and positive skin prick tests in some pollen allergic individuals. However, when oral challenges were performed in 12 individuals showing IgE reactivity to the nut extracts *in vitro*, two individuals showed a positive open challenge which could not be confirmed when a double blind placebo-controlled food challenge, the 'gold standard' of allergy diagnosis, was performed.

An extensive comparison of the allergenic properties of wild-type and glyphosate-tolerant GM soya has also been described, which utilized a combination of *in vitro* analysis of IgE binding, skin prick testing and histamine release together with food challenge (Sten *et al.*, 2004). Due to problems in identifying genuinely soy allergic individuals the study made use of the fact that many people with peanut allergy have IgE to soy, although they can usually tolerate consumption of soy. This comprehensive analysis showed no differences in the allergenic potency of GM and wild-type soy.

Application of GM Technology to Plant Allergens

Although there is some concern that GM technology will result in the introduction of new allergens into the food chain, the risk assessment procedures discussed above mean that this is unlikely. However, it is worth noting in this respect that many biologically active proteins which have potential applications in protection against pests and pathogens (Shewry and Lucas, 1997) are also either allergens or belong to protein families which contain allergenic proteins such as nsLTPs. Thus, the knowledge of protein families outlined above will prove to be valuable in selecting appropriate proteins for expression in transgenic plants.

The greatest impact of GM on allergenicity should be a positive one, by providing tools to reduce or eliminate allergenic proteins or epitopes. However, in practice this may be difficult to achieve, as many foods which cause severe reactions contain multiple allergenic proteins that may belong to separate families. Good examples of this are provided by peanut and soybean, which are related leguminous species. The major storage proteins in these species are 7S and 11S globulins, and in both cases these include major allergens (Ara h 1 in peanut is 7S, and Ara h 3 and 4 are 11S). In addition, peanut contains allergenic 2S albumin storage proteins (Ara h 2, 6 and 7) and a profilin-related protein (Ara h 5) while the soybean allergens include a cysteine proteinase (P34). Furthermore, most of these proteins have also been demonstrated to possess multiple IgE binding sites: at least 23 on Ara h 1, 16 on P34 and 11 on the soybean 11S globulin (see Mills *et al.*, 2004b for full details and references).

Some success has been achieved in downregulating the expression of genes encoding allergenic proteins using 'antisense' technology in transgenic plants. The best documented of these concerns the rice α-amylase inhibitors, which are major dietary allergens. These comprise a number of subunits with M_r about 14 000–16 000, with at least seven immunologically cross-reactive proteins and more than 10 related cDNAs being reported (Alvarez *et al.*, 1995; Matsuda *et al.*, 1991; Nakase *et al.*, 1996). Despite this complexity the antisense expression of a single sequence resulted in reductions in some transcript and protein levels to about 20 % of those in the non-transformed controls (Tada *et al.*, 1996). Although the transgenic rice is 'hypoallergenic', it still contains substantive amounts of allergenic proteins and clearly cannot be regarded as safe for those suffering from dietary allergy to rice.

Transgene-induced gene silencing has also been used to completely suppress the expression of Gly m Bd 30k (Herman, 2003a,b), a major allergen of soybean seeds, which is related to the allergenic cysteine proteinases present in papaya, kiwi fruit, pineapple and fig (see Shewry *et al.*, 2004). The authors reported that the transgenic plants showed no differences to control plants and suggested that the transgenic lines could be combined with naturally occurring mutants which lack other major allergens (Ogawa, Samoto and Takahashi, 2000; Samoto *et al.*, 1997).

Effective suppression of allergen gene expression might be achieved in the future using the new RNAi technology (Baulcombe, 2004), or by manipulating transcription factors to co-ordinately downregulate multigene families. The latter depends on the fact that many multigene families have highly conserved regulatory sequences, which interact with the same transcription factors. This includes most families of seed storage protein genes, which have highly conserved enhancers upstream of their coding sequences (see Nielsen, Bassüner and Beaman, 1997; Shewry, Halford and Lafiandra, 2003). For example, the *opaque-2* gene of maize encodes a transcription factor of the bZIP class, which regulates the expression of a set of genes encoding a family of storage protein genes, the Z22 α-zeins (Schmidt *et al.*, 1990). Consequently, the amounts of these proteins are greatly reduced in mutant *opaque-2* grains (Coleman, Dannenhoffer and Larkins, 1997). Similar downregulation of suites of genes encoding protein allergens, such as the rice α-amylase inhibitors, could possibly be achieved by identifying and manipulating the relevant transcription factors. However, replacement of the genes encoding the allergens may ultimately be required to eliminate any possibility of reversal of the suppression. The technology for this is not currently available in plants.

Antisense technology has also been used to downregulate the production of the major allergens in the pollen of ryegrass: Lol p 1, Lol p 2 (Petrovska et al., 2005) and Lol p 5 (Bhalla, Swoboda and Singh, 1999), leading to the possibility of releasing hypoallergenic grass lines to alleviate hayfever.

Downregulation of families of allergens is clearly feasible technically, although complete removal may not be achievable with current technology. However, this approach may not be applicable where the allergenic proteins have functional or biological properties which must be retained. For example, removal of a substantial proportion of the gluten proteins of wheat would clearly have adverse effects on the processing properties while making bread and other foods (see Chapter 2.4). In such cases removal of IgE binding sites may be possible using an *in vivo* mutagenesis technology called chimeraplasty (Zhu *et al.*, 1999). However, chimer-aplasty has not yet been applied routinely to crop plants and would only be feasible if the mutagenesis of one or a small number of IgE binding sites is sufficient to eliminate the allergenicity.

Nevertheless, the feasibility of this mutagenesis approach has been demonstrated by protein engineering studies of the major allergens of apple (Mal a 1) and cherry (Pru a 1) (both members of the Bet v 1 family) expressed in *E. coli* (Scheurer *et al.*, 1999; Son *et al.*, 1999).

To summarize, genetic engineering provides exciting prospects for reducing the prevalence of allergy to plant proteins but the technology for the complete removal of either allergenic proteins or IgE binding sites from allergenic proteins is still not available for routine use in crop plants. Nevertheless, the partial removal of allergens has already been achieved using antisense technology to provide hypoallergenic lines, which may have applications where complete removal is not essential, for example to reduce the severity of hayfever to pollen of meadow grasses or trees.

Conclusions

The identification of potential allergens, particularly those which have no structural or biological relationships with known allergens, is an uncertain process at present. It is clear that not all foods are equally allergenic, yet the precise molecular basis for these differences remains unclear. There is little evidence that particular protein sequences or structures promote the development of IgE responses, although food allergens appear to be good antigens (i.e. able to stimulate an antibody response). There is some evidence from animal studies that, based on a limited number of allergens tested so far, allergens are intrinsically able to elicit an IgE-response preferentially in experimental animals (Dearman and Kimber, 2001; Dearman et al., 2001). Protein stability and resistance to digestion may also play a role in determining the allergenic potential of proteins when sensitizing via the GI tract. However, only when we have a greater understanding of the molecular pathology of food allergy which embraces both the aspects of food (allergen structure, interactions with the food matrix and the potential adjuvant effects of other food components) coupled with the genetic factors in man which predispose some individuals to becoming atopic, can a completely reliable prediction of novel protein allergenicity be made. Until this information is available, allergen identification will continue to be an uncertain process, requiring a great deal of information on many different aspects of protein structure and properties and relying on comparisons with known food allergens.

Acknowledgment

The authors acknowledge the support of BBSRC through competitive strategic grants to IFR and Rothamsted Research.

References

Adachi, M., Takenaka, Y., Gidamis, A.B., Mikami, B. and Utsumi, S. (2001) Crystal structure of soybean proglycinin A1aB1b homotrimer. *Journal of Molecular Biology* **305**, 291–305.

Alvarez, A., Adachi, T., Nakase, M., Aoki, N., Nakamura, R. and Matsuda, T. (1995) Classification of rice allergenic protein cDNAs belonging to the α-amylase/trypsin inhibitor gene family. *Biochimica et Biophysica Acta* **1251**, 201–204.

Asero, R., Mistrello, G., Roncarolo, D., de Vries, S.C., Gautier, M.F., Ciurana, C.L., Verbeek, E., Mohammadi, T., Knul-Brettlova, V., Akkerdaas, J.H., Bulder, I., Aalberse, R.C. and van Ree, R. (2000) Lipid transfer protein: a pan-allergen in plant-derived foods that is highly resistant to pepsin digestion. *International Archives of Allergy and Immunology* **122**, 20–32.

Astwood, J.D., Leach, J.N. and Fuchs, R.L. (1996) Stability of food allergens to digestion *in vitro. Nature Biotechnology* **14**, 1269–1273.

Baderschneider, B., Crevel, R.W., Earl, L.K., Lalljie, A., Sanders, D.J. and Sanders, I.J. (2002) Sequence analysis and resistance to pepsin hydrolysis as part of an assessment of the potential allergenicity of ice structuring protein type III HPLC 12. *Food Chemistry and Toxicology* **40**, 965–978.

Bannon, G.A., Li, X.F., Rabjohn, P., Stanley, J.S., Burks, A.W., Huang, S.K. and Sampson, H.A. (1997) Ara h 3, a peanut allergen identified by using peanut sensitive patient sera absorbed with soy proteins. *Journal of Allergy and Clinical Immunology* **99**, 568.

Baulcombe, D. (2004) RNA silencing in plants. *Nature* **43**, 356–363.

Becker, W.M. and Reese, G. (2001) Immunological identification and characterization of individual food allergens. *Journal of Chromatography B: Biomedical Applications* **756**, 131–140.

Beyer, K., Bardina, L., Grishina, G. and Sampson, H.A. (2002a) Identification of sesame seed allergens by 2-dimensional proteomics and Edman sequencing: seed storage proteins as common food allergens. *Journal of Allergy and Clinical Immunology* **110**, 154–159.

Beyer, K., Grishina, G., Bardina, L., Grishin, A. and Sampson, H.A. (2002b) Identification of an 11S globulin as a major hazelnut food allergen in hazelnut-induced systemic reactions. *Journal of Allergy and Clinical Immunology* **110**, 517–523.

Bindeslev-Jensen, C., Sten, E., Earl, L.K., Crevel, R.W., Bindslev-Jensen, U., Hansen, T.K., Stahl Skov, P. and Poulsen, L.K. (2003) Assessment of the potential allergenicity of ice structuring protein type III HPLC 12 using the FAO/WHO 2001 decision tree for novel foods. *Food Chemistry and Toxicology* **41**, 81–87.

Bhalla, P.L., Swoboda, I. and Singh, M.B. (1999) Antisense-mediated silencing of a gene encoding a major ryegrass pollen allergen. *Proceedings of the National Academy of Sciences USA* **96**, 11676–11680.

Breiteneder, H. (2004) Thaumatin-like proteins—a new family of pollen and fruit allergens. *Allergy* **59**, 479–481.

Breinteneder, H. and Ebner, C. (2000) Molecular and biochemical classification of plant-derived food allergens. *Journal of Allergy and Clinical Immunology* **106**, 27–36.

Breinteneder, H. and Redauer, C. (2004) A classification of plant food allergens. *Journal of Allergy and Clinical Immunology* **113**, 821–830.

Burks, A.W., Brooks, J.R. and Sampson, H.A. (1988) Allergenicity of major component proteins of soybean determined by enzyme-linked immunosorbent assay (ELISA) and immunoblotting in children with atopic dermatitis and positive soy challenge. *Journal of Allergy and Clinical Immunology* **81**, 1135–1142.

Burks, A.W., Cockrell, G., Connaughton, C., Guin, J., Allen, W. and Helm, R.M. (1994) Identification of peanut agglutinin and soybean trypsin inhibitor as minor legume allergens. *International Archives of Allergy and Immunology* **105**, 143–149.

Burks, A.W., Helm, R.M., Cockrell, G., Bannon, G.A., Stanley, J.S., Shin, D.S. *et al.* (1999) 'Tertiary structure of peanut allergen Ara h 1', PTC Patent Application, Board of Trustees of the University of Arkansas.

Burks, A.W., Williams, L.W., Connaughton, C., Cockrell, G., O'Brien, T.J. and Helm, R.M. (1992b) Identification and characterization of a second major peanut allergen, Ara h II, with use of the sera of patients with atopic dermatitis and positive peanut challenge. *Journal of Allergy and Clinical Immunology* **90**, 962–969.

Burks, A.W., Williams, L.W., Helm, R.M., Connaughton, C., Cockrell, G. and O'Brien, T. (1991) Identification of a major peanut allergen, Ara h I, in patients with atopic dermatitis and positive peanut challenges. *Journal of Allergy and Clinical Immunology* **88**, 172–179.

Bush, R.K. and Hefle, S.L. (1996) Food allergens. *Critical Reviews in Food Science and Nutrition* **36** (Suppl), S119–S163.

Coleman, C.E., Dannenhoffer, J.M. and Larkins, B.A. (1997) The prolamin proteins of maize, sorghum and *Coix*, in *Cellular and Molecular Biology of Plant Seed Development* (eds B.A. Larkins and I.K. Vasil), Kluwer Academic Publishers, The Netherlands, pp. 257–288.

Curioni, A., Santucci, B., Cristado, A., Canistraci, C., Peitravalle, M., Simonato, B. and Giannattasio, M. (1999) Urticaria from beer: an immediate hypersensitivity reaction due to a 10 kDa protein derived from beer. *Clinical and Experimental Allergy* **29**, 407–413.

Dearman, R.J. and Kimber, I. (2001) Determination of protein allergenicity: studies in mice. *Toxicology Letters* **120**, 181–186.

Dearman, R.J., Caddick, H., Stone, S., Basketter, D.A. and Kimber, I. (2001) Characterisation of antibody responses to food proteins: influence of route of exposure. *Toxicology* **167**, 217–231.

Diaz-Perales, A., Sanchez-Monge, R., Blanco, C., Lombardero, M., Carillo, T. and Salcedo, G. (2002a) What is the role of the hevein-like domain of fruit class I chitinases in their allergenic capacity? *Clinical and Experimental Allergy* **32**, 448–454.

Diaz-Perales, A., Tabar, A.I., Sanchez-Monge, R., Garcia, B.E., Gomez, B., Barber, D. and Salcedo, G. (2002b) Characterization of asparagus allergens: a relevant role of lipid transfer proteins. *Journal of Allergy and Clinical Immunology* **110**, 790–796.

Dunn, B.M. and Hung, S-H. (2000) The two sides of enzyme-substrate specificity: lessons from the aspartic proteinases. *Biochimica et Biophysica Acta* **1477**, 231–240.

Dunwell, J.M., Purvis, A. and Khuri, S. (2004) Cupins: the most functionally diverse protein superfamily? *Phytochemistry* **65**, 7–17.

Duro, G., Colombo, P., Costa, M.A., Izzo, V., Porcasi, R., Di Fiore, R., Locorotondo, G., Mirisola, M.G., Cocchiara, R. and Geraci, D. (1996) cDNA cloning, sequence analysis and allergological characterization of Par j 2.0101, a new major allergen of the *Parietaria judaica* pollen. *FEBS Letters* **399**, 295.

Fahlbusch, B., Rudeschko, O., Schumann, C., Steurich, R., Henzgen, M., Schlenvoigt, G. and Jager, L. (1998) Further characterisation of IgE-binding antigens in kiwi with particular emphasis on glycoprotein allergens. *Journal of Investigational Allergology and Clinical Immunology* **8**, 325–332.

FAO-WHO (2001) Evaluation of Allergenicity of Genetically Modified Foods, Report of the joint FAO/WHO expert consultation on allergenicity of foods derived from Biotechnology, 22–25th January 2001, pp. 1–26. http://www.who.int/fsf/Gmfood/Consultation-Jan2001/report20.pdf.

FAO-WHO (2002) Codex Alimentarius Commission Ad Hoc Intergovernmental task force on Foods Derived from Biotechnology (ALINORM 03/34) March 2002. ftp://ftp.fao.org/codex/alinorm03/Al03_34e.pdf.

FAO-WHO (1999) Codex General Standard for the Labelling of Prepackaged Foods *CODEX STAN 1-1985 (Rev. 1-1991).* http://www.fao.org/DOCREP/005/Y2770E/y2770e02.htm#bm02

Fu, T-J., Abbott, U.R. and Hatzos, C. (2002) Digestibility of food allergens and nonallergenic proteins in simulated gastric fluid and simulated intestinal fluid—a comparative study. *Journal of Agricultural and Food Chemistry* **50**, 7154–7160.

Gajhede, M., Osmark, P., Poulsen, F.M., Ipsen, H., Larsen, J.N., Joost van Neerven, R.J., Schou, C., Lowenstein, H. and Spangfort, M.D. (1996) X-ray and NMR structure of Bet v 1, the origin of birch pollen allergy. *Nature Structural Biology* **3**, 1040–1045.

Gall, H., Kalveram, K.J., Forck, G. and Sterry, W. (1994) Kiwi fruit allergy—a new birch pollen-associated food allergy. *Journal of Allergy and Clinical Immunology* **94**, 70–76.

Gavrovic-Jankulovic, M., Irkovic, T., Vuckovic, O., Atanaskovic-Markovic, M., Petersen, A., Gojgic, G., Burazer, L. and Jankov, R.M. (2002) Isolation and biochemical characterization of a thaumatin-like kiwi allergen. *Journal of Allergy and Clinical Immunology* **110**, 805–810.

Gendel, S. (1998) The use of amino acid sequence alignments to assess potential allergenicity of proteins using genetically modified foods. *Advances in Food and Nutrition Research* **42**, 45–62.

Gendel, S.M. (2002) Sequence analysis for assessing potential allergenicity. *Annals of the New York Academy of Sciences* **964**, 87–98.

Guénard-Bilbault, L., Kanny, G. and Moneret-Vautrin, D. (1999) L'allergie alimentaire à la farine de blé. *Allergie et Immunologie* **31**, 22–25.

Helm, R., Cockrell, G., Connaughton, C., Sampson, H.A., Bannon, G.A., Beilinson, V., Livingstone, D., Nielsen, N.C. and Burks, A.W. (2000) A soybean G2 glycinin allergen. 1. Identification and characterization. *International Archives of Allergy and Immunology* **123**, 205–212.

Herman, E.M. (2003a) Genetically modified soybeans and food allergies. *Journal of Experimental Botany*, **54**, 1317–1319.

Herman, E.M. (2003b) Genetic modification removed an immunodominant allergen from soybean. *Plant Physiology* **132**, 36–43.

Hileman, R.E., Silvanovich, A., Goodman, R.E., Rice, E.A., Holleschak, G., Astwood, J.D. and Hefle, S. (2002) Bioinformatic methods for allergenicity assessment using a comprehensive allergen database. *International Archives of Allergy and Immunology* **128**, 280–291.

Hoffman, D., Lowenstein, H., Marsh, D.G., Platts-Mills, T.A.E. and Thomas, W. (1994) *Bulletin of the World Health Organization* **72**, 796–806.

Hoffmann-Sommergruber, K. (2002) Pathogenesis-related (PR)-proteins identified as allergens. *Biochemical Society Transactions* **30**, 930–935.

Inschlag, C., Hoffmann-Sommergruber, K., O'Riordain, G., Ahorn, H., Ebner, C., Scheiner, O. and Breiteneder, H. (1998) Biochemical characterization of Pru a 2, a 23-kDa thaumatin-like protein representing a potential major allergen in cherry (*Prunus avium*). *International Archives of Allergy and Immunology* **116**, 22–28.

Ivanciuc, O., Mathura, V., Midoro-Horiuti, T., Braun, W., Goldblum, R.M. and Schein, C.H. (2003) Detecting potential IgE-reactive sites on food proteins using a sequence and structure database, SDAP-food. *Journal of Agricultural and Food Chemistry* **51**, 4830–4837.

James, J.M., Sixbey, J.P., Helm, R.M., Bannon, G.A. and Burks, A.W. (1997) Wheat α-amylase inhibitor: a second route of allergic sensitisation. *Journal of Allergy and Clinical Immunology* **99**, 239–244.

Jankiewicz, A., Baltes, W., Bögl, K.W., Dehne, L.I., Jamin, A., Hoffmann, A., Haustein, D. and Vieths, S. (1997) Influence of food processing on the immunochemical stability of celery allergens. *Journal of the Science of Food and Agriculture* **75**, 359–370.

Kalinski, A., Weisemann, J.M., Matthews, B.F. and Herman, E.M. (1990) Molecular cloning of a protein associated with soybean seed oil bodies that is similar to thiol proteases of the papain family. *Journal of Biological Chemistry* **265**, 13843–13848.

Kelly, J.D., Hlywka, J.J. and Hefle, S.L. (2000) Identification of sunflower seed IgE-binding proteins. *International Archives of Allergy and Immunology* **121**, 19–24.

Kleber, J.T., Crameri, R., Appenzeller, U., Schlaak, M. and Becker, W.M. (1999) Selective cloning of peanut allergens, including profilin and 2S albumins, by phage display technology. *International Archives of Allergy and Immunology* **119**, 265–274.

Kleine-Tebbe, J., Vogel, L., Crowell, D.N., Haustein, U.F. and Vieths, S. (2002) Severe oral allergy syndrome and anaphylactic reactions caused by a Bet v 1-related PR-10 protein in soybean, SAM22. *Journal of Allergy and Clinical Immunology* **110**, 797–804.

Krebitz, M., Wagner, B., Ferreira, F., Peterbauer, C., Campillo, N., Witty, M., Kolarich, D., Steinkellner, H., Scheiner, O. and Breiteneder, H. (2003) Plant-based heterologous expression of Mal d 2, a thaumatin-like protein and allergen of apple (*Malus domestica*), and its characterization as an antifungal protein. *Journal of Molecular Biology* **329**, 721–730.

Kreis, M., Forde, B.G., Rahman, S., Miflin, B.J. and Shewry, P.R. (1985) Molecular evolution of the seed storage proteins of barley, rye and wheat. *Journal of Molecular Biology* **183**, 499–502.

Kusaba-Nakayama, M., Ki, M., Iwamoto, M., Shibata, R., Sato, M. and Imaizumi, K. (2000) CM3, one of the wheat alpha-amylase inhibitor subunits, and binding of IgE in sera from Japanese with atopic dermatitis related to wheat. *Food Chemistry and Toxicology* **38**, 179–185.

Larson, J.N. and Lowenstein, H. (1996) Allergen nomenclature. *Journal of Allergy and Clinical Immunology* **97**, 577–578.

Leitner, A., Jensen-Jarolim, E., Grimm, R., Wuthrich, B., Ebner, H., Scheiner, O., Kraft, D. and Ebner, C. (1998) Allergens in pepper and paprika. Immunological investigation of the celery-birch-mugwort-spice syndrome. *Allergy* **53**, 36–41.

Lopez-Torrejon, G., Salcedo, G., Martin-Esteban, M., Diaz-Perales, A., Pascual, C.Y. and Sanchez-Monge, R. (2003) Len c 1, a major allergen and vicilin from lentil seeds: protein isolation and cDNA cloning. *Journal of Allergy and Clinical Immunology* **112**, 1208–1215.

Markovic-Housley, Z., Degano, M., Lamba, D., von Roepenack-Lahaye, E., Clemens, S., Susani, M., Ferreira, F., Scheiner, O. and Breiteneder, H. (2003) Crystal structure of a hypoallergenic isoform of the major birch pollen allergen Bet v 1 and its likely biological function as a plant steroid carrier. *Journal of Molecular Biology* **325**, 123–133.

Maruyama, N., Ichise, K., Katsube, T., Kishimoto, T., Kawase, S.-I., Matsumura, Y., Takeuchi, Y., Sawada, T. and Utsumi, S. (1998) Identification of major wheat allergens by means of the *Escherichia* coli expression system. *European Journal of Biochemistry* **255**, 739–745.

Matsuda, T., Nomura, R., Sugiyama, M. and Nakamura, R. (1991) Immunochemical studies on rice allergenic proteins. *Agricultural and Biological Chemistry* **55**, 509–513.

Menedez-Arias, L., Moneo, I., Dominguez, J. and Rodriguez, R. (1988) Primary structure of the major allergen of yellow mustard (*Sinapis alba* L.) seed, Sin a I. *European Journal of Biochemistry* **177**, 159–166.

Metcalfe, D.D., Astwood, J.D., Townsend, R., Sampson, H.A., Taylor, S.L. and Fuchs, R.L. (1996) Assessment of the allergenic potential of foods derived from genetically engineered crop plants. *Critical Reviews in Food Science and Nutrition* **36** (Suppl), S165–S186.

Mills, E.N.C., Madsen, C., Shewry, P.R. and Wichers, H.J. (2003) Food allergens of plant origin— the relationship between allergenic potential and biological activity (PROTALL). *Trends in Food Science and Technology* **14**, 145–156.

Mills, E.N.C., Jenkins, J., Alcocer, M.J.C. and Shewry, P.R. (2004a) Structural, biological and evolutionary relationships of plant food allergens sensitising *via* the gastrointestinal tract. *Critical Reviews in Food Science and Nutrition* **44**, 379–407.

Mills, E.N.C., Jenkins, J.A. and Bannon, G.A. (2004b) Plant seed globulin allergens, in *Plant Food Allergens* (eds E.N.C. Mills and P.R. Shewry), Blackwell Science, Oxford, UK, p. 141.

Mittag, D., Vieths, S., Vogel, L., Becker, W.M., Rihs, H.P., Helbling, A., Wuthrich, B. and Ballmer-Weber, B.K. (2004) Soybean allergy in patients allergic to birch pollen: clinical investigation and molecular characterization of allergens. *Journal of Allergy and Clinical Immunology* **113**, 148–154.

Möller, M., Pashcke, A., Vieluf, D., Kayma, M., Vieths, S. and Steinhart, H. (1997) Character- ization of allergens in kiwi fruit and detection of cross-reactivities with allergens of birch pollen and related fruits. *Food and Agricultural Immunology* **9**, 107–121.

Monsalve, R.I., Gonzalez de la Pena, M.A., Menezes-Arias, L., Lopez-Otin, C. Villalba, M. and Rodriguez, R. (1993) Characterization of a new oriental-mustard (*Brassica-juncea*) allergen Bra j I: detection of an allergenic epitope. *Biochemistry Journal* **293**, 625–632.

Nakase, M., Adachi, T., Urisu, A., Miyashita, T., Alvarez, A.M., Nagasaka, S., Aoki, N., Nakamura, R. and Matsuda, T. (1996) Rice (*Oryza sativa* L.) α-amylase inhibitors of 14–16 kDa are potential allergens and products of a multigene family. *Journal of Agricultural and Food Chemistry* **44**, 2624–2628.

Nielsen, N.C., Bassüner, R. and Beaman T. (1997) The biochemistry and cell biology of embryo storage proteins, in *Cellular and Molecular Biology of Plant Seed Development* (eds B.A. Larkins and I.K. Vasil), Kluwer Academic Publishers, The Netherlands. pp. 151–220.

Ogawa, T., Tsuji, H., Bando, N., Kitamura, K., Zhu, Y.-L., Hirano, H. and Nishikawa, K. (1993) Identification of the soyabean allergenic proteins, Gly m Bd 30K, with the soybean seed 34 kDa oil-body-associated protein. *Bioscience, Biotechnology and Biochemistry* **57**, 1030–1033.

Ogawa, T., Samoto, M. and Takahashi, K. (2000) Soybean allergens and hypoallergenic soybean products. *Journal of Nutritional Science and Vitaminology* **46**, 271–279.

Palosuo, K., Alenius, H., Varjonen, E., Koivuluhta, M., Mikkola, J., Keskinen, H., Kalkkinen, N. and Reunala, T. (1999) A novel wheat gliadin as a cause of exercise-induced anaphylaxis. *Journal of Allergy and Clinical Immunology* **103**, 912–917.

Palosuo, K., Alenius, H., Varjonen, E., Kalkkinen, N. and Reunala, T. (2001a) Rye γ-70 and γ-35 secalins and barley γ-3 hordein cross-react with ω-5 gliadin, a major allergen in wheat-dependent, exercise-induced anaphylaxis. *Clinical and Experimental Allergy* **31**, 466–473.

Palosuo, K., Varjonen, E., Kekki, O.M., Klemola, T., Kalkkinen, N., Alenius, H. and Reunala, T. (2001b) Wheat ω-5 gliadin is a major allergen in children with immediate allergy to ingested wheat. *Journal of Allergy and Clinical Immunology* **108**, 634–638.

Pastorello, E.A., Conti, A., Pravettoni, V., Farioli, L., Rivolta, F., Ansaloni, R., Ispano, M., Incorvaia, C., Giuffrida, M.G. and Ortolani, C. (1998a) Identification of actinidin as the major allergen of kiwi fruit. *Journal of Allergy and Clinical Immunology* **101**, 531–537.

Pastorello, E.A., D'Ambrosio, F.P., Pravettoni, V., Farioli, L., Giuffrida, G., Monza, M., Ansaloni, R., Fortunato, D., Scibola, E., Rivolta, F., Incorvaia, C., Bengtsson, A., Conti, A. and Ortolani, C. (2000a) Evidence for a lipid transfer protein as the major allergen of apricot. *Journal of Allergy and Clinical Immunology* **105**, 371–377.

Pastorello, E.A., Farioli, L., Pravettoni, V., Ispano, M., Conti, A., Ansaloni, R., Rotondo, F., Incorvaia, C., Bengtsson, A., Rivolta, F., Trambaioli, C., Previdi, M. and Ortolani, C. (1998b) Sensitization to the major allergen of Brazil nut is correlated with the clinical expression of allergy. *Journal of Allergy and Clinical Immunology* **102**, 1021–1027.

Pastorello, E.A., Pravettoni, V., Farioli, L., Ispano, M., Fortunato, D., Monza, M., Giuffrida, M.G., Rivolta, F., Scibola, E., Ansaloni, R., Incorvaia, C., Conti, A. and Ortolani, C. (1999b) Clinical role of a lipid transfer protein that acts as a new apple-specific allergen. *Journal of Allergy and Clinical Immunology* **104**, 1099–1106.

Pastorello, E.A., Farioli, L., Pravettoni, V., Ispano, M., Scibola, E., Trambaioli, C., Giuffrida, M.G., Ansaloni, R., Godovac-Zimmermann, J., Conti, A., Fortunato, D. and Ortolani, C. (2000b) The maize major allergen, which is responsible for food-induced allergic reactions, is a lipid transfer protein. *Journal of Allergy and Clinical Immunology* **106**, 744–751.

Pastorello, E.A., Farioli, L., Pravettoni, V., Ortolani, C., Fortunato, D., Giuffrida, M.G., Perono Garoffo, L., Calamari, A.M., Brenna, O. and Conti, A. (2003) Identification of grape and wine allergens as an endochitinase 4, a lipid-transfer protein, and a thaumatin. *Journal of Allergy and Clinical Immunology* **111**, 350–359.

Pastorello, E.A., Farioli, L., Pravettoni, V., Ortolani, C., Ispano, M., Monza, M., Baroglio, C., Scibola, E., Ansaloni, R., Incorvaia, C. and Conti, A. (1999a) The major allergen of peach (*Prunus persica*) is a lipid transfer protein. *Journal of Allergy and Clinical Immunology* **103**, 520–526.

Pastorello, E.A., Farioli, L., Robino, A.M., Trambaioli, C., Conti, A. and Pravettoni, V. (2001a) A lipid transfer protein involved in occupational sensitization to spelt. *Journal of Allergy and Clinical Immunology* **108**, 145–146.

Pastorello, E.A., Varin, E., Farioli, L., Pravettoni, V., Ortolani, C., Trambaioli, C., Fortunato, D., Giuffrida, M.G., Rivolta, F., Robino, A., Calamari, A.M., Lacava, L. and Conti, A. (2001b) The major allergen of sesame seeds (*Sesamum indicum*) is a 2S albumin. *Journal of Chromatography B: Biomedical Applications* **756**, 85–93.

Pastorello, E.A., Vieths, S., Pravettoni, V., Farioli, L., Trambaioli, C., Fortunato, D., Luttkopf, D., Calamari, M., Ansaloni, R., Scibilia, J., Ballmer-Weber, B.K., Poulsen, L.K., Wutrich, B., Hansen, K.S., Robino, A.M., Ortolani, C. and Conti, A. (2002) Identification of hazelnut major allergens in sensitive patients with positive double-blind, placebo-controlled food challenge results. *Journal of Allergy and Clinical Immunology* **109**, 563–570.

Penninks, A.H. and Knippels, L.M.J. (2001) Determination of protein allergenicity: studies in rats. *Toxicology* **120**, 171–180.

Petrovska, N., Wu, X., Donato, R., Wang, Z., Ong, E-K., Jones, E., Forster, J., Emmerling, M., Sidoli, A., O'Hehir, R. and Spangenberg, G. (2004) 'Transenic ryegrasses (*Lolium* spp.) with down-regulation of main pollen allergens', *Molecular Breeding*, **14**, 489–501.

Poltronieri, P., Cappello, M.S., Dohmae, N., Conti, A., Fortunato, D., Pastorello, E.A., Ortolani, C. and Zacheo, G. (2002) Identification and characterisation of the IgE-binding proteins 2S albumin and conglutin gamma in almond (*Prunus dulcis*) seeds. *International Archives of Allergy and Immunology* **128**, 97–104.

Posch, A., Wheeler, C.H., Chen, Z., Flagge, A., Dunn, M.J., Papenfuss, F., Raulf-Heimsoth, M. and Baur, X. (1999) Class I endochitinase containing a hevein domain is the causative allergen in latex-associated avocado allergy. *Clinical and Experimental Allergy* **29**, 667–672.

Redauer, C. and Hoffman-Sommergruber, K. (2003) Profilins, in *Plant Food Allergens* (eds E.N.C. Mills and P.R. Shewry), Blackwells, Oxford, UK, pp. 105–124.

Roux, K.H., Teuber, S.S., Robotham, J.M. and Sathe, S.K. (2001) Detection and stability of the major almond allergen in foods. *Journal of Agricultural and Food Chemistry* **49**, 2131–2136.

Samoto, M., Fukuda, Y., Takahashi, K., Tabuchi, K., Hiemori, M., Tsuji, H., Ogawa, T. and Kawamura, Y. (1997) Substantially complete removal of three major allergenic soybean proteins (Gly m Bd 30k, Gly m Bd 28k and the alpha-subunit of conglycinin) from soy protein by using a mutant soybean, Tohoku 124. *Bioscience, Biotechnology and Biochemistry* **61**, 2148–2150.

Sánchez-Monge, R., Lombardero, M., Garcia-Sellés, F.J., Barber, D. and Salcedo, G. (1999) Lipid-transfer proteins are relevant proteins in fruit allergy. *Journal of Allergy and Clinical Immunology* **103**, 514–519.

Sánchez-Monge, R., Diaz-Perales, A., Blanco, C. and Salcedo, G. (2004) Latex allergy and plant chitinases, in *Plant Food Allergens* (eds E.N.C. Mills and P.R. Shewry), Blackwells, Oxford, UK, pp. 87–104.

Sánchez-Monge, R., Pascual, C.Y., Diaz-Perales, A., Fernandez-Crespo, J., Martin-Esteban, M. and Salcedo, G. (2000) Isolation and characterisation of relevant allergens from boiled lentils. *Journal of Allergy and Clinical Immunology* **106**, 955–961.

Scheurer, S., Son, D.Y., Boehm, M., Karamloo, F., Franke, S., Hoffmann, A., Haustein, D. and Vieths, S. (1999) Cross-reactivity and epitope analysis of Pru a 1, the major cherry allergen. *Molecular Immunology* **36**, 155–167.

Scheurer, S., Wangorsch, A., Haustein, D. and Vieths, S. (2000) Cloning of the minor allergen Api g 4 profilin from celery (*Apium graveolens*) and its cross-reactivity with birch pollen profilin Bet v 2. *Clinical and Experimental Allergy* **30**, 962–971.

Scheurer, S., Pastorello, E.A., Wangorsch, A., Kastner, M., Haustein, D. and Vieths, S. (2001) Recombinant allergens Pru av 1 and Pru av 4 and a newly identified lipid transfer protein in the *in vitro* diagnosis of cherry allergy. *Journal of Allergy and Clinical Immunology* **107**, 724–731.

Schmidt, R.J., Burr, F.A., Aukerman, M.K. and Burr, B. (1990) Maize regulatory gene opaque-2 encodes a protein with a "leucine-zipper" motif that binds to zein DNA. *Proceedings of the National Academy of Sciences USA* **87**, 46–50.

Shewry, P.R. and Lucas, J.A. (1997) Plant proteins that confer resistance to pests and pathogens, in *Advances in Botanical Research*, Volume 26 (ed J. Callow), Academic Press, pp. 135–192.

Shewry, P.R., Halford, N.G. and Lafiandra, D. (2003) Genetics of wheat gluten proteins, in *Advances in Genetics*, Volume 49, Elsevier Science, USA pp. 111–184.

Shewry, P.R., Jenkins, J., Beaudoin, F. and Mills, E.N.C. (2004) The classification, functions and evolutionary relationships of plant proteins in relation to food allergens, in *Plant Food Allergens* (eds E.N.C. Mills and P.R. Shewry), Blackwells, Oxford, UK, pp. 24–41.

Shibasaki, M., Suzuki, S., Tajima, S., Nemoto, H. and Kuroume, T. (1980) Allergenicity of major component proteins of soybean. *International Archives of Allergy and Applied Immunology* **61**, 441–448.

Simonato, B., Pasini, G., Giannattasio, M., Peruffo, A.D., De Lazzari, F. and Curioni, A. (2001a) Food allergy to wheat products: the effect of bread baking and *in vitro* digestion on wheat allergenic proteins. A study with bread dough, crumb, and crust. *Journal of Agricultural and Food Chemistry* **49**, 5668–5673.

Simonato, B., De Lazzari, F., Pasini, G., Polato, F., Giannattasio, M., Gemignani, C., Peruffo, A.D., Santucci, B., Plebani, M. and Curioni, A. (2001b) IgE binding to soluble and insoluble wheat flour proteins in atopic and non-atopic patients suffering from gastrointestinal symptoms after wheat ingestion. *Clinical and Experimental Allergy* **31**, 1771–1778.

Soeria-Atmadja, D., Zorzet, A., Gustafsson, M.G. and Hammerling, U. (2004) Statistical evaluation of local alignment features predicting allergenicity using supervised classification algorithms. *International Archives of Allergy and Immunology* **133**, 101–112.

Son, D.Y., Scheurer, S., Hoffmann, A., Haustein, D. and Vieths, S. (1999) Pollen-related food allergy: cloning and immunological analysis of isoforms and mutants of Mal d 1, the major apple allergen, and Bet v 1, the major birch pollen allergen. *European Journal of Nutrition* **38**, 201–215.

Stadler, M.B. and Stadler, B.M. (2003) Allergenicity prediction by protein sequence. *FASEB Journal* **17**, 1141–1143.

Sten, E., Stahl Skov, P., Andersen, S.B., Torp, A.M., Olesen, A., Bindslev-Jensen, U., Poulsen, L.K. and Bindslev-Jensen, C. (2002) Allergenic components of a novel food, Micronesian nut Nangai (*Canarium indicum*), shows IgE cross-reactivity in pollen allergic patients. *Allergy* **57**, 398–404.

Sten, E., Skov, P.S., Andersen, S.B., Torp, A.M., Olesen, A., Bindslev-Jensen, U., Poulsen, L.K. and Bindslev-Jensen, C. (2004) A comparative study of the allergenic potency of wild-type and glyphosate-tolerant gene-modified soybean cultivars. *APMIS* **112**, 21–28.

Tanabe, S., Arai, S., Yanagihara, Y., Takahashi, K. and Watanabe, M. (1996) A major wheat allergen has a Gln-Gln-Gln-Pro-Pro motif identified as an IgE-binding epitope. *Biochemical and Biophysical Research Communications* **219**, 290–293.

Tada, Y., Nakase, M., Adachi, T., Nakamura, R., Shimada, H., Takahashi, M., Fujimura, T. and Matsuda, T. (1996) Reduction of 14–16 kDa allergenic proteins in transgenic rice plants by antisense gene. *FEBS Letters* **391**, 341–345.

Tejera, M., Villalba, M., Batanero, E. and Rodriguez, R. (1999) Identification, isolation and characterisation of Ole e 7, a new allergen of olive tree pollen. *Journal of Allergy and Clinical Immunology* **104**, 797–802.

Teuber, S.S., Dandekar, A.M., Peterson, W.R. and Sellers, C.L. (1998) Cloning and sequencing of a gene enconding a 2S albumin seed storage protein precursor from English walnut (*Juglans regia*), a major food allergen. *Journal of Allergy and Clinical Immunology* **101**, 807–814.

Teuber, S.S. and Petersen, W.R. (1999) Systemic allergic reaction to coconut (*Cocos nucifera)* in 2 subjects with hypersensitivity to tree nut and demonstration of cross-reactivity to legumin-like seed storage proteins: new coconut and walnut food allergens. *Journal of Allergy and Clinical Immunology* **103**, 1180–1185.

Teuber, S., Jarvis, K.C., Dandekar, A.M., Peterson, W.R. and Ansari, A. (1999) Identification and cloning of a complimentary DNA encoding a vicilin-like proprotein, Jug r 2, from English walnut kernel (*Juglans regia*), a major food allergen. *Journal of Allergy and Clinical Immunology* **104**, 1111–1120.

van Loon, L.C. and van Strien, E.A. (1999) The families of pathogenesis-related proteins, their activities and comparative analysis of PR-1 type proteins. *Physiological and Molecular Plant Pathology* **55**, 85–97.

Varjonen, E., Vainio, E. and Kalimo, K. (2000) Antigliadin IgE—indicator of wheat allergy in atopic dermatitis. *Allergy* **55**, 386–391.

Varjonen, E., Vainio, E. and Kalimo, K. (1997) Life-threatening, recurrent anaphylaxis caused by allergy to gliadin and exercise. *Clinical and Experimental Allergy* **27**, 162–166.

Varjonen, E., Vainio, E., Kalimo, K., Juntunen-Backman, K. and Savolainen, J. (1995) Skin-prick test and RAST responses to cereals in children with atopic dermatitis. Characterization of IgE-binding components in wheat and oats by an immunoblotting method. *Clinical and Experimental Allergy* **25**, 1100–1107.

van Ree, R., Fernandez-Rivas, M., Cuevas, M., van Wijngaarden, M. and Aalberse, R.C. (1995) Pollen-related allergy to peach and apple: an important role for profilin. *Journal of Allergy and Clinical Immunology* **95**, 726–734.

Vioque, J., Sanchez-Vioque, R., Clemente, A., Pedroche, J., Bautista, J. and Millan, F. (1999) Purification and partial characterization of chickpea 2S albumin. *Journal of Agricultural and Food Chemistry* **47**, 1405–1409.

Wang, F., Robotham, J.M., Teuber, S.S., Tawde, P., Sathe, S.K. and Roux, K.H. (2002) Ana o 1, a cashew (*Anacardium occidental*) allergen of the vicilin seed storage protein family. *Journal of Allergy and Clinical Immunology* **110**, 160–166.

Watanabe, M., Tanabe, S., Suzuki, T., Ikezawa, Z. and Arai, S. (1995) Primary structure of an allergenic peptide occurring in the chymotryptic hydrolysate of gluten. *Bioscience, Biotechnology and Biochemistry* **59**, 1596–1597.

Wensing, M., Akkerdaas, J.H., van Leeuwen, W.A., Stapel, S.O., Bruijnzeel-Koomen, C.A., Aalberse, R.C., Bast, B.J., Knulst, A.C. and van Ree, R. (2002) IgE to Bet v 1 and profilin: cross-reactivity patterns and clinical relevance. *Journal of Allergy and Clinical Immunology* **110**, 435–442.

Wensing, M., Knulst, A.C., Piersma, S., O'Kane, F., Knol, E.F. and Koppelman, S.J. (2003) Patients with anaphylaxis to pea can have peanut allergy caused by cross-reactive IgE to vicilin (Ara h 1). *Journal of Allergy and Clinical Immunology* **111**, 420–424.

Yagami, T., Haishima, Y., Nakamura, A., Osuna, H. and Ikezawa, Z. (2000) Digestibility of allergens extracterd from natural rubber latex and vegetable foods. *Journal of Allergy and Clinical Immunology* **106**, 752–762.

Yagami, T., Osuna, H., Kouno, M., Haishima, Y., Nakamura, A. and Ikezawa, Z. (2002) Significance of carbohydrate epitopes in a latex allergen with beta-1,3-glucanase activity. *International Archives of Allergy and Immunology* **129**, 27–37.

Zhu, T., Peterson, D.J., Tagliani, L., St Clair, G., Baszczynski, C.L. and Bowen, B. (1999) Targeted manipulation of maize genes *in vivo* using chimeric RNA/DNA oligonucleotides. *Proceedings of the National Academy of Sciences USA* **96**, 8768–8773.

Zorzet, A., Gustafsson, M. and Hammerling, U. (2002) Prediction of food protein allergenicity: a bioinformatic learnining systems approach. *In Silico Biology* **2**, 48–58.

3.2

Environmental Impact and Gene Flow

P.J.W. Lutman and K. Berry

Rothamsted Research, Harpenden, Hertfordshire, AL5 2JQ, United Kingdom

Introduction

The impact that transgenic crops have on the environment has been one of the main political issues surrounding their commercialization in Europe, along with potential effects on human health. Those groups opposing the technology, such as Greenpeace and Friends of the Earth, have strongly emphasized the potential environmental impacts of GM crops (Hill, 1995; Treu and Emberlin, 2000). It has become a much greater political issue in Europe than it has been in N. America for a number of reasons, some totally unrelated to the technology itself. For example, following the outbreak of bovine spongiform encephalopathy (BSE), consumers in the UK and continental Europe are reluctant to believe scientific information, especially if transmitted via a politician. This makes it extremely difficult to persuade consumers that novel technologies are safe. This concern is exaggerated by the fact that the current commercialized GM crops only offer direct benefit to the grower, and linked suppliers, and offer no direct value to the consumer. The risks associated with mobile phone use are arguably higher than they are for the current commercialized GM crops, yet because of the convenience that mobile phones provide to the user, they are much more acceptable to consumers.

The main point that has to be made in this introduction about the environmental and gene flow impacts of GM crops is that most effects are specific to the crop and transgene under investigation. An environmental risk assessment could come up with totally different conclusions when, for example, assessing a maize (*Zea mays* L.) containing Bt genes, making it toxic to some insects, compared to an enhanced oil cultivar of oilseed rape (*Brassica*

Plant Biotechnology. Edited by Nigel Halford.

napus L.). There is a tendency in the popular press to talk about 'hazards from GM crops'. Such a generalization is inappropriate.

Although much of this chapter will discuss potential negative impacts on the environment (e.g. gene flow to wild relatives, effects on non-target organisms), it must be emphasized that the technology can provide positive impacts too (e.g. increase use of minimum cultivation prior to planting herbicide-tolerant (HT) crops, reduction in insecticide use in Bt cotton).

Environmental Impact of GM Crops

All agricultural activities have an impact on the natural environment; arable agriculture is based on completely unecological activities. Nature does not usually create large areas of plant monocultures and such areas only exist as a result of man's intervention aimed at providing adequate food for the human population. There is evidence in the UK that the increasing intensity of agriculture over the last 50 years has had an increasingly negative impact on the arable ecosystem as a whole by decreasing the number of species of plants and animals that co-exist with arable farming (Chamberlain *et al.*, 2000; Preston, Pearman and Dines, 2002). With this background of concern about the impact of agriculture on the natural environment it was not surprising that questions were asked about the environmental impact of the HT and insecticidal (Bt) GM crops, which happened to be in the first 'cohort' of GM crops to be commercialized.

Before considering the environmental effects of GM crops in more detail, it is helpful to establish a framework for assessment of the potential impacts.

- First, the GM crop itself may pose a threat to the natural environment as a result of acquiring greater 'fitness' and an ability to colonize natural habitats and replace indigenous species. As mentioned in the first paragraph, crops generally require man's intervention to survive and do not become invasive weeds. However, it is theoretically possible for crops normally unadapted to natural ecosystems to acquire genes by genetic transformation such that they become invasive.
- Second, the transformation may impact on the behavior of non-target organisms that exist in association with the GM crop in the arable ecosystem. This can be a direct or an indirect effect. An obvious example would be the introduction of Bt maize to control corn borer, which also could have a negative effect on other Lepidoptera susceptible to the Bt toxin. Another theoretical example would be a change in quality of the leaves of a crop associated with a desired change in crop marketability or disease resistance that could impact on the palatability of the crop to invertebrates and birds (either positively or negatively).
- Third, changes in management associated with the GM crop may have positive or negative effects on other components of the arable ecosystem. The introduction of HT canola in Canada has raised the overall level of weed control. This could be perceived as a negative effect on the arable ecosystem but it has also resulted in a rapid increase in 'no-tillage' establishment practices, which are known to have environmental benefits, such as reduction in soil erosion.
- Finally, the introduction of GM crops can cause changes in cropping patterns, which in turn result in positive or negative effects on the overall arable environment. This

change is not an attribute of GM crops in particular, but is a reflection of the environment's response to changed cropping.

Effects of Genetic Modification on 'Fitness' of Transformed Plants

It is possible that a transgenic crop plant could possess increased 'weedy' characteristics that would result in it becoming invasive. 'Volunteer' crops have been serious weeds in arable fields for many years (Cussans, 1978) but are rare in natural ecosystems. Invasive weeds cause serious problems in many countries around the world. These invasive plants are generally not crop species but wild plants (weeds) of natural ecosystems. It also needs to be noted that the potential for a crop to become weedy depends on the attributes of the crop itself. In the UK, oilseed rape is a common arable weed because of the persistence of its seeds (Lutman, Freeman and Pekrun, 2003). Conversely, maize does not establish volunteer populations in the UK because the seeds and plants are susceptible to cold winter weather. However, volunteer maize can pose weed problems in warmer climates in the first year after a maize harvest.

While it is possible that a transformed crop species could become invasive, this would depend on the attribute that was being introduced and is not a generic attribute of all GM crops. Current commercialized GM crops such as those that tolerate herbicides have been studied to assess whether the transgene affects the plant's fitness. Evidence from work by Fredshavn and Poulsen (1996) and from Crawley *et al.* (1993; 2001) indicates that GM HT oilseed rape is no fitter than conventional varieties. Similarly, Crawley *et al.* (2001) could find no evidence of increased fitness with four crops that were transformed for either herbicide or insect resistance. However, it cannot be assumed that the same is true for all crops that have been transformed, especially those delivering, for example, pest resistance or drought tolerance. Each transformation needs to be assessed on its own 'merits'.

A good example of such an effect has been identified with Bt sunflower. Volunteers from this crop appear to be less susceptible to grazing by insects and consequently produce more seeds (Snow *et al.*, 2003). Whether this will have a major impact on the ecology of volunteer and wild sunflowers remains to be established.

Direct and Indirect Effects of Genetic Modification of Crops on Non-Target Organisms

The potential of a GM crop to have direct effects on non-target organisms is one that has been addressed by a considerable number of researchers, especially those concerned with insect-resistant (e.g. Bt) crops. However, the potential indirect effects are more difficult to establish and involve considerable debate as to the relevant comparator. These issues are of particular relevance to 'input' traits, such as herbicide or insecticide resistance, but are much less relevant to transformations that alter crop quality traits. GM crops containing a Bt gene aimed at controlling Lepidopteran pests could have adverse effects on non-target arthropods. These effects can be direct, such as the death of non-target Lepidoptera, or indirect, through tritrophic interactions. The latter is an important issue as biological control of insect pests by their predators and parasitoids plays an appreciable role in the containment of pest populations.

The potential direct effects of Bt genes on non-target organisms received considerable public discussion and media attention when an American paper was published in Nature in 1999 (Losey *et al.,* 1999). This reported some laboratory studies indicating that the talismanic Monarch butterfly (*Danaus plexippus*) might be adversely affected by Bt corn. Several million dollars were spent subsequently to clarify whether this laboratory effect was replicated in the field. The work showed that the potential impact of Bt corn on this species in the field was negligible compared with other mortality factors (Stanley-Horn *et al.*, 2001).

A considerable amount of work has also explored the tritrophic effects of the Bt gene, particularly Bt corn resistant to the European corn borer (*Ostrinia nubilalis*) on insect predators and parasitoids. Although laboratory studies have shown potential effects of Bt corn on corn borer predators, such work has not been confirmed in the field. Nevertheless, there is some evidence from the field that parasitoids can be adversely affected (work reviewed by Obrycki *et al.*, 2001). These types of study do highlight that caution is needed when new transgenic crops are approved for commercialization. It has also been suggested that GM crops resistant to insect attack will produce pollen that will damage honey bees (*Apis mellifera*). However, field evidence to date suggests that the current insect-resistant crops do not affect bees (Connor *et al.*, 2003). It must also be realized that non-Bt crops often receive appreciably more quantities of toxic broad-spectrum insecticides than do Bt crops, so non-GM crops may have an overall more detrimental environmental profile. It has been estimated that the planting of GM soybeans, oilseed rape, cotton and maize reduced insecticide and herbicide use worldwide in 2000 by 22.3 million kg of product (Phipps and Park, 2002). All these inter-related issues need to be considered when assessing environmental impacts.

Similar concerns have been expressed about indirect effects associated with herbicide use on HT crops. There is no evidence that the transgenic HT crop itself has direct adverse effects on the environment, therefore research has focussed on the impact that the changes in herbicide practice associated with the GMHT crop could have on the environment. Between 1999 and 2003, 197 sites were used to investigate the indirect effects of HT GM oilseed rape, maize and sugar/fodder beet in the UK. This work concentrated on identifying whether the herbicides used in glyphosate- and glufosinate-tolerant GM crops had a greater or lesser effect on plants and invertebrates in arable fields than the 'conventional' products normally used by farmers. This work said nothing about the transgene itself and examined purely the impact of changed herbicide practice. The results showed that HT beet and spring rape had fewer surviving plants and invertebrates than 'conventional' crops, whereas there were more of these species in HT maize (Figure 3.2.1) (Heard *et al.*, 2003). On this basis the UK government was prepared to consider the commercialization of GM maize but not rape and beet. This concern about the indirect effects of farming on diversity in rural ecosystems, highlighted by these 'Farm Scale Evaluations' clearly has much wider implications than simply whether an HT crop can be grown or not. All farming operations have indirect effects on the ecosystem, but the issue of the impact of GMHT crops has set a precedent in the UK that any new farming practice that increases the decline in species abundance in arable ecosystems should not be approved. The discussion on this paradigm continues!

The Farm Scale Evaluations only examined one aspect of the potential changes in crop agronomy associated with HT crops that could impact on diversity, the direct effects of changed herbicide practice. It is clear from countries where the use of HT crops is

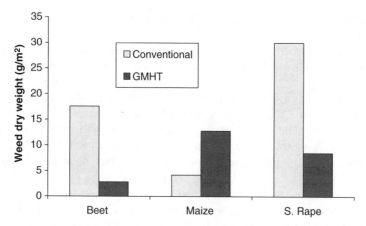

Figure 3.2.1   *Comparison of the impact of conventional and HT crops on the weight of weeds present in the summer (means of 62 beet, 40 maize and 62 spring rape split-field comparisons) (Heard et al., 2003) Reproduced by the permission of the Royal Society.*

widespread that the indirect effects can be far more complex. In Canada, the uptake of GMHT oilseed rape (canola) has been considerable and in 2003 nearly 90 % of all crops were HT. Weed control in the HT crops tends to be more effective, cheaper and simpler than the conventional systems that they replace. From the conclusions in the previous paragraph, such increased weed control should cause a decline in arable diversity. This indeed may be the case, but this decline from a Canadian viewpoint is actually better for the indigenous flora, as over 80 % of species in the Canadian arable weed flora are imported alien species from Europe. Even though weed control may be higher, the overall environmental profile of HT crops can be better than its conventional alternatives as the products used on the HT crop have a much lower impact on all components of the ecosystem (Altieri, 2000; Glick, 2001).

A further complication when trying to assess the biodiversity consequences of GMHT crops relates to other changes in crop production practices resulting from the planting of these crops. As weed control is simpler and more effective with GMHT crops, an appreciable number of Canadian farmers are exploiting this benefit by planting their HT rape using minimum tillage techniques (Devine and Buth, 2001). Minimum tillage has much less of an adverse effect on soil-living invertebrates than ploughing, the technique that was used previously. Thus, the overall environmental impacts of GM crops can be difficult to estimate, especially if the technology also induces other changes in agricultural practice. As with the direct effects of GM crops, indirect effects must be considered on a case-by-case basis.

The overintensive planting of GMHT crops can nullify any environmental benefits. For example, the widespread planting of glyphosate-tolerant GM soybeans in the USA and the consequent mono-usage of this herbicide has resulted in the development of a weed flora that avoids the product by emerging after treatment or which has developed resistance. Such 'weed shifts' and resistance problems are not surprising because, since the development of pesticides, mono-usage of single active ingredients has frequently resulted in the evolution of resistant populations of weeds, pests and diseases. As a result of these changes in the weed flora, mixtures of products are now being used to control weeds, reducing some of the environmental benefits associated with glyphosate.

A further impact of GM crops that is hard to quantify lies in the response of farmers to the technology. A GM crop that is widely adopted could cause a change in crop rotations, which could have positive or negative effects on the arable ecosystem. If the GM crop replaces a crop that has a severely negative impact on diversity, then the use of GM crops will be advantageous. Conversely, if the GM crop replaces less environmentally damaging crops then the overall impact of the GM crop will be negative. When the GM crop affects input traits, it is possible that the GM crop will have the potential to change inputs in subsequent years. For example, it has been postulated that the availability of GMHT crops could reduce herbicide use in the previous and subsequent crops. In the former case there is no particular need for the target crop to be preceded by a clean, weed-free crop. Similarly, as weed control in the GMHT crop is likely to be high, the seed return for the next crop will be low and so weed control will not be so vital. Again this issue has nothing directly to do with the GM crop itself but relates to how the technology fits in the overall context of farming practice. As such, the same questions can be asked of any new farming technology. For example, is the widespread cultivation of coppice willow for renewable energy likely to have an adverse effect on biodiversity? The questions being asked by government about the environmental impact of GM crops are often equally relevant to other agricultural technologies.

Appreciable amounts of research have been done over the last 10 years to address the potential impact of GM crops on the wider environment. Each transgene needs to be considered individually. Conner, Glare and Nap (2003), in their review of the environmental impact of GM crops, conclude that putative threats to biodiversity would seem to be largely hypothetical and marginal, especially when put in the context of the impact of native land conversion and habitat fragmentation. They state that 'given the multidimensional complexity of the biodiversity concept, assessing the impact of a technological development such as GM crops is far from straightforward', a sentiment that we would support.

Gene Flow and Co-existence of GM and Conventional Crops

One of the main issues currently being discussed in Europe in relation to the approval of the planting of GM crops is the issue of co-existence. If one farmer wishes to grow a GM crop and his neighbor does not, what precautions are needed to ensure that the GM crop does not affect the neighboring non-GM crop? The answers to this question lie in a sound understanding of the mechanisms of gene flow, which covers more issues than whether two farmers can successfully grow GM and non-GM crops. The following section will consider both crop to crop and crop to wild relative gene flow and the respective roles of pollen and seeds.

What is gene flow via pollen? It is a combination of the potential of pollen to carry genes from one plant to another, plus the ability of that pollen to fertilize the ovule in the receptor plant to produce a viable seed, such that the gene becomes established in a new population. Many discussions in the press assume that pollen flow and gene flow are the same. They are not. The latter includes many aspects not included in the former. Once the

Table 3.2.1 *Wild relatives which are known to hybridize with the 14 most important crop species (Ellstrand et al., 1999).* Reprinted, with permission, from the *Annual Review of Ecology and Systematics*, Volume 30 © 1999 by Annual Reviews   www.annualreviews.org

Crop species		Wild relative
Wheat	*Triticum aestivum*	Wild *T. aestivum, T. turgidum, Aegilops* spp.
Rice	*Oryza sativa*	*O. sativa* f. *spontanea*
Maize	*Zea mays*	The teosintes, e.g. *Zea mays* ssp. *mexicana*
Soybean	*Glycine max*	*G. soja, G. gracilis*
Barley	*Hordeum vulgare*	*H. spontaneum* wild *H. vulgare*
Cotton seed	*Gossypium hirsutum*	Wild and feral *G. hirsutum, G. barbadense*
Sorghum	*Sorghum bicolor*	*S. halepense* and wild *S. bicolor*
Millet	*Eleusine coracana*	*E. coracana* ssp. *africana*
Beans	*Phaseolus vulgaris*	*P. aborigineus, P. mexicanus*
Rapeseed	*Brassica napus*	*B. campestris/rapa, B. juncea*
Groundnut	*Arachis hypogaea*	*A. monticola*
Sunflower seed	*Helianthus annuus*	Wild *H. annuus*
Sugar beet	*Beta vulgaris*	Wild annual *B. vulgaris, B. vulgaris* ssp. *maritima*
Sugarcane	*Saccharum officinarum*	Wild *Saccharum* spp.

pollen has arrived on the receptive flower it has to compete with pollen from the local plants and fertilize the seed. Any seed resulting from cross-fertilisation must then produce a healthy seedling which in turn must produce viable pollen and seeds.

Gene Flow to Wild Relatives

Most crops have closely related species somewhere in the world where the possibility of cross-fertilization could occur. Traditional plant breeding depends on the ability of the breeder to find related species with desirable genes and then cross-fertilize the crop with these desirable related species. Ellstrand, Prentice and Hancock (1999) reviewed the risks for 13 of the world's major crops (Table 3.2.1), and there is evidence that wild relatives exist for 44 crop species. How much of a risk there is will depend on where in the world the GM crop is being grown. It should be noted that many of the wild species that could cross-fertilize with the crop are actually the wild progenitors of the crop.

Chevre *et al.* (1999) listed the following aspects that need to be studied to determine the probability of inter-specific hybridization and introgression:

 (i) the existence of close relatives growing in sympatry, with overlapping flowering times
 (ii) a mating system allowing pollen flow and allogamy
(iii) the production and survival of fertile F_1 inter-specific hybrids
(iv) fertile plants produced in successive generations
 (v) the possibility of chromosome recombination and spread of the gene in populations of the wild species (this is affected by factors including chromosomal location, adaptation encoded for and wild population size)

This list shows that crop to wild relative gene flow is not always simple and in many cases is extremely unlikely, for a range of reasons. One further issue that needs to be considered is the impact of the transgene on the fitness of the new hybrid. If new hybrid plants have no selective advantage they are likely to be eliminated from the population, unless hybridization results in the production of huge numbers of seeds. With herbicide tolerance transgenes, the fitness of the hybrid is not likely to be enhanced except in the presence of the herbicide. As hybrids will probably arise in semi-natural habitats and not in fields, they will not be exposed to the herbicide and so will have no advantage. The subsequent paragraphs will consider the risks of crop to wild relative gene flow for the main European GM crops that are in development: oilseed rape (*B. napus*), sugar beet (*Beta vulgaris* L.) and maize (*Z. mays*).

Maize. This crop is simply resolved as although there are wild *Teosinte* species in N. America, none are present in Europe. Consequently, there is no risk of gene flow.

Sugar beet. Sugar beet has a limited number of related species, the most important being wild beet (annual *B. vulgaris* ssp. *vulgaris*) and sea beet (*B. vulgaris* ssp. *maritima*). The latter tends to be restricted to coastal areas, whereas the former is widely distributed as a result of contamination of sugar beet seed in earlier decades. Both can inter-breed freely with the crop plants if the latter are permitted to flower. The commercial sugar beet crop is a biennial producing a root in the first year and flowering in the second. Thus, provided the seedlings are not vernalized, the crop should not flower. The main areas at risk from gene flow would be the sugar beet seed-producing regions of northern Italy and southern France. There is evidence that gene flow between the beet seed crops and wild beets has occurred in the past and the risks of it occurring with GM beets are appreciable (Bartsch *et al.*, 1999).

Oilseed rape. Oilseed rape has been created from hybrids between *B. rapa* L. (*B. campestris*), *B. oleracea* L. and *B. nigra* (L.) Koch, and the tetraploids derived from them, *B. juncea* (L.) Czern. and Coss, *B. carinata* Braun and *B. napus* L. Consequently, it shares aspects of its genome with these other species and so cross-fertilization is possible. In Europe, several of these wild species have been investigated for potential hybridization to oilseed rape. The most studied species at present are *B. rapa*, *B. juncea* and *Raphanus raphanistrum* L. Table 3.2.2 lists the main wild relatives studied and their hybridization and back-crossing potential with oilseed rape.

The overall conclusion from the studies of potential hybridization between oilseed rape and all of these related members of the Brassicaceae is that, for most combinations, although hybrids can be formed they are difficult to create, rare to occur and even rarer to establish. In some cases, such as with *Sinapis arvensis* L., although it is theoretically possible to create hybrids in the laboratory, it is very difficult or impossible for them to be created in the field and even more difficult for them to establish successfully. As shown by Gueritaine *et al.* (2003a,b) with hybrids between rape and *R. raphanistrum*, the fitness of the hybrids can often be lower than that of their parents. A related aspect of the creation of these hybrids was that the frequency of the herbicide tolerance gene (*BAR*) declined in the population in the absence of selective pressure from glufosinate.

It is clear from some of the more recent work that the potential to hybridize varies within rape cultivars and within populations of the wild species (Gueritaine and Darmency, 2001). The weed that is most likely to hybridize with oilseed rape in the UK is *B. rapa*. Work in Denmark (Hauser *et al.*, 1998a,b; Jorgensen *et al.*, 1996) and more recently in the UK (Wilkinson *et al.*, 2003) has shown that hybrids are created

Table 3.2.2 *Hybridization and backcrossing potential between oilseed rape (B. napus) and the listed wild relatives (based on Scheffler and Dale, 1994 and Berry et al., 2004).*

Wild relative		Hybridization		Backcross to weed (BC₁)	Selected references
Latin name	Common name	Crop ♂	Crop ♀	(BC₁)	Selected references
Brassica spp.					
Brassica rapa	wild turnip, turnip rape	✓	✓	✓	Mikkelsen *et al.* (1996), Hauser *et al.* (1998a,b), Snow, Anderson and Jorgensen (1999)
B. carinata	brown mustard	✓	✓	✓	Frello *et al.* (1995), Jorgensen *et al.* (1996)
B. oleracea	wild cabbage	?✓	?✓	?✓*	Scheffler and Dale (1994), Eastham and Sweet (2002)
B. carinata	Ethiopian mustard	?✓	?✓	?✓	Scheffler and Dale (1994)
B. nigra	black mustard	?✓	?✓	?✓	Scheffler and Dale (1994), Estham and Sweet (2002)
Other genera					
Raphanus raphanistrum	wild radish, runch	✓	✓	✓	Chevre *et al.* (2000), Gueritaine *et al.* (2002)
Sinapis arvensis	charlock, wild mustard	?✓	?✓	?✓	Lefol, Danielou and Darmency (1996a), Moyes *et al.* (2002)
Hirschfeldia incana	hoary mustard	✓	✓	✓*	Lefol, Fleury and Darmency (1996b), Darmency and Fleury (2000)
Erucastrum gallicum	dog mustard	?✓	?✓	?✓	Lefol, Seguin-Swartz and Downey (1997)

✓ When a hybrid is produced by open pollination in field or glasshouse (even with low fertility).
?✓ When a hybrid is produced using controlled methods (including emasculation and hand pollination)
* BC hybrids became non-viable.

relatively easily and have intermediate fitness between the two parents. Wilkinson's work explored the sympatry between the two species and their distributions, calculating that 49 000 hybrid seeds could be produced in a year. The fate of these seeds is unknown but clearly would be an issue in the context of gene flow from GM rape to wild *B. rapa*.

The previous paragraphs have discussed the potential for pollen to transfer genes from crop to related species but, with the exception of the work of Wilkinson *et al.* (2003), have not addressed all the other related issues associated with the creation of hybrids. For example, are the two species flowering simultaneously (sympatric)? How big an area and how far away is the donor crop? How big is the receptor population and thus its pollen cloud? All these issues will influence the potential frequency of hybrid production.

Crop to Crop Gene Flow

The potential for cross-fertilization is much higher for crop to crop gene flow than it is for crop to wild relatives. There are many genetic barriers to the latter, as discussed above, but provided the crops are sympatric there are none for crop to crop gene flow. The level of cross-fertilization that occurs will be affected by the factors mentioned in the previous paragraph (size (area) of donor and receptor populations, distance apart, etc.). Clearly, gene flow between adjacent crops has always been of concern and plant breeders take this into account when planting seed crops that need to have a defined level of seed purity. The issue of crop to crop gene flow will vary according to the crop concerned, depending on the degree of out-crossing of the crop, the nature of the pollen and the level of production. Plant breeders routinely use isolation distances of 50–200 m, depending on the crop and the required purity of the seed.

In the UK, the crop of most concern in the context of gene flow and GM crops is oilseed rape as, although it is able to self-fertilize, pollination and seed-set can be enhanced by external pollen. Indeed, with varietal associations and hybrids, cross-fertilization is an important contributor to creating a well-fertilized crop. Clearly there is potentially a need for some spatial separation, if it is required that attributes of one cultivar are not to be transferred to adjacent crops. An obvious example of this would be the debate about the degree of separation required between GM and organic crops. The separation required will depend on the threshold of cross-fertilization that is acceptable in the 'receptor' crop. In the case of GM crops in Europe, the European Commission has proposed that produce that contains less than 0.9 % GM material does not have to be labeled as containing GM. Lower levels of 0.3 % and 0.5 % have been proposed for crop seed production systems so that the commercial crops can meet this 0.9 % limit. These levels are being contested by organic growers who are advocating a 0 % tolerance level for organic crops. Consequently, there has been much research and modeling on the separation distance required to meet these limits.

It is clear that the level of cross-fertilization declines rapidly with increasing separation distance between the source and receptor crops. With oilseed rape, out-crossing rates are variable but are estimated at <1 % at 100 m from the common field border. The rate continues to decrease beyond 100 m but is still present at 800 m or more (Beckie *et al.*, 2003; Downey, 1999; Eastham and Sweet, 2002; Rieger *et al.*, 2002). The factors influencing the low levels of cross-fertilization that can occur at considerable distances from source oilseed rape crops are subject to current debate and research. The distributions of cross-fertilized plants at distance from the source crop show no clear pattern (e.g. Rieger *et al.*, 2002), with no evidence of greater frequency in the side of the field closest to the source of the pollen, suggesting that wind movement of pollen is not the main cause. This and other work suggests that this low-level gene movement may be due to insect-mediated pollinations. A study using male-sterile bait plants has found evidence of gene flow several kilometres from the source (Ramsay *et al.*, 2003).

Such studies show potential gene flow but do not reflect 'real' cropping situations, as in a normal crop this incoming pollen would have to compete with indigenous pollen to fertilize the seeds. Modeling by Perry (2002) suggested that if insect-mediated pollen was important in long-distance gene flow, exponential decline curves for calculating the relationship between distance from the source and pollen density or cross-fertilization events were not appropriate. The general consensus is that there is a very rapid decline in cross-fertilization in the first 50 m from the pollen source and further decline thereafter, but the tail of the curve is longer and more variable than would be expected from a normal exponential decline. Although rare cross-pollination events can occur more than 1 km from the source crop, the standard plant breeding separation distance of 100–200 m would ensure that cross-pollination did not result in a non-GM crop breeching the EU GM admixture limit (0.9 %). Greater isolation might be needed where the acceptable threshold of admixture was lower (e.g. seed crops) and where the 'receptor' crop was a varietal association, with its greater sensitivity to incoming pollen.

The previous paragraph on cross-fertilization was written in the context of gene flow from a GM crop causing a non-GM crop to exceed the EU threshold. A second aspect of this subject is the role of gene flow in creating gene stacking. For example, pollen from a glyphosate-tolerant rape cultivar could transfer genes to a glufosinate-tolerant one, creating plants with both tolerance traits. Such gene stacking has occurred quite frequently in Canada where canola varieties tolerant of glyphosate, glufosinate and imidazolinone herbicides are widely grown. Indeed, in one well-documented example plants that tolerated all three groups of herbicides were created in two seasons (Hall *et al.*, 2000). This creation of multiple HT plants may or may not pose a management problem, depending on the cropping practices involved, but does highlight that good management is needed to ensure that unwanted gene flow is minimized.

Role of Seed Persistence in Gene Flow

The preceding section has been devoted to pollen movement and consequent spatial gene flow. Although the movement of genes from field to field is of concern and has been much publicized in the debates about GM technology in the popular press, of probably greater practical significance to farmers is gene flow in time through the GM crop seeds left in the field after harvest. No harvesting operation removes all seeds (or tubers etc.) from the field and approximately 5 % of seeds of grain/seed crops are left after harvest. For some species these seeds do not survive (e.g. maize seeds in the UK), but for others (e.g. oilseed rape) a persistent seed bank can be created. Plants from this seed bank emerge over a number of years, providing a source of the genes from the original crop. This becomes an issue if the farmer decides not to grow a GMHT crop, having grown a GM type of the same crop earlier in the rotation.

As with pollen flow, the most problematic UK GM crop would be oilseed rape. This crop forms persistent seed banks (Lutman, 2003; Lutman *et al.*, 2003) and, because the crop produces huge numbers of small seeds, seed banks of several thousand seeds/m^2 can easily be created. Additionally, feral populations of oilseed rape can be established in field margins and road-side verges as a result of seeds 'escaping' from transport vehicles, providing another source of GM genes for adjacent crops. The appearance of GM volunteer rape plants in subsequent non-GM rape crops could cause the latter to exceed the EU GM threshold of 0.9 %. The problem with oilseed rape can be minimized with careful harvesting and appropriate post-harvest management, such as avoidance of ploughing immediately after the rape harvest (Pekrun, Hewitt and Lutman, 1998). Similar problems can arise with seeds of sugar beet and with grain crops such as wheat, though seed persistence is shorter for the grain crops (Lutman, 2003).

Conclusions

The first generic point to be reiterated here is that the environmental impact of GM crops must be considered on a case-by-case basis. Any environmental effects are not linked to the technology of genetic modification itself, but are associated with the transgenes incorporated into the crop's genome. Additionally, any environmental impact depends on the interaction of the new GM crop with the existing farm management practices in which it will be grown. In the case of HT crops, changes in production systems through a complete rotation resulting from the introduction of the technology have potentially both positive and negative effects on biodiversity. Additionally, indirect effects may be of greater significance than direct effects. Thus, the effects of Bt maize on predators and parasitoids are probably more significant than direct toxicity on non-target species. Similarly, the indirect impact of a GM HT crop on levels of invertebrates is possibly more environmentally significant than the direct effects of the herbicide treatments on the weeds. The environmental effects of GM crops with input traits, such as herbicide tolerance, are likely to be higher than those associated with altered quality traits (e.g. oilseed rape oil quality). But each new transgenic crop must be assessed for its potential impact.

As the EU member states are being required to develop national co-existence guidelines, the question of co-existence of GM and conventional crops in a farmed landscape is currently being seriously debated. Such guidelines focus strongly on the issues of gene flow resulting from both pollen movement and seed persistence. For the development of appropriate codes of practice, agreement has to be reached on the acceptable thresholds of admixture of GM with conventional and organic crops. The EU proposed limit of 0.9 % in food is rejected by some organic growers' organizations who have proposed a zero threshold. Such a threshold is a surrogate for banning the technology, as no biological system can guarantee that cross-fertilization will not occur. Now that the 0.9 % food limit and associated lower limits of 0.3 % and 0.5 % for seed crops are finally approved, it is possible to consider the spatial and temporal isolation, combined with production hygiene systems required to avoid problems arising after the crop has been harvested, that are needed to ensure the approved thresholds are not exceeded. Denmark published a report in 2003 that explores the problems and sets out appropriate control measures for a range of actual and potential GM crops (Tolstrup *et al.*, 2003).

It is clear that pollen movement between crops requires appropriate spatial isolation, as described in the previous sections of this chapter. It would be possible to isolate fields to restrict the level of cross-fertilization so that less than 0.3–0.9 % of the harvested seed contained the GM transgene. However, lower thresholds for some crops, especially oilseed rape, would be hard to meet. New work on gene flow at the landscape scale, as has been started in the GENESYS model (Colbach *et al.*, 2001), will endeavor to indicate how appropriate isolation can be achieved in a mixed arable landscape. Similarly, appropriate temporal isolation may also be needed to avoid contamination problems arising from persisting crop seeds from crops earlier in the rotation. Both the issues of pollen flow and seed persistence are particularly significant for oilseed rape. The associated isolation requirements for this crop are likely to be more rigorous than for crops where pollen production is less of an issue (e.g. GM sugar beet) and seed persistence is much lower (e.g. cereals).

There may also be technological solutions to the problems of spatial and temporal isolation. If the transgenic crop seed is not viable, through the introduction of genes that prevent germination, problems arising from seed persistence are no longer an issue. When Monsanto acquired a patent for such a technology (using a so-called terminator gene) in the 1990s it was vilified by Greenpeace and other NGOs as jeopardizing the livelihood of small-scale farmers who traditionally saved their own seeds. However, it is still a viable solution for farming systems in the developed world. Similarly, the problems arising from pollen flow could be minimized if the introduced transgenes were not expressed in pollen. Thus, there could be technological solutions to gene flow as well as agronomic/managerial solutions

The approval of GM crops for widespread cultivation in Europe still seems to be as far in the future as it was in the late 1990s, when the areas of the first HT and Bt crops were expanding rapidly in N. America. In Europe, it seems very likely that the socio-economic and political issues associated with GM crops, rather than scientific assessment, will drive decisions as to the commercial acceptability of these crops.

References

Altieri, M.A. (2000) The ecological impacts of transgenic crops on agroecosystem health. *Ecosystem Health* **6**, 13–23.

Bartsch, D., Lehnen, M., Clegg, J., Pohl-Orf, M., Schuphan, I. and Ellstrand, N.C. (1999) Impact of gene flow from cultivated beet on genetic diversity of wild sea beet populations. *Molecular Ecology* **8**, 1733–1741.

Beckie, H.J., Warwick, S.I., Nair, H. and Seguin-Swartz, G. (2003) Gene flow in commercial fields of herbicide-resistant canola (*Brassica napus*). *Ecological applications* **13**, 1276–1294.

Berry, K., Lutman, P.J.W. Lotz, L.A.P. and Kempenaar, C. (2004) Genetically modified herbicide-tolerant crops—a European perspective with a United Kingdom emphasis, in *Transgenic Crop Protection: Concepts and Strategies* (eds O. Koul and G.S. Dhaliwal) Science Publishers Inc., USA, pp. 117–156.

Chamberlain, D.E., Fuller, R.J., Bunce, R.G.H., Duckworth, J.C. and Shrubb, M. (2000) Changes in the abundance of farmland birds in relation to the timing of agricultural intensification in England and Wales. *Journal of Applied Ecology* **37**, 771–788.

Chevre, A.M., Eber, F., Darmency, H., Fleury, A., Picault, H., Letanneur, J.C. and Renard, M. (2000) Assessment of interspecific hybridization between transgenic oilseed rape and wild radish under normal agronomic conditions. *Theoretical and Applied Genetics* **100**, 1233–1239.

Chevre, A.M., Eber, F., Renard, M. and Darmency, H. (1999) Gene flow from oilseed rape to weeds, in *BCPC Symposium Proceedings No 72: Gene Flow and Agriculture—Relevance for Transgenic Crops* (ed P.J.W. Lutman), British Crop Protection Council, Farnham, Surrey, UK, pp. 125–130.

Colbach, N., Molinari, N. and Meynard, J.M. (2001) GENESYS: a model of the influence of cropping system on gene escape from herbicide-tolerant rapeseed crops to rape volunteers. *Agricultural Ecosystems and Environment* **83**, 235–270.

Conner, A.J., Glare, T.R. and Nap, J.-P. (2003) The release of genetically modified crops into the environment: part II. Overview of ecological risk assessment. *The Plant Journal* **33**, 19–46.

Crawley, M.J., Hails, R.S., Rees, M., Kohn, D. and Buxton, J. (1993) Ecology of transgenic oilseed rape in natural habitats. *Nature* **363**, 620–623.

Crawley, M.J., Brown, S.L., Hails, R.S., Kohn, D. and Rees, M. (2001) Transgenic crops in natural habitats. *Nature* **409**, 682–683.

Cussans, G.W. (1978) The problems of volunteer crops and some possible means of their control, in *Proceedings British Crop Protection Conference—Weeds*, BCPC, Farnham, Surrey, pp. 915–921.

Darmency, H. and Fleury, A. (2000) Mating system in *Hirschfeldia incana* and hybridization to oilseed rape. *Weed Research* **40**, 231–238.

Devine, M.D. and Buth, J.L. (2001) Advantages of genetically modified canola—a Canadian perspective. Proceedings the BCPC conference—Weeds, BCPC, Alton, UK, pp. 367–372.

Downey, K. (1999) Gene flow and rape—the Canadian experience, in *BCPC Symposium Proceedings No. 72: Gene Flow and Agriculture—Relevance for Transgenic Crops* (ed P.J.W. Lutman), British Crop Protection Council, Farnham, Surrey, UK, pp. 109–116.

Eastham, K. and Sweet, J.B. (2002) *Genetically Modified Organisms (GMOs): The Significance of Gene Flow Through Pollen Transfer*, European Environment Agency Report 28, Luxembourg.

Ellstrand, N.C., Prentice, H.C. and Hancock, J.F. (1999) Gene flow and introgression from domesticated plants into their wild relatives. *Annual Review of Ecology and Systematics* **30**, 539–563.

Fredshavn, J.R. and Poulsen, G.S. (1996) Growth behaviour and competitive ability of transgenic crops. *Field Crops Research* **45**, 11–18.

Frello, S., Hansen, K., Jensen, J. and Jorgensen, R. (1995) Inheritance of rapeseed (*Brassica napus*) specific RAPD markers and a transgene in the cross *B. juncea* × (*B. juncea* × *B. napus*). *Theoretical and Applied Genetics* **91**, 236–241.

Glick, H.L. (2001) Herbicide-tolerant crops: a review of agronomic, economic and environmental impacts. *Brighton Crop Protection Conference—Weeds* **1**, 359–366.

Gueritaine G. and Darmency, H. (2001) Polymorphism for interspecific hybridisation within a population of wild radish (*Raphanus raphanistrum*) pollinated by oilseed rape (*Brassica napus*). *Sexual Plant Reproduction* **14**, 169–172.

Gueritaine G., Bazot, S. and Darmency, H. (2003a) Emergence and growth of hybrids between *Brassica napus* and *Raphanus raphanistrum*. *New Phytologist* **158**, 561–567.

Gueritaine, G., Sester, M., Eber, F., Chevre, A.M. and Darmency, H. (2002) Fitness of backcross six of hybrids between transgenic oilseed rape (*Brassica napus*) and wild radish (*Raphanus raphanistrum*). *Molecular Ecology* **11**, 1419–1426.

Gueritaine, G., Sester, M., Eber, F., Chevre, A.M. and Darmency, H. (2003b) Fitness of backcross six of hybrids between transgenic oilseed rape (*Brassica napus*) and wild radish (*Raphanus raphanistrum*). *Molecular Ecology* **11**, 1419–1426.

Hall, L., Topinka, K., Huffman, J., Davis, L. and Good, A. (2000) Pollen flow between herbicide-resistant *Brassica napus* is the cause of multiple-resistant *B. napus* volunteers. *Weed Science* **48**, 688–694.

Hauser, T.P., Jorgensen, R.B. and Ostergard, H. (1998a) Fitness of backcross and F2 hybrids between weedy *Brassica rapa* and oilseed rape (*B. napus*). *Heredity* **81**, 436–443.

Hauser, T.P., Shaw, R.G. and Ostergard, H. (1998b) Fitness of F-1 hybrids between weedy *Brassica rapa* and oilseed rape (*B. napus*). *Heredity* **81**, 429–435.

Heard M.S., Hawes, C., Champion, G.T., Clark, S.J., Firbank, L.G., Haughton, A.J., Parish, A.M., Perry, J.N., Rothery, P., Scott, R.J., Skellern, M.P., Squire, G.R. and Hill, M.O. (2003) Weeds in fields with contrasting conventional and genetically modified herbicide-tolerant crops 1. Effects on abundance and diversity. *Philosophical Transactions of the Royal Society B* **358**, 1819–1832.

Hill, J.E. (1995) Herbicide-tolerant crops: environmentalists, concerns and regulatory responses, in *Brighton Crop Protection Conference—Weeds*, Volume 3, BCPC, Brighton, pp. 1027–1034.

Jorgensen, R.B., Andersen, B., Landbo, L. and Mikkelsen, T.R. (1996) Spontaneous hybridisation between oilseed rape (*Brassica napus*) and weedy relatives. *Acta Horticulturae* **407**, 193–200.

Lefol, E., Danielou, V. and Darmency, H. (1996a) Predicting hybridization between transgenic oilseed rape and wild mustard. *Field Crops Research* **45**, 153–161.

Lefol, E., Fleury, A. and Darmency, H. (1996b) Gene dispersal from transgenic crops 2. Hybridization between oilseed rape and the wild hoary mustard. *Sexual Plant Reproduction* **9**, 189–196.

Lefol, E., Seguin-Swartz, G. and Downey, K. (1997) Sexual hybridisation of cultivated *Brassica* species with the crucifers *Erucastrum gallicum* and *Raphanus raphanistrum*: potential for gene introgression. *Euphytica* **95**, 127–139.

Losey, J.E., Raynor, L.S. and Carter, M.E. (1999) Transgenic pollen harms monarch larvae. *Nature* **399**, 214.

Lutman, P.J.W. (2003) Coexistence of conventional, organic and GM crops—role of temporal and spatial behaviour of seeds. *Proceedings Conference GMCC-03, GM Crops and Coexistence*, DIAS, Slagelse, Denmark, pp. 33–42.

Lutman, P.J.W., Freeman, S.E. and Pekrun, C. (2003) The long-term persistence of seeds of oilseed rape (*Brassica napus*) in arable fields. *Journal of Agricultural Science* **141**, 231–240.

Mikkelsen, T.R., Jensen, J. and Jorgensen, R.B. (1996) Inheritance of oilseed rape (*Brassica napus*) RAPD markers in a backcross progeny with *Brassica campestris*. *Theoretical and Applied Genetics* **92**, 492–497.

Moyes, C.L., Lilley, J.M., Casais, C.A., Cole, S.G., Haeger, P.D. and Dale, P.J. (2002) Barriers to gene flow from oilseed rape (*Brassica napus*) into populations of *Sinapis arvensis*. *Molecular Ecology* **11**, 103–112.

Obrycki, J.J., Losey, J.E., Taylor, O.R. and Jesse, L.C. H. (2001) Transgenic insecticidal corn: beyond insecticidal toxicity to ecological complexity. *BioScience* **51**, 353–361.

Pekrun, C., Hewitt, J.D.J. and Lutman, P.J.W. (1998) Cultural control of volunteer oilseed rape (*Brassica napus*). *Journal of Agricultural Science* **130**, 155–163.

Perry, J.N. (2002) Sensitive dependencies and separation distances for genetically modified herbicide-tolerant crops. *Proceedings of the Royal Academy of London, Series-B—Biological Sciences* **269**, 1173–1176.

Phipps R.H. and Park J.R. (2002) Environmental benefits of genetically modified crops: global and European perspectives on their ability to reduce pesticide use. *Journal of Animal and Feed Sciences* **11**, 1–18.

Preston, C.D., Pearman, D.A. and Dines, T.D. (2002) *New Atlas of the British and Irish Flora*, Oxford University Press: Oxford.

Ramsay, G., Squire, G.R., Thompson, C.E., Cullen, D. Anderson, J.N. and Gordon, S.C. (2003) Understanding and predicting landscape scale gene flow in oilseed rape. *Proceedings Conference GMCC-03, GM Crops and Coexistence*, DIAS, Slagelse, Denmark, pp. 102–104.

Rieger, M.A., Lamond, M., Preston, C., Powles, S.B. and Roush, R.T. (2002) Pollen-mediated movement of herbicide resistance between commercial canola fields. *Science* **296**, 2386–2388.

Scheffler, J.A. and Dale, P.J. (1994) Opportunities for gene transfer from transgenic oilseed rape (*Brassica napus*) to related species. *Transgenic Research* **3**, 263–278.

Snow, A.A., Andersen, B. and Jorgensen, R.B. (1999) Costs of transgenic herbicide resistance introgressed from *Brassica napus* into weedy *B-rapa*. *Molecular Ecology* **8**, 605–615.

Snow, A.A., Pilson, D., Rieseberg, L.H., Paulsen, M.J., Pleskac, N., Reagon, M.R., Wolf, D.E. and Selbo, S.M. (2003) A Bt transgene reduces herbivory and enhances fecundity in wild sunflowers. *Ecological Applications* **13**, 279–286.

Stanley-Horn, D.E., Dively, G.P., Hellmich, R.L., Mattila, H.R., Seears, H.R., Rose, R., Jesse, L.C., Losey, J.E., Obrycki, J.J. and Lewis, L. (2001) Assessing the impact of Cry1Ab-expressing corn pollen on monarch butterfly larvae in the field studies. *Proceedings National Academy of Science USA* **98**, 11931–11936.

Tolstrup, K., Andersen S.B., Boelt, B., Buus, M., Gylling, M., Holm, P.B., Kjellsson, G., Pedersen, S., Østergård, H. and Mikkelsen, S. (2003) Report from the Danish working group on the co-existence of genetically modified crops with conventional and organic crops. *DIAS Report Plant Production* (94), p. 275.

Treu, R. and Emberlin, J. (2000) *Pollen Dispersal in the Crops Maize (Zea mays), Oilseed Rape (Brassica napus ssp. oleifera), Potatoes (Solanum, tuberosum), Sugar Beet (Beta vulgaris ssp. vulgaris) and Wheat (Triticum aestivum)*, Soil Association, Bristol, UK.

Wilkinson, M.J. Elliott, L.J., Allainguillaume, J. Shaw, M., Norris, C., Welters, R., Alexander, M., Sweet, J. and Mason, D.C. (2003) Hybridization between *Brassica napus* and *B. rapa* on a national scale in the United Kingdom. *Nature* **302**, 457–459.

3.3

Risk Assessment, Regulation and Labeling

Nigel G. Halford

Crop Performance and Improvement, Rothamsted Research, Harpenden, Hertfordshire, AL5 2JQ, United Kingdom

Introduction

Governments can deal with perceived risk by either banning a technology or regulating it. Different governments around the world have responded to the advent of plant biotechnology in different ways but all have felt the need to adopt regulations to control its use. At the heart of these regulatory processes is risk assessment. Put simply, how 'risky' something is can be defined as the chance of it occurring combined with its potential consequences. In the case of GM crops it may be their effect on human health, their impact on the environment or both that are important. Each new GM variety will pose a different risk and has to be assessed independently. Of course, exactly the same could be said of new varieties produced by other methods in plant breeding; there is no scientific basis for regarding genetic modification as inherently more risky than mutagenesis or crossing. Nevertheless, a raft of regulations has been applied specifically to GM crops and has to be dealt with by the industry.

The development of a new GM variety begins in the laboratory, progresses to growth in a containment glasshouse, field trials with a degree of containment and then release for commercial cultivation. Clearly, the chance of the new variety escaping into the environment or entering the food chain increases through this progression while uncertainty regarding unforeseen effects of the modification that has been made diminishes with study and analysis. It is not possible to describe how GM crops are regulated through this

Plant Biotechnology. Edited by Nigel Halford.

progression in every country. However, GM crops are regulated most tightly in Europe, so I will describe the regulations faced by scientists and the biotechnology industry within the United Kingdom, and how the United Kingdom and European Union regulations interact. This is relevant everywhere: first because all countries have adopted some of these regulations, and second because GM produce from outside the EU is subjected to the same degree of scrutiny before it is approved for import.

Regulations Covering the Growth of GM Plants in the United Kingdom

GM Plants Grown in Containment

The production and use of genetically modified organisms (GMOs) in containment is covered by an act of government: the Genetically Modified Organisms (Contained Use) Regulations 2000 (GMO(CU)) and its amendment, the Genetically Modified Organisms (Contained Use) (Amendment) Regulations 2002. The legislation is enforced by a government regulatory body called the Health and Safety Executive (HSE), without whose permission it is illegal for any organization or individual within the United Kingdom to produce or hold GMOs of any kind, plant, animal or microbe.

In summary, the regulations require that activities involving GMOs in containment are assessed for risk to human health and the environment, and assigned on the basis of this assessment to a category of containment (the single exception to this is humans; people who have been genetically modified through receiving gene therapy are not covered by the legislation). An organism is said to be in containment if physical, chemical or biological barriers are used to limit contact between it and other organisms (particularly humans) or the environment. Biological containment is often achieved with bacteria by using strains that are genetically disabled and therefore unable to survive outside the laboratory. It can be applied to plants by using species that do not survive in the local environment and do not cross with native wild plant species (wheat, for example, does not survive outside agriculture and has no wild relatives in the UK), or by adopting measures such as the removal of flowers before pollen is shed.

An example of a chemical barrier to the spread of a GM plant is the use of a herbicide, although it would be more likely to be applied in the post-harvest treatment of a field test site than under strict containment conditions. Physical containment involves the use of specially designed laboratories, plant growth rooms and glasshouses.

There are four categories of containment for GM microorganisms and two for GM plants and animals (Table 3.3.1). In the case of GM plants, the legislation only covers their use in the laboratory and glasshouse; field releases are covered separately. The legislation also requires that the HSE is notified in advance that a premises is to be used for genetic modification activities; HSE inspectors then assess the site to ensure that it is suitable. The inspectors must satisfy themselves that any organization that proposes working on GMOs has the facilities required and has staff who are trained and experienced in the handling and disposal of the organisms and contaminated waste. Even facilities that handle GMOs that are considered to be of no or negligible risk to human health or the environment must meet basic standards of containment and cleanliness. For example, access to the laboratory must be restricted to trained staff,

Table 3.3.1   *Risk categories of GM micro-organisms, plants and animals kept in containment in the UK.*

	GM microorganisms				GM plants and animals	
	Class 1	Class 2	Class 3	Class 4	A	B
Risk to human health or environment	No or negligible	Low	Medium	High	Less risky than non-GM parent	More risky than non-GM parent
Notification requirement	HSE must be notified that GM work is to be undertaken	HSE must be notified of each activity in these categories			HSE must be notified that GM work is to be undertaken	HSE must be notified of each activity
Containment and safety measures	Increasing requirement for measures to contain organism and protect handler				Increasing requirement for containment measures	

the laboratory must be easy to clean and the work surfaces and floor must be sealed. The work surfaces must be cleaned after use, and spillages must be dealt with immediately with disinfectant that must be ready to hand if required. Hand-washing facilities must be provided by the exit and basic protective clothing, such as a lab-coat and disposable gloves, must be worn and removed before leaving the laboratory. A safety cabinet must be available for all procedures that might produce contaminated airborne droplets and if the laboratory is mechanically ventilated the air flow must be inwards. All contaminated glassware must be sterilized and all contaminated waste must be stored safely in a designated bin and autoclaved (sterilized by heat and pressure) before disposal. Once they have left the laboratory, GM plants must be grown in designated glasshouses or controlled environment rooms with filtered, negative air pressure ventilation, sealed drains and a chlorination treatment system for drainage water to ensure that no viable plant material escapes into the environment.

Risk assessments of specific projects are carried out by the project-leading scientists and then considered by an internal Genetic Modification Safety Committee. Any risk associated with GM plants will depend on the host plant species and the nature of the gene or genes being inserted. The project leader has to consider any possible induction of or increase in toxicity and/or allergenicity compared with the parent plant and the risk of accumulation of toxicity through food chains. Any possible risk to the environment must also be considered, particularly the potential of the GM plants to be more 'weedy' than the parent plant. This assessment will include factors such as colonization ability, seed dispersal mechanisms, resistance to control measures such as herbicides and increased toxicity to insects and other grazers. The potential for and consequences of sexual transfer of nucleic acids between the GM and other plants of the same or compatible species must be considered; this is particularly important if the plants have the ability to transfer novel genetic material to wild UK plant species. The likelihood and consequences of horizontal gene transfer to unrelated species, for example by a virus, bacterium or other vector, also has to be taken into account. Finally, an assessment has to be made of the potential of the GM plants to cause harm to animals or beneficial microorganisms. The proper risk assessment of experiments involving GMOs and their safe handling and containment are legal requirements.

The Field Release of GM Plants

Regulations covering the field release of GM plants were included in the Environmental Protection Act of 1990, and a statutory advisory committee called the Advisory Committee for Releases into the Environment (ACRE) was set up to provide advice to government on the matter. The regulations were updated through the GMO Deliberate Release Regulations 2002, which implemented the European Union's directive on the deliberate release of GMOs, Directive 2001/18/EC. The UK government minister responsible for overseeing GM crop releases is the Secretary of State for Environment, Food and Rural Affairs. ACRE also advises ministers within the devolved administrations of Scotland and Wales and the Department of the Environment in Northern Ireland.

ACRE is made up predominantly of academics, with expertise in subjects such as agronomy, biodiversity, conservation, ecology, entomology, genetics, microbiology, molecular biology, plant breeding, plant physiology, rural affairs, virology and weed ecology. It also includes two farmers. Members are appointed for a three-year term.

A field trial (also known as a Part B release) of a GM plant in the UK can only go ahead after permission is granted by the Department for Environment, Food and Rural Affairs (DEFRA) and permission is only granted after a detailed risk assessment has been considered by ACRE. A typical risk assessment will include:

- Information on the host plant species:
  - The full name of the plant species and the breeding line used.
  - Details of the sexual reproduction of the plant, generation time and the sexual compatibility of the plant with other cultivated or wild plant species.
  - Information concerning the survivability and dissemination of the plant and the geographical distribution of the species.
- A description of the methods used for the genetic modification and the nature and source of the vector used to modify the plant.
- Information on the nature of the genetic modification:
  - Details of the novel genes introduced into the plant, including size, intended function and the organisms from which they originated.
  - Information on when and where in the plant the novel gene or genes is/are active and the methods used for finding out.
  - Information on the location of the inserted novel DNA in the plant cells and the number of copies of the novel gene or genes that are present.
  - The size and function of any region of the host plant genome that was deleted as a result of the genetic modification.
  - An analysis of the genetic stability of the novel gene or genes.
- An assessment of the GM plants:
  - A description of the trait or traits and characteristics of the GM plant, which have been introduced or modified.

- o An assessment of any differences between the GM plant and its parent in respect of methods and rates of reproduction, dissemination and survivability.
- o A description of detection and identification techniques for the GM plants.
- An assessment of potential risks posed to health and/or the environment:
  - o An assessment of any potential toxic effects on humans, animals and other organisms.
  - o An assessment of the likelihood of the GM plant becoming more persistent than the recipient or parental plants in agricultural habitats or more invasive in natural habitats.
  - o A description of the mechanism of interaction between the GM plant and target organisms (e.g. if the plant has been engineered to be resistant to insects) and any potentially significant interactions with non-target organisms.
  - o An assessment of the potential environmental impact of the interaction between the GM plant and target or non-target organisms.
  - o An assessment of the potential for transfer of genetic material from the GM plants to other organisms.
  - o Any selective advantage or disadvantage conferred to other species, which may result from genetic transfer from the GM plant.
  - o Information about previous releases of the GM plant.
- Information on the release site:
  - o The location and size of the release site or sites.
  - o A description of the release site ecosystem, including climate, flora and fauna.
  - o Details of any sexually compatible wild relatives or cultivated plant species present at the release sites.
  - o The proximity of the release sites to officially recognized protected areas that may be affected.
- A description of the management of the field trial:
  - o The purpose of the trial.
  - o The foreseen dates and duration of the trial.
  - o The method by which the GM plants will be released.
  - o The method for preparing and managing the release site, including cultivation practices and harvesting methods.
  - o The approximate number of GM plants (or plants per m^2) to be released.
- A description of containment measures to be adopted during and after the field trial.
  - o A description of any precautions to maintain the GM plant at a distance from sexually compatible plant species and to minimize or prevent pollen or seed dispersal.
  - o A description of the methods for post-release treatment of the site. These are likely to include ploughing of the site, irrigation to encourage germination of any seed in the soil, the removal of any plants that sprout by spraying with an appropriate total herbicide, and a period of 1–2 years when the site is kept fallow and monitored.
  - o A description of how the GM plant material will be disposed of.
  - o Details of how the site is to be monitored after the trial is over.
  - o A description of emergency plans in the event that an undesirable effect occurs during the trial or that the plants spread.

- An assessment of the likelihood and consequences of theft of GM material from the trial, vandalism of the trial, or accidental movement of GM material off the trial site by any means (e.g. on field machinery).

Clearly, providing all of this information is not a trivial or inexpensive matter and applicants also have to pay a fee of several thousand pounds. Typically, new technologies are subjected to tight regulation and control, but these restrictions are gradually relaxed as evidence builds that the technology is safe; this is exactly what has happened with GM plant field experiments in the USA, where several tens of thousands of GM plant field experiments have now been undertaken. The opposite has applied to GM field trials in the UK, with increasingly tight restrictions being applied despite GM field experiments having an excellent safety record over nearly two decades. This has undoubtedly hampered researchers and the plant biotechnology industry.

Assessing the Safety of GM Crops for Commercial Use

There are two aspects to the assessment of GM crops for commercial use: potential risks to the environment and potential risks to human and animal health. Within the UK, the potential environmental risks posed by the commercial cultivation of a new GM crop variety are assessed by ACRE, and the questions that have to be answered are essentially the same as those applied to field trials except that there is now no way in which the crop can be contained. The committee must also consider the fact that possibly a large area of land will be planted with the new crop; any negative impact will be multiplied many times over compared with that of a field trial. However, the committee will have more data to go on because the crop will have been studied under trial conditions; the point of a progression from laboratory to glasshouse to field trial before commercial release is that uncertainties should be dealt with along the way.

The safety of foods derived from GM crops is assessed by another committee, the Advisory Committee on Novel Foods and Processes (ACNFP). ACNFP was established in 1988 to advise the responsible authorities in the United Kingdom on any matters relating to novel foods and novel food processes. Currently it comprises 14 academic members with expertise in allergenicity, genetics, immunology, microbiology, nutrition and food toxins as well as a consumer representative and an ethicist. It comes under the jurisdiction of the Food Standards Agency (FSA). The opinion of a similar committee, the Advisory Committee for Animal Feedingstuffs (ACAF), is consulted if the GM crop is to be used for animal feed.

The assessments made by the ACNFP and ACAF essentially follow guidelines endorsed by the World Health Organization (WHO), at the heart of which is the concept of substantial equivalence, in which a GM plant or food is compared with its non-GM counterpart. This concept arose from the recognition that traditional toxicological testing and risk assessment procedures cannot be applied to whole foods, GM or otherwise, in the way that they are to single compounds (such as food additives). It is therefore impossible to establish absolute safety of a whole food; the fact is that very few foods consumed today have been subject to any toxicological studies. The aim of the substantial equivalence approach is to consider whether the GM food is as safe as its traditional counterpart, where such a counterpart exists.

Establishing substantial equivalence involves a comprehensive biochemical and molecular comparison of a GM food and its conventional equivalent and a detailed analysis of any differences. These differences then become the focus of the safety assessment and, if necessary, further investigation. Nevertheless, the risk assessment of the GM food will also consider other factors, including the following:

- The identity and source of novel genes (in particular whether the source is a well-characterized food or is entirely new to the food chain).
- The methods used to make the GM plant.
- The stability and potential for transfer of the novel gene or genes.
- The nature of the proteins encoded by the novel gene or genes, including potential toxicity and allergenicity.
- Potential changes in function of novel genes and proteins (in other words could they adopt a different function in the host plant from that in their species of origin).
- Possible secondary effects arising from the genetic modification event. These may include effects of disrupting a gene in the host plant, knock-on effects on metabolic pathways, or changes in the production of nutrients, antinutrients, toxins, allergens and physiologically active substances.
- The effects of processing or cooking.
- How much of the GM food a person is likely to consume.

Technologies for the analysis of plant and food material are advancing at a rapid rate through the development of genomics (the analysis of the entire gene complement of an organism), transcriptomics (determining the levels of activity of tens of thousands of genes in a single experiment), proteomics (the single-experiment analysis of the entire protein complement of a plant or food) and metabolomics (the single-experiment analysis of all the compounds present in a plant or food). However, they are expensive and the insistence that safety testing be undertaken for novel plants and foods produced by genetic modification but not those produced by other methods discriminates against biotechnology; there is no scientific basis for making this distinction.

One method that is not regarded as particularly suitable for the safety assessment of GM plants and foods is animal feeding experiments. The WHO considered this matter in the light of experience gained in the testing of irradiated food in the early 1990s. While animal studies are useful in the safety assessment of individual compounds such as pesticides, pharmaceuticals, industrial chemicals and food additives, they tell little about the safety of the complex mixtures of compounds that make up whole plants and foods. Toxicology testing requires that animals be exposed to the test substance at far higher levels than humans would be exposed to, so that any detrimental effects on health can be identified. This may be appropriate if the plant has been modified to produce a protein or metabolite that has never been in the food chain before and can be undertaken before the GM plant is made by engineering the gene into a microorganism to produce lots of protein for study. However, it is impossible with whole foods because animals cannot be persuaded to eat orders of magnitude more than humans can. Furthermore, feeding only one type of food to an animal almost inevitably reduces the nutritional value of the diet, causing adverse effects that are not related directly to the material itself. For these reasons, relating any effects on the welfare of an animal to a particular genetic modification can be extremely difficult.

Despite these reservations about the usefulness of animal feeding trials, they are now routinely performed with new GM crops before they are released for commercial cultivation in Europe. Many people question the ethics of animal testing at all; it is certainly questionable to undertake studies on animals if the results are unlikely to be meaningful (animal feeding experiments end with the slaughter and dissection of the animal).

Regulations Governing the Approval of GM Crops for Commercial Use

A centralized system for assessing and authorizing new GM crop varieties for commercial use in the European Union was introduced in April 2004 with the adoption of the GM Food and Feed Regulation (EC) no. 1829/2003. Each new crop is assessed by the European Food Safety Authority with responsibility for the final decision resting with the European Commission. The new regulations allow for a single application to cover the use of GMOs, including GM crops, for food, animal feed and cultivation (environmental release). Authorization is given in what is called a Part C consent and applies across the whole European Union; a member state can obtain an exemption only if it can demonstrate that the crop would present a specific threat to the local environment in that country. GM crops that have been issued with a Part C consent are listed in Table 3.3.2.

Although the European Commission has the final say on the matter, applications are voted on first by the Union's member states at the Council of the European Union using a qualified majority voting system (QMV). The way that the UK votes depends on the opinion of ACRE on the potential environmental risks of the crop and of the FSA (taking advice from ACNFP and ACAF) on any potential risks of the crop in food and animal feed.

In the QMV system, each member state has a fixed number of votes, between two and ten, depending on its population. Up to November 2004, any decision taken had to gain 88 votes of 124 to be carried (so only 37 were needed to block it). The number of votes assigned to each country changed in November 2004 to take into account the fact that ten new member states joined the EU on the 1st May 2004 and will be simplified if the 'Treaty Establishing a Constitution for Europe' comes into force. Under the simplified rules, a decision will require a 'double majority' of 55 % of member states representing 65 % of the population.

The QMV system is important because it enabled six member states, France, Italy, Denmark, Greece, Austria and Luxembourg, to block every application for approval of a new GM crop from 1998 onwards. In cases where a QMV is not obtained, the decision should be referred to the Council of Ministers, which can overturn the Commission's recommendation but only by a unanimous vote; otherwise, the Commission will grant the Part C consent. However, this did not happen between 1998 and 2004, during which time there was a *de facto* moratorium on the authorization of new GM crop varieties for import or for cultivation in Europe, a state of affairs that undoubtedly damaged trade relations and stymied the development of the plant biotechnology industry. The impasse was finally broken in May 2004 when the Commission authorized Syngenta Bt-11 sweetcorn, an insect-resistant and gluphosinate-tolerant variety, for import and food use. Later in 2004 it extended the consent for Mon810, an insect-resistant variety that had been

Table 3.3.2 (a) *GM crops currently holding Part C product approval for importation and processing for food and animal feed use (not for cultivation).*

Company	Crop	GM trait	Consent issued
Monsanto	Soybeans	Herbicide tolerance	May 1996
AgrEvo/Aventis (now Bayer)	Oilseed rape	Herbicide tolerance	June 1998
Northrup King (Now Syngenta)	Maize	Insect resistance (Bt) and herbicide tolerance	June 1998
Syngenta	Maize (sweet corn)	Insect resistance and herbicide tolerance	May 2004
Monsanto	Maize	Herbicide tolerant	July 2004 (For animal feed only)

(b) *GM crops holding Part C consents for cultivation and food use in the EU.*

Company	Crop	GM trait	Date consent issued
SNETA	Tobacco	Herbicide tolerance	June 1994
PGS/Aventis (now Bayer)	Oilseed rape	Herbicide tolerance	February 1996 (seed production) June 1997 (cultivation)
Bejo Zaden	Chicory	Herbicide tolerance	August 96 (seed production only)
Ciba-Geigy/Novartis (Now Syngenta)	Maize	Insect resistance and herbicide tolerance	February 1997
Florigene	Carnation	Flower color	December 1997
AgrEvo/Aventis (now Bayer)	Maize	Herbicide tolerance	August 1998
Monsanto	Maize	Insect resistance	August 1998 for cultivation in Spain and France and consumption across EU. September 2004 for cultivation across EU.
Florigene	Carnation	Increased vase-life	October 98

authorized for food use throughout the EU and for cultivation in France and Spain since 1998. The extension allowed for cultivation of this variety, which has been quite successful in Spain, throughout the EU.

Labeling and Traceability Regulations

The first GM plant product to come onto the market in the United Kingdom was paste made from slow-ripening tomatoes. This product was approved for food use in the UK by ACNFP in 1995, but the product has not been available since 1999. The paste was sold through two major food retailers, Sainsbury and Safeway, clearly marked with a large label stating that it was made from GM tomatoes. The UK food retail industry intended to pursue this policy for all foods derived from GM plants, at least until consumers were familiar with the new technology. However, these plans were thrown into confusion in late 1996 when shipments of that year's harvest of soybean and maize imported from the USA arrived. Both contained at that time approximately 2 % GM material with the GM and non-GM all mixed together. With no legislation covering the labeling of GM foods, no agreement with the Americans to supply segregated GM and non-GM produce, and insufficient time to identify alternative suppliers (the industry appeared to have been taken by surprise), retailers had a choice: either label everything containing US soybean and maize as potentially containing GM material or abandon the labeling policy. It chose the latter, a decision that, with hindsight, was undoubtedly a mistake because consumers felt that GM food was being introduced behind their backs.

Labeling controls covering GM foods were finally introduced in Europe in 1997, with further updates in 1998 and 2000. Essentially the regulations required that any food-containing material from GM crops had to be labeled with the following exemptions: refined vegetable oils, sugar and other products that did not contain DNA or protein; foods that contained small amounts (below 1 %) of GM material as a result of accidental mixing; food sold in restaurants and other catering outlets (the UK opted out of this exemption and required caterers to provide written or verbal information covering GM foods to their customers); and animal feed.

The regulations have been updated with two new directives, one on the regulation of GM food and feed (1829/2003), and the other on the traceability and labeling of GMOs (1830/2003). The new EU regulations came into force in April 2004, although UK domestic legislation providing penalties for their infringement were still being drafted in late 2004. The food and feed regulations require the labeling of all food and animal feed containing GMOs or GMO-derived ingredients. The threshold for labeling has been reduced to 0.9 % for the accidental presence of GM material; 0.5 % if the GM material has not been approved for use in Europe. The traceability and labeling regulations require a system for tracking GM products through the supply chain from seed company to farm, processors, distributors and retailers. The exemptions for animal feed and refined products no longer apply. A full list of products that have to be labeled is given in Table 3.3.3.

It is not clear how the regulations covering refined products will be enforced because the products are identical, whether they come from a GM or non-GM source, and without the presence of DNA or protein there is no way of confirming a GM origin. Given the current market advantage in claiming that products come from a non-GM source, this may lead to widespread fraud.

Although the stringent labeling laws undoubtedly make GM crops less attractive for European farmers and the food supply chain, they may have the effect of forcing the food industry to face the reality that major commodity-exporting countries have adopted GM crops widely. In 2003, the UK imported approximately 21.3 million tonnes of soybeans and soybean meal from the USA, Canada, Brazil and Argentina, and approximately three quarters of a million tonnes of maize gluten feed from the USA, plus smaller quantities of rapeseed and cotton meal from other parts of the world. The latest figures for planting in 2004 indicate that over 85 % of US soybean and 30 % of US maize plantings were GM (US Department of Agriculture (USDA)), while plantings of soybeans in Argentina are almost 100 % GM. The UK food and feed industry has claimed for several years to

Table 3.3.3   *GMOs and GM foods that do or do not have to be labeled under EU regulations 1829/2003 and 1830/2003 that came into force in April 2004 (Source: Food Standards Agency)*

GMO or GM food	Labeling required?
GM plant or seed	Yes
Whole foods derived from a GM plant source	Yes
Refined products such as oil and sugar from a GM plant source	Yes
Foods containing GM microorganisms	Yes
Animal feed containing GM crop products, including derivatives such as oil and gluten	Yes
Fermentation products produced for food or feed use	Yes
Food from animals fed GM animal feed	No
Food made with an enzyme produced in a GM microorganism (e.g. some cheeses)	No
Products containing GM enzymes where the enzyme is acting as an additive or performing a technical function	Yes
Flavorings and other food and feed additives produced from GMOs	Yes
Food containing GM ingredients that are sold in catering establishments	Open to interpretation! FSA opinion is yes.

be obtaining GM-free soybeans from Brazil (which begs the question of what has the 21.3 million tonnes of largely GM soybean from the USA, Argentina and elsewhere been used for). However, in 2003 the Brazilian government finally admitted what many had known for some years, that GM soybean was used widely in the main soybean-growing areas, and issued a decree to legalize the practice. The proportion of Brazilian soybean that is GM is estimated to be between 20 % and 30 % with no segregation of GM and non-GM product.

Imported soybean, maize and oilseed rape are all important raw materials for the food and, along with cotton meal, the animal feed industries and it is doubtful that they could be replaced. Clearly the food industry will have to start labeling its products as containing GM materials. Perhaps the appearance of GM labels on a wide range of familiar foods will reduce the consumer fears that anti-GM pressure groups have played on so effectively.

Safety Assessment and Labeling Requirements in the USA

New GM crop varieties have to undergo field trials in the USA in the same way that they do in Europe. However, much less detail is required in the risk assessment of the variety before the trial can go ahead and the procedure has been relaxed as more field trials have been undertaken and no problems have ensued. The total number of field trials of GM crops that have been run in the USA now runs into the tens of thousands, covering a variety of crop species and traits.

Test and commercial releases of GM crops in the USA are controlled by the Animal and Plant Health Inspection Service (APHIS) within the USDA, with the exception of crops that produce their own insecticide, such as Bt, which are controlled by the Environmental Protection Agency (EPA). The safety of foods derived from GM crops is assessed by the Food and Drug Administration (FDA). This is not a legal requirement but all companies seek the approval of the FDA before marketing a novel food. The FDA issues an Advisory Opinion on a specific characteristic of a product or its suitability as a food but liability for problems that arise after marketing rests with the company involved.

This system was first tested with the development of Calgene's Flavr Savr tomato (a GM variety with improved shelf-life). Calgene undertook four years of pre-market testing and submitted the data to the FDA and the USDA, at the same time publishing the data for public comment. The company requested two separate Advisory Opinions from the FDA, one on the use of the kanamycin resistance marker gene, the other on the status of the Flavr Savr tomato as a food. The FDA issued a preliminary report that all relevant safety questions about the Flavr Savr tomato had been resolved (a standard FDA position) and this was ratified by its Food Advisory Committee in a public meeting in April 1994. It announced its findings that the Flavr Savr tomato was as safe as non-GM tomatoes in May 1994 and the product was marketed shortly after.

The labeling laws covering GM foods in the USA are quite different from those in Europe in that products derived from GM plants do not have to be labeled unless they are significantly different from their conventional counterpart. In other words, products are labeled according to their properties, not how they were made.

The UK's Farm Scale Evaluations (FSE) Program

By the year 2000 a gluphosinate-tolerant maize variety, Chardon LL from Bayer, had been granted Part C consent for cultivation throughout Europe and gluphosinate-tolerant oilseed rape and glyphosate-tolerant sugar beet were expected to be granted consent shortly after. In the United Kingdom, GM crops that are approved for commercial release still have to undergo standard new variety trials carried out by the National Institute of Agricultural Botany (NIAB). Nevertheless, the UK government was faced with the prospect of the first GM crops becoming available for commercial cultivation while the public remained fearful and hostile. It therefore placed yet another barrier in the way of commercial use of GM crops in the UK by persuading the biotechnology industry to agree not to commercialize GM crops before 2003–2004. In the meantime, farm scale trials of the GM varieties in question would be carried out in order to determine what effect they, together with their appropriate herbicide regime, would have on the environment. This agreement was called SCIMAC (Supply Chain Initiative for Modified Agricultural Crops).

The results of these studies were published in a special edition of the Philosophical Transactions: Biological Sciences of the Royal Society (volume 358, number 1439, November 29th 2003). In summary, it was found that for the sugar beet and oilseed rape varieties the numbers of weeds and consequently insects in the GM crop were lower than those in the non-GM equivalent. This was widely reported as showing that GM crops were 'bad for the environment'. In reality it meant that the herbicide regime used with the GM varieties did what it was supposed to do.

Unfortunately, the management regimes that were examined in the studies were extremely narrow; no attempt was made to assess no-till systems, for example, even though these have been used extensively with herbicide-tolerant GM varieties in the USA and have reportedly brought substantial environmental benefits. The possible advantages to the farmer of growing herbicide-tolerant GM crops (e.g. May (2003) estimated that sugar beet farmers could save £150 per hectare per year) were not considered and different ways of using the crops and the herbicide were not included in the study. As a result, the government put off approval of these varieties until the companies involved, Monsanto and Bayer, could show that they could be used in an 'environment-friendly' manner. However, there is no indication that Monsanto and Bayer are interested in persevering with the issue.

In contrast to the oilseed rape and sugar beet varieties the GM maize variety with the associated gluphosinate treatment gave relatively poor weed control and was therefore deemed 'better' for the environment than the non-GM variety. In fact, this poor weed control was a concern to some farmers participating in the study. However, the herbicide used on the non-GM maize in the study and used by most maize farmers in the UK was atrazine, which was banned for use within the EU in 2004, because of its toxicity. The GM maize/gluphosinate combination therefore represented a possible alternative.

In March 2004, the UK Government announced that it agreed in principle to the commercial cultivation of Chardon LL maize but that a number of constraints would be placed on its use. In April 2004, Bayer announced that, in view of the fact that details of these constraints had still not been made available, resulting in another period of delay, it was not worth proceeding with commercialization.

Ironically, both GM and non-GM maize were found to harbor fewer weeds and insects than the other crops studied (maize is taller and weed growth is restricted by shading). However, there is no indication that maize cultivation will be banned because of its adverse effect on the environment.

Quite where all this leaves the legal situation with GM crops in the UK is not clear, since approval for GM crop cultivation is granted at the EU, not national government level. Presumably the UK government could use the results of the farm scale evaluations, despite their narrow remit, to claim that it has demonstrated that the oilseed rape and sugar beet varieties that were studied could pose a threat to the environment in the UK, thereby excluding the UK from a Part C consent, if it were granted. This may be a mute point if the companies are no longer interested in marketing these varieties.

In the meantime, varieties derived from Mon810, a Monsanto insect-resistant (Bt) maize line, have been granted Part C consent for cultivation in Europe. Up to now these varieties have only been grown in Spain, but they can now be grown commercially throughout the EU, including the UK. Maize is not affected by Bt-controlled pests in the UK, so there would be no obvious commercial advantage to growing these varieties. Nevertheless, in theory, there would be nothing to stop a farmer doing so if the seed were available.

No industry would operate in the sort of confused legal situation that the plant biotechnology industry finds itself in the UK and not surprisingly the industry has turned its back on the UK market.

Reference List and Useful Web Sites

European Food Safety Authority (EFSA): www.efsa.eu.int

Health and Safety Executive: Genetically Modified Organisms (Contained Use). www.hse.gov.uk/infection/gmo/index.htm

May, M.L. (2003) Economic consequences for UK farmers growing GM herbicide-tolerant sugar beet. *Annals of Applied Biology* **142**, 41–48.

Philosophical Transactions: Biological Sciences of the Royal Society (volume 358, number 1439, November 29th 2003) (and papers therein).

United Kingdom Advisory Committee on Novel Foods and Processes (ACNFP): www.food.gov.uk/science/ouradvisors/novelfood/

United Kingdom Advisory Committee on Animal Feedingstuffs (ACAF): www.food.gov.uk/science/ouradvisors/animalfeedingstuffs/

United Kingdom Department for Environment, Food and Rural Affairs (DEFRA): www.defra.gov.uk/

United Kingdom Food Standards Agency: www.food.gov.uk

United Kingdom Food Standards Agency position on GM foods: www.food.gov.uk/gmfoods/

United States Department of Agriculture Animal and Plant Health Inspection Service: www.aphis.usda.gov

United States Environmental Protection Agency: www.epa.gov

United States Food and Drug Administration, Center for Food Safety and Applied Nutrition Q and A Sheet: June 1992: FDA's Statement of policy; foods derived from new plant varieties. www.cfsan.fda.gov/~lrd/bioqa.html

United States Food and Drug Administration (FDA): www.fda.gov

World Health Organization: www.who.int

Index

Plant Biotechnology. Edited by Nigel Halford.
© 2006 John Wiley & Sons Ltd.